SILENT INVASION

SILENT INVASION

The Untold Story of the
Trump Administration,
Covid-19, and Preventing
the Next Pandemic
Before It's Too Late

DR. DEBORAH BIRX

HARPER
An Imprint of HarperCollins*Publishers*

HarperCollins books may be purchased for educational, business, or sales promotional use. For information, please email the Special Markets Department at SPsales@harpercollins.com.

FIRST EDITION

Library of Congress Cataloging-in-Publication Data has been applied for.

ISBN 978-0-06-320423-2

22 23 24 25 26 LSC 10 9 8 7 6 5 4 3 2 1

To my daughters,
Danielle and Devynn,
your creativity, passion for discovery, and
compassion for others bring me joy and hope

Contents

Preface

More than two years ago, I first learned of a viral illness originating in China. Since then, the Covid-19 pandemic has greatly occupied our thoughts, has altered to one degree or another our way of life, and most regrettably, might have even cost the lives of people we cared about deeply. Whatever your experience has been, I'm sorry for your loss and the catastrophic loss of life around the world.

Globally, humans have suffered great pandemics before. In some cases, we've learned the lessons from the previous one to reduce the damage in the future and bring about public health reforms. More often though, either we haven't learned those lessons, or those lessons faded quickly with time, leaving us vulnerable to the next pandemic—because as human history tells us, there will always be a next one. That's the reason why I wrote this book.

In it, I share with you my insights while serving at the highest level of the Trump administration's response to the SARS-CoV-2 virus pandemic, and as a private citizen still working in the background as the Biden administration has overseen our public health efforts. As the White House Coronavirus Task Force Coordinator, it was my responsibility, as the job title says, to coordinate efforts across a broad range of federal agencies to deal with the enormously complex problems this pandemic brought about. This pandemic has presented greater challenges than any other in my lifetime, in your lifetime, and maybe even in all of human history. As society has evolved, as technology and medicine have advanced, we've done wonders. Yet, in the face of what some scientists refer to as "organisms at the edge of life," many of us, as well as all of our advances, have been pushed to the edge.

It's tempting when faced with something as complex as a viral pandemic to find comfort in simplicity. By nature and by discipline, I don't do that. I'm not tempted to do that here in this book, either, because that reductive approach doesn't reflect reality and will make it too easy to dismiss what's happened these past two years. That approach will not help us learn the lessons we need to, and would allow us to go into the next one as equally unprepared as we were for this one. I can't let that happen.

Pandemics don't lend themselves to simplicity. There are too many interwoven layers: the politics are too intrinsic, the science sometimes seems too difficult to convey, and the cost of mistakes is so high that determining who is accountable is cast in doubt. In short, pandemics are hard to get right, and more than two years into this one, we can safely say that no one got this one right—not the Trump administration and not the Biden administration. Yet we are not alone in this track record. As you look around the globe, this imperfection has been replicated everywhere during the Covid-19 pandemic. No country has been completely right in its handling, and the few that have come close have only done so with extreme measures that are hard to implement in most places. The reality is that countries everywhere have experienced victories and also defeats against this virus. There have been moments of celebration as well as horror in equal measure as the virus has appeared to retreat and then returned with a whole new extreme of savagery.

This is not a book about Donald Trump alone. It is not a book that portrays the failures that occurred solely through the complexities of his character. In our imagination and our politics, Donald Trump looms large to be sure, but the scale of what occurred in 2020 was far greater than even him. Of course, he is a part of this story, but he is just that—a part. There is no one scapegoat for the greater than 950,000 Americans dead—as much as we might want there to be. That number is too vast, the damage far too great. To point the finger exclusively at any one group or individual misses the larger point: There is plenty of blame to go around. There have been lots of errors made by many people and institutions that have gotten us to this point. It may make you feel better to think of this as the result of one man, but you'll still be infected with another figurative pathogen that will make it more difficult for all of us collectively to see, to understand, to evaluate, and to do better next time.

Conversely, this is also a story with heroes and victories, with people who, through their attempts to save the lives of those around them, ushered in a broader sense of what was possible, of what we need to do differently now and into the future. People quietly doing their own part in their corner of the country, changing their behaviors and caring for their neighbors and families. People who recognized this pandemic for what it was: an opportunity to provide help in all the ways it was needed—help that continues to reverberate today.

The truth is, this is a story where multiple levels of behavior and decisions—both good and bad—compound one another. Here in 2022, we have seen how this virus has evolved, and our understanding of this pandemic's history must evolve as well. As a people, we've made errors and we've made good choices. Our leaders have done the same. We must learn from all of these choices, so that in the future we can make different ones.

I believe that we have to continue to learn from what has worked and what hasn't. I believe that we should appreciate the successes and acknowledge the people and organizations who contributed to them. I also believe that we should hold accountable those groups and individuals who contributed to the problems that plagued the response and ensure corrective actions. Sometimes they exist on both sides of that ledger. In a rush to judgment, it's too easy to forget this is frequently the case in all of life. That it is true in this situation, one of such historical significance, demands that we thoroughly examine both right and wrong, good and bad, and all the points in between those extremes.

My purpose isn't simply to condemn or commend, but to recommend. I am most frequently an optimist, sometimes a pessimist, but most consistently I'm a data-driven realist and that is the perspective that is most reflected in the pages that follow. We have solutions to the problems this pandemic presents. I have recommendations for how we can be better prepared for the next one. In both cases, we have to break the cycle of dysfunction that has produced so much devastation. If there was one thread that ran through my experiences in the White House and in many of our states, it was that where a spirit of community and cooperation thrived, we achieved the most gains. Where and when this spirit did not exist, we lost the most ground. When minds and hearts worked together, we were better.

We have to do better. We can do better. So much depends on it, as we continue to face this crisis and, I hope, as we plan better for the next one.

—Dr. Deborah Birx, February 2022

SILENTINVASION

Prologue

My back is against the wall. Literally.

I've just been ushered into the president's private dining room. A too large table and a credenza are crammed into the tiny twelve-by-twelve space. Seated at the head of a table, the president dominates this space. His head and chest are visible above stacks and stacks of newspapers. It's as if he's behind sandbags in the trenches.

The president barely acknowledges my presence. Staffers hover around him, taking turns leaning in to speak to him. His eyes briefly dart from side to side, then are drawn again to the large TV screen opposite him. My eyes follow his and take in flashes from the four major cable news networks packaged into smaller viewing areas on the screen. He regards them briefly and then picks up his phone to speak with someone. He ends a phone call mid-sentence, asking to be connected to someone else. Not sure where to look, I scan the room for the next few minutes. Oddly, the sound of ice swirling in a glass cuts through the cacophony and draws me back to the president.

It is March 2, 2020. I've just flown in overnight from South Africa to take on the role of response coordinator for the White House Coronavirus Task Force, a job I didn't seek but felt compelled to accept. I'm physically tired but mentally alert. After weeks of urging from Matthew Pottinger—President Trump's deputy national security advisor, a task force member himself, and the husband of a former colleague and friend of mine—I finally gave in to Matt's request that I come on board to help with the response to the coronavirus outbreak. I trusted his inside assessment that the administration's response to date had been meandering and flawed, putting the American people at potential risk.

It wasn't easy for me to leave my full-time position as the U.S. global AIDS coordinator. I am now dual-hatted, working for both the State Department, continuing my oversight of the President's Emergency Plan for AIDS Relief (PEPFAR) program, and the White House. Stepping away from the daily execution of PEPFAR at a critical juncture was extremely difficult. Still, I had to go where I was most needed, and this looming public health crisis requires my attention.

I'm not anxious about meeting the president. In my forty years as a civil servant in various capacities in public health and the military, I've met and worked with other U.S. presidents, heads of state across the globe, generals, and chief executive officers. I am used to working behind the scenes with critical decision makers. I know how important policies are to public health, and seated across from me is the most important policy maker of the moment.

For more than a month, I've been sending back-channel communications to the task force through Matt. In doing so, I hoped to incite a course correction—but those efforts have produced few immediate results. Matt has become increasingly frustrated by the lack of action across the federal government. But I don't know where the individual members of the task force are compared to where I am. From thousands of miles away, I thought the task force and coordinating agencies were like a duck on water, calm on the surface but paddling furiously underneath—building testing capacity and ensuring that PPE, ventilators, and other equipment were adequately stockpiled. Most important, I believed they were creating the essential data streams to determine where the virus was and where it was going. Matt's urgent tone made me less certain.

Today I have two primary goals: I want to gauge the president's sense of urgency and convey to him how strong mine is. I can wait patiently in this meeting, but in a larger sense, none of us can waste one more second. This isn't PEPFAR, where we have had months, and sometimes years, to move political leaders to enact the policies needed to save lives. Now days, even hours, matter.

I've got one shot at this; it had better be a direct hit. I've been sharing data about the burgeoning epidemic for weeks with the White House, through Matt. The message hasn't been getting through. I believe that the president, as a businessman, will be persuaded by the figures, will appreciate a bottom-line number. But will he be able to comprehend the numbers I am seeing now? They are at odds with what most others are forecasting. Things are much worse than he likely believes. This administration has been trying to downplay the seriousness of the virus. I wonder what the president has been told.

Has he been briefed on the level of asymptomatic silent spread? Does he understand that the majority of cases can be detected only by testing and not by symptoms? Has it been made clear to him that the virus is undoubtedly already circulating widely, below the radar, in the United States?

I've been reading and hearing messages coming out of the White House and the federal health agencies that the virus can be contained, that there are systems in place to isolate it and prevent it from spreading. The bearers of these messages seem adamant that this virus's spread will be visible, that they will be able to contain the virus by merely identifying those people with symptoms and isolating them and those to whom they've been exposed. The belief is that if they do these things, the disease itself will be, in their estimation, no more a threat than the annual version of the flu.

They are wrong.

This virus is not flu-like. It's not behaving or spreading in ways that mimic the seasonal flu. This virus cannot be tracked and traced only through finding those with symptoms. That may have worked for the seasonal flu or even pandemic flu, but it will not work for this virus, one that is clearly much more deadly than the seasonal flu.

If the Centers for Disease Control and Prevention, our premier public health agency, seemingly hasn't been able to grasp this concept of silent spread, how can a layman, a real estate developer by trade, be expected to? In part, I get it. I've seen this cocktail of confusion, bias, and denial in the public health agencies before. I've also seen the devastation that viruses mete out. HIV, SARS-CoV-1, MERS-CoV, Ebolavirus—I've been on the front lines and have worked with many other experts in the field as the world navigated these public health crises. Handling infectious diseases ranging from tuberculosis to AIDS has been inherently political, throughout history and now.

As president, you don't have the luxury of focusing on only one thing at a time. Yet, the presidents I served under before, George W. Bush and Barack Obama, had the ability to shift gears and direct their focused attention in a way President Trump has not.

I'm not going to get him to change.

I have to change my approach.

Experience has taught me that you have to meet people where they are. It helps if they're willing to greet you, but I'm being held at a distance, even in this small room. I can feel it. It is a palpable presence overcrowding an already crowded, noisy room. What's the one thing I can say that will span this distance, be heard above the buzz and hum of this private dining room?

Then the president is making a few brief remarks to the room, welcoming me. Finally, I get a chance to speak with him.

"Mr. President. This is not the flu. This is far more serious than the flu. We have to shape our response differently."

He holds up his hand. He smiles that glib grimace of a smile.

I stop speaking. This is my commander in chief.

"Well, the people I'm talking to say that this isn't going to be any worse than the flu," he says.

"Mr. President, I don't know who you are speaking with, but I have evidence to fully support the conclusion that this outbreak is not going to be like the seasonal flu or even pandemic flu. This virus is very deadly."

"Well, these are good people. Smart people. I trust these people. They know what they're saying."

I reiterate my position. He nods. His eyes return to his television screens. He reaches for a remote control, and the voice of someone at Fox News enters what passed for a conversation between us.

Sources tell us that . . .

I don't hear the rest. Someone takes a few steps toward me and gestures toward the door. I've had less than thirty seconds to speak with the president.

My mind shifts gears to the afternoon meeting ahead, which the president will be attending. I'll be there with select task force members and, critically, the biotech and pharmaceutical companies. This will be my first opportunity to see where we are in the development of effective treatments and vaccines.

Will they be given short shrift like I was? "Short shrift." Historically, the expression refers to the time between confession and execution.

It's as fitting an expression as any for what I'm up against—indeed, what we all are.

PART I

CHAPTER 1

Tracking a Mystery

I can still see the words splashed across my computer screen in the early morning hours of January 3. Though we were barely into 2020, I was stuck in an old routine, waking well before dawn and scanning news headlines online. On the BBC's site, one caught my attention: "China Pneumonia Outbreak: Mystery Virus Probed in Wuhan."

Anytime I see a phrase like "mystery virus" my antennae go up. Anytime that mystery virus is in China, I am even more concerned. I've worked my entire life, in one capacity or another, in the fields of immunology, infectious diseases, and public health. My medical specialty is in immunology, and epidemics and pandemics have played a large role in shaping my career—from HIV to avian flu. I was doing work in Asia back in 2002 when the sudden acute respiratory syndrome (SARS) outbreak began. Visceral recollections of the fear that gripped the region and the public health community still haunt me. The numbers of those infected with the virus SARS-CoV-1 weren't extraordinarily high, as pandemics go, but the rate at which it killed those infected was.

Numbers alone don't tell the story of a virus. Every viral outbreak is unique and requires some variation of measures in the handbook to overcome it. When it came to SARS-CoV-1 and China, I recalled the outrage the medical and public health fields felt, and continued to feel. China violated one of the most fundamental principles of managing any infectious disease: to share information early and to share everything you know about a new pathogen. China did neither. With SARS-CoV-1, China may have taken the right actions locally, but it certainly did not do so regionally or globally. Even back then, the world was too small for parochial interests to have outweighed our common interest.

During that outbreak, I sat in coach class on one of my frequent flights to Asia. I had empty row after empty row on which to stretch out and sleep. That I had that choice did more to prevent me from resting well than being crammed shoulder to shoulder would have. Nature abhors a vacuum, and those empty seats were, for me, filled with the specter of the virus and its victims. The flight attendants, airport staff, and the few Asians on my flight all wore masks, adding to the ghostly effect.

The phrase "mystery virus" goes back even further for me. In the early 1980s, as an active-duty Reserve officer in the U.S. Army while serving as a medical doctor at Walter Reed Army Medical Center, I treated U.S. soldiers suffering from another mysterious illness. Early on, we knew it as adult respiratory distress syndrome (ARDS). As was the case for SARS, we didn't know what was causing patients to die from what started as an atypical respiratory infection. We could see their immune systems were being destroyed, but we still didn't understand why or by what.

We faced a heartrending set of circumstances, seeing previously vibrant, healthy young men being killed by inexplicable, unrelenting immune system deficiencies. With ARDS, too many soldiers died the most terrifying deaths. Their eyes would grow wide as they struggled to breathe. Later, with the soldiers unconscious and medicated to minimize their pain, we could only sit beside them holding their hands and watch as their faces twisted into a rictus of suffering and, despite all our efforts, they essentially drowned in their beds as pneumonia filled their lungs with fluid and starved them of oxygen.

Normally, we could have treated the root cause of the immune deficiency, but for these young men, we had no answers. We saw the evidence of an invasion crippling their body's immune system with one rare infection after another, but we didn't know its cause. We were desperate, and we were humbled. Our patients went from one bad moment to the next. Our interventions were temporary. We lived in a world of so many unknowns but one—that these young men were going to die and we couldn't do anything to prevent that.

Returning to the BBC article about this latest "mystery virus," I noted that the piece focused on two areas: what little was known about the spectrum of the disease (that is, the progression of its symptoms) and how the Chinese citizens and public health officials were responding to it. It was already the annual influenza season in the northern latitudes. If the Chinese had started tracking this outbreak based solely on symptoms and not on a definitive laboratory diagnosis, their initial presumption was that the virus was a seasonal flu variety. That it was now a "mystery virus" meant it could have been circulating for quite some time already. Were we seeing only the tip of the iceberg?

Using Google Translate, I read Chinese social media entries expressing fear that the new illness could be linked to a SARS outbreak. It was easy to see why some online chatter had made the SARS connection both in China and across Asia. Likely, many of those posting had lived through or lost someone to that earlier SARS and Middle East respiratory syndrome (MERS) crisis. Government officials and citizens across Asia knew both the pervasive fear and the personal response that had worked before to mitigate the loss of life and the economic damage wrought by SARS and MERS. They wore masks. They decreased the frequency and size of social gatherings. Crucially, based on their recent experience, the entire citizenry and local doctors were ringing alarm bells loudly and early. Lives were on the line—lots of them. They knew what had worked before, and they would do it again.

To that end, my heart sank a bit as I read on. The BBC article reported that Wuhan police had already cracked down on those who were "publishing or forwarding false information on the Internet without verification." Would data be withheld again as it was in 2003? Certainly, I recognized the possibility that some of those reporting could be alarmists. But many of them were equally, if not more likely, truth tellers. People risking jail to share information likely meant one thing: the situation was worse than Chinese authorities were reporting.

I hoped this wasn't the case. In the two decades since SARS, officials around the world, including in China, had agreed to focus on global health security and ensure transparency and information sharing early, even when the data points were incomplete. Based on what I was reading, though, the outbreak in China was not just worse than Chinese officials said, but it had likely started earlier. This meant that the virus had already had the opportunity to spread widely before they enacted any measures to contain it. This had implications for the rest of the world.

When it comes to handling emerging pathogens, the Chinese government is not alone in being motivated by self-interest. Economic and reputational concerns lead to a pattern of denial and downplaying. I saw this happen during other outbreaks, ranging from Ebola in West Africa in 2014/16; to the 2014 MERS, which originated in Saudi Arabia; to the latest instance of the Zika virus's spread in the Americas in 2017. Governments always believe they can contain a virus and prevent it from spreading widely in their country and to others. But viruses can change rapidly, viruses move rapidly across borders, and humans are by nature slow and often too arrogant to act when they should, convinced they have the power to control and contain viruses with technology and win.

THE NEXT FEW DAYS after that first BBC story, my early morning internet-browsing sessions quickly turned into my taking a few moments regularly throughout the day—some might say excessively—to check where this new virus was and where it was going. Viral outbreaks evolve quickly, so I'd scour the internet between meetings. I'd use different search terms. I'd integrate data points in my head, turning single-source reports into a two-dimensional picture of the new virus on the move.

Much of my career has been shaped by the desire to be of service in the most effective way possible. In college, I chose medicine over my first loves, physical chemistry and math, for precisely this reason. Helping a company like Kodak develop a new and better green dye that didn't turn photo paper yellow over time certainly would have made me money, but it wouldn't have helped change the world.

I switched to medicine when medical research and understanding of our immune system were expanding at an explosive rate. The immune system fascinated me because I saw it as a very sophisticated mathematical equation. It has to strike a very delicate balance and stay within a critical window where it can fight off pathogens while not going too far and destroying the very body it's meant to defend. That the same system contained both the ability to kill us and to save us enthralled and challenged me.

Throughout my medical/research career—which has taken me from the dawn of broad immunological study to places and organizations like

Walter Reed, the National Institutes of Health (NIH), and the CDC; to my role as ambassador-at-large and global AIDS coordinator as part of PEPFAR, at the State Department—the immune system and its role in fighting disease has been at the forefront of my work to mitigate the effects of infectious spread. For many years, this meant the AIDS pandemic, but also other diseases, like tuberculosis.

It has been enormously rewarding work, and I was looking forward to ending my tenure in 2021. I had decided that four decades in public service was a good, round number, and I planned to move on to a second career. I wasn't quite there yet, and had entered into a very busy time with PEPFAR, when I began to read those first accounts coming out of Wuhan, China. I had no special access to other inside information, just a long-held need to keep informed.

Along with monitoring publicly available sources and journalism, I also watched the information coming from the World Health Organization's situation reports. As the mystery evolved, I dug deeper, moving beyond mainstream news organizations to new websites and online posts tracking the virus across the globe. Clearly, and as I expected, others saw the need to probe more fully into this developing story. Many in the field of public health began to mobilize, putting together what little data we had to create a more complete picture of what was happening.

On January 6, I read a *New York Times* story about the same clusters of illnesses the BBC had reported on. It confirmed much of the earlier reporting on them, but also included details about a suspected source. The Huanan Seafood Market in Wuhan had been shut down and decontaminated. The virus that caused SARS and the H7N9 strain of bird flu, which had caused five epidemics of avian flu between 2013 and 2017, had been traced back to similar markets. Close interactions between humans and animals can lead to a virus jumping from one species to another. When animal-to-human transference takes place, and the virus adapts to infect human hosts, we call the resulting virus "zoonotic."

Novel (new) zoonotic viruses are particularly alarming. All zoonotic viruses are worrying, and the fact that Ebola, SARS, MERS, avian flu, and AIDS were all caused by zoonotic viruses is especially worrisome. (Basically, epidemiology is a field that requires a strong stomach for worrying.) Why? With a zoonotic virus, human beings usually don't have any preexisting immunity to the pathogen that arises in and adapts to a specific animal. The more rapidly it adapts to the human host and can spread from human to human, especially through the air, the more easily the human population can become relatively easy pickings. This is especially true if we are otherwise immunocompromised (as in the case of people who are HIV-positive) or already have other conditions (known as comorbidities) that lessen our ability to produce a fully effective response to pathogens. Airborne pathogens are also particularly dangerous because of the ease with which they are transmitted. As horrific as Ebola and AIDS are, they are not as easily spread, as they are transmitted through the exchange of bodily fluids.

Along with sharing this detail about the market, the *Times* piece also confirmed two early suspicions I had, both quite troubling. The WHO had received notification from Chinese authorities about a pneumonia-like cluster of infections on December 31, 2019. Yet, by that date, Chinese authorities already had tracking and containment plans in place at airports outside Wuhan to monitor airline passengers arriving from that provincial capital. For the virus to have risen to a level of infection that necessitated these state actions meant it had likely been spreading for weeks, and the Chinese were now aggressively acting to control the spread within their borders while underplaying the outbreak globally. If the virus had in fact been spreading for weeks, it meant the Chinese were also behind in seeing and responding to the outbreak and their containment efforts would fail.

Furthermore, the Chinese were claiming to the WHO that there had been no human-to-human transmission. The only ones infected, they said, had had direct contact with animals at the Huanan wet market. If there was no human-to-human transmission, then the number of victims of the disease would be very small, restricted to those who had been to that single wet market or other wet markets.

Whether the officials in Wuhan or higher up on the chain of command had delayed the release of information by days or weeks was impossible to know at this point. But I did know that any delay could prove deadly. Whatever lessons the Chinese authorities had learned from SARS, they apparently weren't frightening enough to inspire a change to full transparency.

The *Times* article also confirmed that the Chinese were on the hunt for those who were already exhibiting signs of infection. To complicate matters, China was in winter, a period when many other respiratory viruses circulate, including influenza. It would be hard to tell, therefore, who had pneumonia-like symptoms that were not related to this novel zoonotic virus—at least, not without a test to detect that particular virus. From past experience with other viral outbreaks, I was dubious about this kind of containment strategy.

In tracking viral outbreaks, it is critical to account for four types of spread. The first is asymptomatic, which applies to people who are infected but, despite not having symptoms such as fever, cough, and nasal discharge, are indeed infectious and able to transmit the virus to others. The second is presymptomatic. Immediately following initial infection and replication of the virus and before exhibiting any signs of the infection, these individuals are infectious and will transmit the virus to others during this window. The third, mildly symptomatic, are those with symptoms so mild and non-febrile that they either ignore them or pass them off as symptoms of allergies or a hangover; nevertheless, these individuals are infectious and can transmit the virus to others. The fourth, the fully symptomatic, are those currently presenting typical signs of the infection and able to transmit the virus to others.

Asymptomatic, presymptomatic, and even mildly symptomatic spread are particularly insidious because, with these, many people don't know they are infected. They may not take precautions or may not practice good hygiene, and they don't isolate. As a consequence, they come in contact with more people than someone who is symptomatic. Sick people with high fevers and body aches often can't physically work and tend to stay at home. As a result, those in the first three categories often infect more people than the fully symptomatic do.

Placing a major emphasis on the fully symptomatic is typical of the containment strategies devised to lessen symptomatic spread. That's step one, but if it's the only step you take, if that's the only type of spread you feel you have to mitigate, then containment will never work. In my experience, from the earliest onset of any cluster of infections, you have to be alert to the possibility of, and account for the first three types of spread. In my mind, this is job one.

Sure, there are other variables that determine the scope of an outbreak. Knowing how the virus is spread—whether it is airborne or passed on by blood or other body fluids—the length of the incubation period after exposure, and how long a person remains infectious to others is also critical information. For example, a person infected with HIV can remain asymptomatic for as many as ten years, which contributes to that virus's being such a difficult one to control. Some viruses, like the one for measles, are more easily transmitted than others through aerosols (that is, fine particles). Some mutate at a greater rate, becoming more adaptable to their new host's changing infectiousness and/or virulence.

However, the one variable that stands out most for me is the type of spread/transmission. After years of experience seeing asymptomatic, presymptomatic, and mildly symptomatic cases being ignored in tabulations, anytime I read a number indicating a confirmed case, I multiply that by a factor of between three and ten. Whatever the number of infected the Chinese had put out, 44 in their first report, I read as between 132 and 440.

The only way to accurately account for all four types of spread is to test as many people as possible early and often. The Chinese weren't doing this—or, if they were, they were far behind where the outbreak actually was. It seemed highly unlikely that they'd developed a test specific to this novel virus yet. If they didn't believe (or didn't want to admit) that human-to-human transmission was going on, and if they weren't accounting for asymptomatic spread, then they wouldn't prioritize test development.

Often in pandemics, we focus significant effort on the development of treatments and vaccines, and we neglect the development of tests. This is a fundamental error. In Africa, we had spent years moving from testing only those with AIDS symptoms to testing everyone independent of perceived risk. We'd saved countless lives through active community testing to determine if individuals had been infected with the virus, ensure access to lifesaving treatment, and prevent unknowing transmission to others. This approach was working for controlling HIV/AIDS community by community, and even though the novel virus was not being transmitted in the same way as HIV, the same model could apply in the case of this outbreak.

Without widespread testing, the Chinese were providing inaccurate data. They may not have been intentionally underreporting, but their numbers were wrong nonetheless. Whether this was an error of commission or omission doesn't really matter. Either way, without testing, you could not see the full extent of the virus and number of people infected.

In the case of this novel virus, I believed that significant asymptomatic and presymptomatic spread (which, together, are also known as "silent spread") was already occurring in China in early January and had likely been happening for weeks. I didn't have to wait long to find further evidence to support my contention. In the days following the original news reports, other provinces within China began identifying similar pneumonia-like cases. I also knew this: If this novel virus was related to SARS, then it was already spreading more rapidly than the late 2002/3 SARS-CoV-1 version had. The puzzle pieces were being laid down to create a frightening picture. I worried about the HIV-positive people we were supporting throughout Asia and Africa and their potential susceptibility to this new virus.

All zoonotic viruses aren't created alike, and they don't act alike, either. They range from extremely and immediately deadly (Ebola) to deadly over time (HIV), with a longer component of silent infections with HIV. SARS and MERS killed many people quickly. Ironically, the more rapidly a person becomes symptomatic and dies from infection, the less community spread there typically is, because there are fewer chances for that person to infect other people. While this may sound counterintuitive, it occurs because those stricken are sick in bed at home or in hospitals and, therefore, not silently and unknowingly spreading the virus in the community.

As deadly as Ebola is, and as large as it looms in our collective imagination, measures to contain it are relatively effective. It spreads through contact with the body fluids of those infected with the virus. You know immediately who has been infected, so tracking and tracing those who have developed, or might develop, the disease are more straightforward. That's not the case with SARS and its variant MERS. These diseases are produced mainly when an infected person coughs or sneezes, emitting droplets containing the virus that another person comes in contact with. You know when you've come in contact with body fluids. You're not always aware when you've come in contact with aerosols and droplets, which remain unseen and are suspended in the air for long periods, allowing us to unknowingly breathe them in. For the medical community, viruses that can spread through aerosols suspended in air (highly infectious) and that are virulent (deadly) are the most concerning.

The SARS-CoV-1 strain of virus that caused the 2003 SARS outbreak is extremely virulent. Its average 10 percent case fatality rate (or CFR, the number of confirmed infected individuals divided by the number of deaths over a specified period of time) is extremely high. Fortunately, the SARS virus is less transmissible, requiring a higher viral exposure for longer periods in order to spread. As a result, even though SARS had a very high CFR, because it wasn't as "contagious" and because it moved less silently and undetected from host to host, the number of total cases, and thus deaths, for the 2003 outbreak was fairly low. That being said, the SARS outbreak killed indiscriminately across age groups, rapidly taking out healthy twenty-somethings, forty-somethings, all the way up to eighty-somethings, and struck fear throughout Asia from the end of 2002 through 2003.

I understood just how fortunate we all were that SARS wasn't both easily transmissible and highly fatal. That was the stuff of nightmares.

WHILE I HAVE LONG been fascinated with the immune system's response to a pathogen, as a public health official, I'm far more concerned about how governments and health agencies respond to the presence of a potentially new and deadly virus. Alerting the public early and being very aggressive can change the course of an outbreak and prevent it from developing into a full-blown epidemic.

On January 3, the same day the BBC piece ran, the Chinese government officially notified the United States of the outbreak. Bob Redfield, the director of the Centers for Disease Control and Prevention, was contacted by his Chinese counterpart, George F. Gao. At the time, this wasn't reported, but later, in April 2020, the *Washington Post* revealed that Alex Azar, the secretary of the U.S. Department of Health and Human Services, directed his chief of staff to notify the National Security Council about the severity of the situation in China. I was unaware of any of these developments.

On January 6, the same day the initial *New York Times* piece was published, the CDC in Atlanta issued a Level 1 travel advisory, the lowest in its three-tier system. It advised those traveling to a specific region of the world to practice the usual precautions. Unfortunately for many, "the usual precautions" meant traveling nonetheless and doing what they would normally do. During the course of a normal year, the CDC and the WHO offices around the world issue dozens of such advisories. The CDC was following standard operating procedures, indicating that it was monitoring the situation. I continued to follow the WHO status updates, but often, to my dismay, they appeared to be focused on merely tracking the evolving situation rather than taking action or alerting others to take action. It's like they were passively watching a movie. When it comes to viruses, tracking without action leads to continued spread.

Regardless of the testing status in China, and regardless of the issues with the Chinese numbers, I presumed that China (and other Asian countries) would likely fare better with this first outbreak than they eventually did. From the nature of the symptoms, I thought this was likely a respiratory infection similar to the one caused by the SARS virus. Having been through a viral outbreak of that type before, the governments across Asia had enacted aggressive changes to public health policy and preparedness. They not only had a plan, but they and their populations knew how to implement it when necessary. Their populations would be more compliant because they understood the benefits of behavioral changes. They had lived that frightening nightmare and, as a consequence, had developed a kind of muscle memory that guided their actions. They immediately knew what to do and did what was necessary to protect themselves and their families. They weren't waiting for guidance from the global health officials.

During the SARS crisis, we in the United States dodged a bullet, mostly because SARS was not as highly transmissible. It also was a fairly "loud" and "visible" disease, meaning it was detectable early, through symptoms that clearly and loudly announced themselves with their severity. This made contact tracing, with isolation (of the infected) and quarantining (of the exposed), more straightforward and possible. Yet, our avoiding that major outbreak also meant that, unlike the Asian populations, we didn't have a cultural understanding of what the impact of a viral outbreak could do to individuals, families, businesses, and our general way of life. We didn't have that shared experience that told us that wearing masks and social distancing were effective actions to take.

I felt this acutely because, for as long as I could remember, I understood that the risk of a virus is borne by all. As an eleven-year-old, my grandmother was infected at school during the deadly 1918 Spanish flu epidemic. Her mother, my young, vibrant great-grandmother Leah, from whom I get my middle name, had just given birth to another daughter when my grandmother transmitted the virus to her. When my great-grandmother died from it, the entire trajectory of my grandmother's life was changed in an instant. For a time, she became the caregiver for her infant sister and the rest of her family. She was racked with guilt until the day she died at age ninety-four.

My grandmother was my touchstone, a woman who believed in me unconditionally. I spent every summer with her growing up, and I have carried her wisdom with me through the years. That one innocent moment of bringing the Spanish flu home changed her life for the next eighty-three years. In a very real way, my life was shaped by a viral outbreak. It was part of my lived experience, and my decision to study immunology and epidemiology was shaped by it. I didn't want anyone to experience what my grandmother had. I didn't want any country to reexperience the devastation of the Spanish flu, SARS, or any other pandemic event. Youth, vibrancy—these things do not always protect us, but more important, they don't protect our families. Nothing about our circumstances can insulate us; only our behavior can.

In the case of the novel virus, on January 13 Thailand reported its first case, then a second case. On January 20, the Korean CDC reported its first case. The infected Korean woman had been in Wuhan but had not visited the suspect market. The medical community knew that it had to have been spread from human to human.

Later in the month, a Japanese businessman who had traveled to the Wuhan area and also hadn't spent any time in the suspected wet market became sick. He was eventually hospitalized and tested positive for a novel coronavirus infection. All this signaled to me, once again, that in both these cases, human-to-human transmission was going on. I couldn't see it any other way.

By mid-January, though, Wuhan's Municipal Health Committee had published a FAQ piece claiming, among other things, that there was no clear evidence of human-to-human transmission. On January 14, the WHO tweeted the same thing: no evidence of human-to-human transmission. Later that day, the WHO held a press briefing during which it was stated that "it is certainly possible that there is limited human-to-human transmission." The spokesperson went on to say that it was important to "ascertain . . . the presence of asymptomatic or mildly symptomatic cases that are undetected." I shook my head when I heard this—I knew that politics was at work, at least in China. There had to be human-to-human transmission to account for the approximately five hundred cases being reported there. Fortunately, on January 12, the WHO reported that the Chinese had isolated a novel coronavirus as the cause of the spreading illness. This meant that tests—the first pillar of any twenty-first-century public health response—could now be developed to detect the presence of that particular virus. Crucially, those tests could detect infection before a person developed symptoms. This new information would instigate the development of treatments (the second pillar) and vaccines (the third pillar of an effective response).

Like the SARS virus, the novel pathogen was from the *Coronaviridae* family of viruses and therefore shared some of its ribonucleic acid, or RNA, with SARS-CoV. Knowing that the genetic sequence of this new RNA virus was related to the SARS virus was worrying—but better the devil you know than the one you don't. The long-term investment in SARS research following the 2003 outbreak had created a strong platform of baseline understanding of this type of virus and would accelerate the research necessary to develop tests, therapies, and vaccines for the new one.

After identifying the type of pathogen, the Chinese did the right things scientifically: On January 12, the China CDC released three genetic sequences for the novel coronavirus. Two others were posted to the Global Initiative on Sharing All Influenza Data, or GISAID, which provides open access to genomic information. This was critical and a good sign, both within and outside China. Scientists could now ramp up their efforts on multiple fronts. But as is often the case with outbreaks, bad news often follows close behind the good.

Despite the Chinese crackdown on the online spread of information, word was getting out. Hospitals in Wuhan were filling up, an incredibly troubling development. If you are, in fact, effectively mitigating, if you are isolating those infected and quarantining those exposed, you should be *preventing* hospitals from becoming overwhelmed—what we call "flattening the curve" of the pandemic. Based on what I was reading, though, this did not appear to be happening in Wuhan. This was a bright red warning light. Even with the rules in place, the Chinese were not containing the virus, and it was spreading quickly through human-to-human transmission. The Chinese were missing a critical part of the spread. With hospitals being filled so suddenly, there simply couldn't have been that many people directly in contact with whatever animal at whatever suspected market was carrying the transmissible virus. I knew then that this virus was worse than SARS. It was spreading faster, and hospitals were filling up quicker than they had in 2002/3.

On January 14, when the WHO had tweeted about the limited possibility of human-to-human transmission, I knew that "limited" wouldn't adequately account for the five hundred cases being reported in Wuhan. The WHO was trying to thread the needle between science and politics, hoping that its words would be seen as a warning without contradicting the Chinese. But using such tentative language to address such an obvious reality was a serious mistake, one that undermined the central tenets of the organization. By this point, we were already way past caution. Public health officials recommend a very different approach to containing a virus that can be spread only from animals to humans, with no human-to-human transmission; the potential for global spread is dramatically different. Warning the world right then and there of the presence of a highly transmissible respiratory virus with a spectrum of disease ranging from asymptomatic to serious illness would have changed the global response and would have spurred the world to greater action and the manufacturing of tests. In hindsight, I see this as an important, early missed opportunity.

On January 14, we also learned that the incubation period (the time between exposure and symptoms being expressed) of the new virus could be as long as fourteen days. But the data was spotty, and there were still many unknowns. Even so, it seemed apparent that whatever we were seeing now was likely to be vastly different (read: worse) in two weeks.

The Chinese data was suspect and scarce. I turned to other sources to find evidence to support my instinct that asymptomatic transmission was also responsible for the spread beyond Wuhan. The website for the University of Minnesota's Center for Infectious Disease Research and Policy cited two previous cases I'd read about—the Japanese businessman and the Korean. I felt certain other scientists saw these as evidence of human-to-human transmission.

On January 17, the CDC announced that it would be screening passengers arriving in the United States on direct or connecting flights originating out of Wuhan. The screening at three airports consisted of a temperature check to identify those who were febrile and a verbal self-report of symptoms. Just as the Chinese had initially, and then the WHO, the CDC was looking for symptomatic people and their close contacts coming from known hot zones. Once these people were found, they would be isolated or quarantined. This was the singular focus—containment through symptoms, not testing. Without testing, this would never be enough. Everyone who was coming from "hot zones," independent of their symptom status, needed to be tested.

On January 20, the CDC reported that the United States had its first confirmed case. Like the infected South Korean woman, the Seattle man who was the first reported and known American infected with the novel coronavirus had neither been to the animal market nor been in contact with anyone who was ill. Clearly, he had been in contact with someone who was carrying the virus, but as far as this first-known infected American knew, he had not been in contact with any symptomatic person in China or upon his return to the United States.

Of the three cases outside China that I was aware of, two fell clearly into this same potential asymptomatic or presymptomatic exposure category. I was feeling more and more confident that there was silent spread at work on some level. Yet, as a trained scientist, I understood that I didn't yet have enough evidence to support what my gut and my head were telling me. I believed that I was ahead of where the CDC and the WHO were, but I wasn't far enough through the curve to be able to provide the abundance of evidence I would need to defend my beliefs. I could hear the virus calling out, but as far as I knew, it remained silent to others. Still, I was confident that our scientists, among the best in the world, were likely seeing what I was seeing and that our country's vast resources would offer us the best protection.

I was hopeful that this wasn't arrogance talking. I trusted that—though the president said presumptuously on January 22, "We have it totally under control," and though Dr. Nancy Messonnier of the CDC had characterized that agency's approach as "cautious"—the United States would recognize the seriousness and rally as needed. So, at that point, my concern loomed larger for Africa, and was amplified further when an online video from a hospital in Wuhan made its way to me.

The video showed a hallway crowded with patients slumped in chairs. Some of the masked people leaned against the wall for support.

The camera didn't pan so much as zigzag while the Chinese doctor maneuvered her smartphone up the narrow corridor. My eye was drawn to two bodies wrapped in sheets lying on the floor amid the cluster of patients and staff. The doctor's colleagues, their face shields and other personal protective equipment in place, barely glanced at the lens as she captured the scene. They looked past her, as if at a harrowing future they could all see and hoped to survive. I tried to increase the volume, but there was no sound. My mind seamlessly filled that void, inserting the sounds from my past, sounds from other wards, other places of great sorrow. I had been here before. I had witnessed scenes like this across the globe, in HIV ravaged communities—when hospitals were full of people dying of AIDS before we had treatment or before we ensured treatment to those who needed it. I had lived this, and it was etched permanently in my brain: the unimaginable, devastating loss of mothers, fathers, children, grandparents, brothers, sisters.

Staring at my computer screen, I was horrified by the images from Wuhan, the suffering they portrayed, but also because they confirmed what I'd suspected for the last three weeks: Not only was the Chinese government underreporting the real numbers of the infected and dying in Wuhan and elsewhere, but the situation was definitely far more dire than most people outside that city realized. Up until now, I'd been only reading or hearing about the virus. Now it had been made visible by a courageous doctor sharing this video online.

The images amplified everything I'd been learning. All the aggregated data—the news articles, the scientific reports, the WHO status updates, the social media entries, my doubts about the reliability of the Chinese authorities' figures—everything coalesced for me, forming an image of suffering, frightened, overworked human beings and their suffering and dying patients. That video drew me in, and I sat there literally moving toward it, trying to make out details I couldn't see from a distance. It put me there, in a place where I didn't want to be and where none of those pictured wanted to be, either. This was just *one* scene in *one* particular location with those *specific* patients and doctors. I wondered how many other scenes in how many other places involving how many other human beings, in Hubei Province and elsewhere in China, were playing out like this—only, without video evidence. Worse, how many more times would all this be replicated elsewhere and multiplied many times over in the weeks and months ahead?

This question was still lingering when, a few days later, I saw images online of a large plot of cleared land. Dotting it were various pieces of earth-moving equipment, enough of them in various shapes and sizes that I briefly wondered if the photograph was of a manufacturing plant where the newly assembled machines were on display. Quickly, I learned that the machines were in Wuhan and that they were handling the first phase of preparatory work for the construction of a one-thousand-bed hospital to be completed in just ten days' time.

This move was straight out of the Chinese SARS playbook, when China implemented the same measure in Beijing. Sitting there watching video evidence of a need for an enormous prefabricated hospital drove home for me just how ominous things really were. The Chinese may not have been giving accurate data about the numbers of cases and deaths, but the rapid spread of this disease could be counted in other ways— including in how many Chinese workers were being employed to build new facilities to relieve the pressure on the existing, and impressive, Wuhan health service centers. You build a thousand-bed hospital in ten days only if you are experiencing unrelenting community spread of a highly contagious virus that has eluded your containment measures and is now causing serious illness on a massive scale.

In other words, you build a thousand-bed hospital in ten days only if you need a thousand-bed hospital *right now*.

I had to make sure Africa was prepared.

CHAPTER 2

Many Hats

For the last two decades, whenever I've heard of a new pathogen emerging, my first thought has been the risk to Africans. In the United States, we have many different layers of public health protection. Americans receive support from the states themselves and from the CDC, with thousands of public health officials in positions to shape and support the public response. Africa, though—whether you're talking about the entire health care system or just the number of health care providers—is in a fundamentally different position.

I love Africa and the people PEPFAR serves, but even with the substantial support that many nations, including the United States, had put into bolstering its health care system, sub-Saharan Africa was one of the most vulnerable parts of the world. Throughout the region, we were still confronting HIV, TB, and malaria, and any new threat to the region was a threat to the progress of our work and the very people we served. I have always believed that, in life and especially in public health, a proactive approach is best. The commerce and travel between China and Africa had increased logarithmically over the past two decades. China was in deep trouble; Africa needed to be prepared for this new threat.

The entire time I was tracking, observing, and calculating the possible dimensions of the novel coronavirus outbreak in China to other parts of the world, I was in the midst of planning for one of the most important events in my yearly calendar: PEPFAR's annual meeting to evaluate the progress in all the African programs we support. It was at this meeting that we'd help plan the effective use of the more than six billion dollars in U.S. taxpayer money we'd receive for the next year. The meeting was to be held over three weeks in Johannesburg from mid-February 2020 to the beginning of March, commencing on February 17. We'd been working long hours in preparation for it since late fall 2019, reviewing data and writing the "Technical Considerations" section we put together each year for our report "Country and Regional Operational Plan Guidance for All PEPFAR Countries," to ensure that the most recent science and data were available across all our programs.

PEPFAR represents what is best about humanity and, in particular, the United States. We enjoy a privileged position in the world and often lead the way in doing good, impactful work in global health. PEPFAR is an expression of the American people's compassion for those less fortunate than us. It is an ambitious and highly successful program. It has to be, to take on the many challenges of HIV/AIDS—its diagnosis, treatment, and, critically, its prevention. Working in deep partnership with impacted and concerned communities and governments, we partnered to take on directly the structural issues of inequality; human rights; gender-based violence; and the access and availability of services for young women, the LGBTQ community, and others marginalized by host governments. All these fall under the umbrella, some at the edges, of public health. At PEPFAR, we're always driving, above all, to make possible what many have viewed as impossible: controlling the HIV pandemic without a vaccine. I liken this to the ethos of the special operations community in the military. Among its slogans is the creed "These things we do, that others may live." You don't work for organizations like PEPFAR because the work is easy and the financial rewards are great. You do it because it is a calling.

I thrive on being around like-minded individuals whose dedication to making a difference is their true north. We all come from different backgrounds, cultures, and training, but we are united in our goals. We share a real sense of community and action. It's all about now, now, now—doing what we can to have an immediate and lasting impact on people's lives. That's been the ethos under my watch and since President George W. Bush and his administration worked to get PEPFAR funded and implemented. The president and the First Lady, Laura, had a shared world vision and a deep understanding that to whom much is given much will be required. This included addressing HIV/AIDS in Africa. At the same time, the legislation that helped create PEPFAR created the position within the State Department of global AIDS coordinator, the other hat I was wearing as ambassador-at-large.

Normally, PEPFAR's annual meeting would have commanded all my attention. But 2020 was only three weeks old, and already it was abnormal. Whenever I wasn't planning for the meeting or evaluating our programs on the ground through intensive data analysis, my mind was squarely on the exploding coronavirus cases in Asia and the implications for Africa.

To make sure that the people of Africa, particularly sub-Saharan Africa, weren't going to be caught flat-footed due to the WHO's lack of urgency on the new virus, I wrote to Erin Walsh at the National Security Council on January 20. Erin was the head of the Africa region at the NSC, but I wasn't looking to get any insight or intel from her; instead, my focus remained on what I knew. The threat of the virus jumping from its home ground in Asia to elsewhere in the world was very, very real.

I based my serious threat assessment on the number of cases and deaths and on another factor—one that was more behavioral than biological. While it was true that in 2002/3 the damage SARS did outside Asia was minimal, we now lived in a vastly different world. The SARS outbreak had originated in China, but it was spread primarily to other countries through outsiders coming into China and then returning to their home countries, bringing the virus with them. Back then, only a trickle of Chinese nationals traveled outside the country. Now, Chinese nationals traveled around the world in the millions. The timing of the outbreak also couldn't have been worse. It coincided with the Lunar New Year, when even more Chinese would be traveling both within and outside their country. The distribution and spread of the virus would be far greater and far quicker due to the undetected silent invasion I fundamentally believed was taking place across the globe.

The collective effort to track the novel coronavirus outbreak was made much easier in the third week of January, when a professor at Johns Hopkins Whiting School of Engineering unveiled a dashboard she and her graduate students had put together that allowed so many of us to track global cases in real time. The dashboard was wonderfully easy and accessible. With a frequency bordering on the compulsive, I would click over to it throughout the day and in the early hours of each morning, when the aggregated data for Asia appeared. It was through this data that I watched the alarming speed of the virus's advance. Watching SARS spread back in 2003 had been like watching a house fire consuming one home and seeing a single ember land on the roof of another house and slowly smolder for a bit. But watching the novel coronavirus outbreak was like watching satellite imagery of many, seemingly unrelated blazes popping up in different areas of the globe independent of an originating source.

I used the Hopkins dashboard's numbers to demonstrate to Erin Walsh why it was important to immediately hold a meeting with all African diplomats in Washington, where I would speak. They needed to be informed of the dangers posed to their areas of concern. I would be in Africa in three weeks, but that would be too late. We needed to put out an alert now, through the African Diplomatic Corps in DC.

Walsh agreed to schedule the meeting.

In that first week, once the Johns Hopkins data went live on January 22, I watched as the cases mounted: 314 . . . 581 . . . 846 . . . 1,320 . . . 2,014 . . . 2,798 . . . 4,593 . . . By the end of the month, in just nine days, they were up to 9,826. Going from 314 to 9,826 reported cases in nine days is a large increase. More worrying was the doubling—from 2,798 to 4,593 to 9,826—every twenty-four to forty-eight hours. And these were the *visible* cases; testing was not widespread yet. So, I believed there weren't nearly 10,000 cases but—based on my silent-spread arithmetic of three to ten times—potentially, *100,000* cases and growing, spreading unrelentingly, community by community.

The number of countries reporting cases also increased to twenty-four, including some in the Middle East, North America, and Europe. Rapid geographic spread was evident, and what had taken SARS weeks and months to travel was taking this novel coronavirus hours and days. In my mind, I kept seeing the bull's-eye on Africa.

BY THE LEAD-UP TO my meeting with the African Diplomatic Corps in DC, the fifteen-member WHO Emergency Committee still, apparently, hadn't seen enough evidence even to declare that the Wuhan cases constituted a public health emergency of international concern. This didn't make any sense—unless you understood the bureaucratic logic of the

World Health Organization. The WHO has historically never wanted to appear wrong or to seem to rush to judgment, an instinct that, in this case, was already costing lives. My responsibility was to the African countries, and by meeting with their ambassadors and having them relay my message to their capitals and on to their public health officials, I'd be doing what the WHO wasn't—giving them adequate time to prepare for a worst-case (but reality-based) scenario.

I wasn't just raising the alert for the African countries; I was providing them with solutions and options. We at PEPFAR had spent the last nearly two decades investing in all aspects of the health systems in sub-Saharan Africa, not only to address HIV/AIDS, TB, and malaria, but also to ensure that those same systems would be there in the event of the next pandemic. I planned to let those critical laboratories, the broad array of health personnel we were funding for the HIV response—everything and everyone at their disposal to assist in the response to this new pathogen. I believed that the United States, as it so often did, could serve as a model for the proper procedures to implement to combat this outbreak. To act as a guide for various public health operations on the African continent, I asked two of the people primarily responsible for shaping the U.S. domestic response, Dr. Anthony Fauci, the director of the National Institute of Allergy and Infectious Diseases (NIAID), and Dr. Robert Redfield, head of the CDC, to share their expertise with the African representatives. Whatever plan was in place for the resource-rich United States, I wanted it replicated in Africa.

Getting Fauci and Redfield to speak was an easy ask. I'd known Tony Fauci for more than three decades. I consider him one of my mentors. More than anything else, I learned from Tony the importance of listening and then adjusting your responses based on listening. He really heard what his patients had to say. I can't tell you how rare that is. He was super smart about disease pathology, but it was his decency and his empathic nature that truly set him apart. He spent many long hours doing research and then evening rounds, and even if we were working a twelve-hour day, a fifteen- or sixteen-hour day, Tony listened, and his patients felt his concern for them. Combine that with his mission focus, and you have as formidable a presence as there could be without his ever coming across as intimidating, except to us junior members of the team. Tony combined the attributes I most admired: He was a brilliant thinker and a highly capable communicator. He could move easily between the world of science, with its many introverts, and the public realm, where extroversion makes it more likely you will be listened to.

Along with being the director of NIAID, Tony was the chief medical advisor to President Trump, but more than that, he was someone I trusted implicitly to help shape America's response to the novel coronavirus. He agreed to present at the African Diplomatic Corps meeting to let the African ambassadors know what the United States was doing to prepare for the novel coronavirus regarding treatments and vaccines.

Bob Redfield, the director of the CDC, also quickly answered my request for help. Bob and I also had a long history of working together. As was the case with Tony, Bob and I both worked at Walter Reed Army Medical Center in the 1980s and with renowned NIH researcher Dr. Robert Gallo, who co-discovered the HIV virus. Like Tony, Bob was an innovative, out-of-the-box thinker—something I had never been before I met them. I like to have data, to question that data, and to push the data. I worked in support of both Tony and Bob, attempting to design clinical trials and evaluate the results of these two men's leaps of intuition. Bob and I worked in the trenches together, and we developed the kind of we've-got-each-other's-back mentality that comes only from being tested under the most challenging circumstances. We were also a good pair—Bob was focused on HIV from a virology perspective, while I focused on the body's response to the virus, the immunity perspective.

For years, long after we spent hours in the lab or on the ward, Bob and I, joined by another colleague, Dr. Craig Wright, would sit in the residents and fellows' room late into the night running through the possible causes of the disease that had suppressed our patients' immune response. Those were the darkest days of my medical career, and I was fortunate to have shared them with Bob and Craig. Not only were we haunted by the gaunt faces and the ravaged bodies of our patients, but we had to deal with the stigma attached to what was then perceived as a "gay disease."

I can still hear a surgeon, the general in charge of Walter Reed, berating us because he believed we ran the risk of having his hospital known as "an AIDS hospital." He was concerned, he said, that active duty soldiers and retirees with health problems wouldn't want to come there because of that reputation and fear of infection. Bob and I were united in the fight against this kind of thinking. That some people stigmatized AIDS patients made all those hours I spent poring over the literature searching for a cause even more worthwhile. Unraveling the mystery and then eradicating the disease became a lifelong passion. For decades, Bob and I would both do anything we could to prevent even one more person from suffering the way so many of our patients already had.

Back then, Bob was willing to be aggressive and take chances. For AIDS in the 1980s, the stakes were high, and the field was relatively wide open. Bob brought both brilliant thinking and fearlessness to problem solving. As for me, I worked as a kind of translator, the strategist, the get-things-done-behind-the-scenes implementer.

Since leaving Walter Reed, Bob and I had remained in contact. Working with him now, I had every confidence, based on past performance, that whatever path the virus took, the United States and the CDC would be on top of the situation. We weren't going to talk about containment. We were going to discuss mitigation efforts—early testing, available therapeutic measures to back up diagnostics, preventative measures. We had one advantage over those in China: They had faced a sneak attack. We had systems in place within our public health care services to deal with viral outbreaks. We had the CDC and the NIH. The scientists working in the United States were among the best in the world.

ON JANUARY 28, AFTER meeting with Erin Walsh to solidify the planning and schedule for the upcoming African Diplomatic Corps State Department meeting, I received a text from Yen Pottinger. Aside from being the wife of my friend Matt, the deputy national security advisor, Yen was also a former colleague at the CDC and a trusted friend and neighbor.

Like me, Yen was among the many outside researchers who were now tracking the virus. A brilliant woman, she had recently played a key role in developing a new assay (test) for diagnosing whether an HIV infection in someone was recent or old. In our three years working together at the CDC, I had marveled at her abilities in the lab. As early as mid-January, Yen and I had been in communication about the outbreak in China. As events unfolded, we shared whatever insights, information, and anxiety we had.

Her husband, Matt Pottinger, was one of the good ones in the Trump White House. A former journalist turned highly-decorated U.S. Ma-

rine who served as an intelligence officer for part of his time, Matt
had deep experience in China (including during the 2002–2003 SARS
outbreak there) and was fluent in Mandarin. Matt took a position in
the National Security Council in the earliest stage of the Trump administration, while still serving in the Marine Reserves. Unlike many in
the White House, particularly in the security services, Matt managed
to survive the rash of dismissals, scandals, and changing of the guard
that took place over the course of the Trump administration, and in
September 2019, he was appointed to the post of U.S. deputy national
security advisor.

In November 2019, shortly after settling into his new role, Matt had
communicated to me that he wanted me to work at the White House in
some capacity as a public health security advisor. I told Matt I appreciated his thinking of me, but I was declining. I needed to remain focused
on controlling HIV around the globe.

Of course, Matt respected my decision, but he reserved the right to
ask me to reconsider.

Since he'd made the job offer, I'd had so much on my plate that I'd
quietly filed it away, never thinking I would reconsider a job inside the
White House. Off and on in early January 2020, I'd share my thoughts
with Matt: about the larger picture, about how the virus response in
the United States should go, and about how the White House could better manage its messaging around the virus—usually mentioning these
things through Yen. She was happy to assist in any way possible.

Yen knew I would be on the White House complex for my meeting
with Erin Walsh, and the text she sent me said that Matt had a "propo-
sition" for me. She didn't know any of the details, but Matt had apolo-
gized for the short notice and said he hoped we could meet face-to-face.
Yen arranged so that I could meet him in the West Wing, and once we
were both there, Matt got to the point quickly.

He offered me the position of White House spokesperson on the
virus.

I told him I wasn't interested because my skill set didn't match the
job title. In truth, I'd never been particularly media savvy—not because
I didn't understand the game, but because I'd never had the patience for
playing it. While I'd done plenty of press events when asked to do so
by leaders in government, I'd never sought out the media, and I didn't
need or want to be out front. I prefer doing things rather than talking

about doing them. I have always felt that if you do the right thing for the right reasons, you don't need to be validated in public.

Of course, I understood quite well the necessity of messaging and how important a role rhetoric plays in shaping and implementing policy, but there were others more suited to that role than I. I offered to provide Matt with the names of better candidates, while continuing to give him some unofficial guidance about this virus through Yen. Matt accepted my declining the post, took me up on the list of other candidates, but he again reserved the right, he said, to ask me to reconsider later.

Along with feeling a mismatch between my skills and the spokesperson role, I had another reason for saying no. For nearly forty years, throughout the course of my career, whether in the U.S. Army or as a public servant, I'd chosen not to align myself with a specific political party. I had been careful to keep my political leanings personal. Instead, as long ago as the Carter presidency, I'd worked as a public servant across Republican and Democratic administrations. I had served the people—not a party or administration. I was not a DC or a White House insider who spent her hours plotting how to ingratiate herself with the right people to make sure she had the president's ear. And I didn't want to spend my time cracking the code to gain entry, only to be dismissed because I'd worked for one administration and not another. I also didn't want to be merely a mouthpiece, having to restate what senior officials in the White House had decided or believed. In doing so, I would be seen, by implication, as affiliated with that party or administration.

What I *wanted* to do was define the actions being taken on the emerging virus based on the data. In my years of working with high-level leaders around the world, I had wielded metrics to move minds and formulate policies, standing behind data to justify the changes and encouraging political figures to make the hard decisions needed to save lives—even if those decisions didn't help them politically. A number of times, I'd been able to move world leaders who didn't have their people's best interests at heart.

I wasn't certain I could move President Trump. It would take someone with much more political savvy than I to do so.

IN MY BACK-CHANNEL COMMUNICATIONS with Matt, I pulled together all the publicly available data I'd been compiling and analyzing, connecting the dots to create a concerning picture, and sent it to Yen to forward to him. For privacy and security reasons, I wasn't ready to use official White House email. I trusted that Matt would share the information with those who needed it and not reveal that I was his source. In communicating with Matt, I had ensured they would have everything I was seeing, to use during White House meetings. I let Yen know that the earliest data available showed that the Wuhan outbreak and subsequent spread would be, at a minimum, ten times what SARS had been.

I also passed along communication strategies. At this early stage in the crisis, communication would be key. If we were truly going to engage in a campaign to mitigate the spread of the virus, and not rely solely on containment, then people were going to have to change their behavior, as they were already doing in the Asia region. Accounting for and then containing those who were symptomatic, rather than definitively determining them through testing, would be inadequate. With so many people already silently infected, it would be nearly impossible to mitigate effectively without other efforts, ones that involved individual behavioral changes.

In any health crisis, it is crucial to work at the personal behavior level. With HIV/AIDS, this meant convincing asymptomatic people to get tested, to seek treatment if they were HIV-positive, and to take preventative measures, including wearing condoms; or to employ other pre-exposure prophylaxis (PrEP) if they were negative. Prevention and knowledge were crucial to slowing the spread of HIV. We expended enormous amounts of time, energy, and money devising campaigns to get this message across.

We also understood that people needed to have information tailored to their age and sex. We knew that the platform used to convey the message, and who delivered that message, was also important to changing behavior and getting people to act. This level of detailed analysis and action-oriented messaging meant relying on those outside the public health field to assist us. In the private sector, companies hire advertising and public relations firms who specialize in communicating effectively to influence behavior (get consumers to purchase goods), and we relied on their expertise to help us get those at risk for HIV to "buy into" our message around safe sex practices and preventing infection. Those public relations and advertising firms taught us crucial lessons. They conducted focus groups that helped us hear what our target audience

thought, believed, and felt. We had to learn to listen better to better address the needs of the people we wanted to help. We were scientists, but
we weren't behavioral scientists, and accommodating human action and
cultural perception is important in all things to do with public health.

Well before I came onto the task force, I knew the government agencies would need to do the same thing to have a similar effect on the spread
of this novel coronavirus. The most obvious parallel with the HIV/AIDS
example was the message of wearing masks. Because the novel coronavirus was airborne, wearing a mask limited the amount of aerosols or
droplets an infected person could spread and reduced the number of
these particles others could inhale. One of the things that had kept the
SARS case fatality rate from being worse was that, in Asia, the population (young and old alike) adopted the wearing of masks routinely, to
protect themselves from air pollution and infections in crowded indoor
and outdoor spaces when social distancing wasn't possible. Masking
was a normal behavior. Masks saved lives. Masks were good.

In the United States, however, we didn't have that same history of
successful mitigation fresh in our minds. Independently from me, Matt
became the self-appointed White House prophet of mask wearing. Having also been in Asia during the SARS epidemic, he'd seen how the
Chinese people and those throughout Asia had adopted wearing masks
as an antidote to the government's initially flat-footed response to the
outbreak. He and I had also seen that, across Asia, N95 respirators
(those masks that form a seal around the nose and mouth) weren't readily available or used outside hospitals. For this reason, people regularly
wore cloth masks. This distinction would loom large later on.

At the White House, Matt's message about wearing masks to prevent
silent spread had fallen on deaf ears. The consensus there, and among
some in the United States, seemed to be that masks weren't necessary
because people were at low risk of getting the disease. The other reason wearing masks didn't gain traction among Americans was that, besides not having a history of success to fall back on, masks required
the wearer to make multiple commitments—to purchase them, to keep
them in various locations, to remember to put them on, to deal with the
physical and mental discomfort, to get over the stigma attached to wearing them. Simply put, wearing a mask required more effort than most
Americans were accustomed to putting in. Change is hard. Behavioral
change, and remembering that change later, is really hard.

Another strategy that suppressed the 2003 SARS outbreak was social distancing guidelines—limiting how close you got to other people, especially indoors, but also how frequently you gathered with others indoors and, critically, reducing the number of people with whom you interacted by reducing the frequency and size of gatherings. Along with wearing masks, these behavioral changes had the greatest effect on mitigating the SARS epidemic by limiting community spread and not letting the virus claim more lives.

In those early days of the novel coronavirus, when few were acknowledging the role of silent spread, I knew it would be extremely difficult to begin a public campaign touting these three measures. With no clear-cut numbers to convince people of an obvious need for them, who would engage in behavioral changes as drastic as wearing masks and reducing the size and frequency of gatherings? This was not new: Back in the 1980s, even when it was clear that many, many people were dying of AIDS, it was still difficult to get the message across about the use of condoms and the other behavioral changes needed to decrease the spread of HIV. Similarly, if you hadn't yet seen anyone in your family or your community getting sick from the coronavirus, or if the number of those infected was very small, it was far too easy to shrug and say, "I don't see the need."

This was a variant of the "Not in My Backyard" phenomenon. Unless this virus was actively affecting people's lives or the life of someone connected to them, getting people to adopt a precaution above the most basic level would always prove to be difficult. Also, here in the United States, we are not particularly attuned to the idea of prevention, especially as it pertains to our own health and even when it comes to a virulent disease. Even "CDC," the abbreviation for our most trusted health agency, the Centers for Disease Control and Prevention, leaves off the *P*, which represents the most crucial word in its name. This says a great deal about what we value. Prevention, in many people's minds, takes far more effort than treatment. While we're sometimes reluctant to do the latter, most people loathe doing the former.

Also, good and effective preventative interventions are often invisible. If you take care of yourself and don't get sick, there are no real markers you can look back on to say, "That was a near miss." It is extremely difficult to prove a negative, to show that a "non-event" happened. Individuals' tendency to place treatment over prevention is one of the greatest challenges our public health officials face all the time.

One way to convince people of the need for behavioral change is to develop a consistent messaging strategy. When public health measures are centered on behavioral change as the primary intervention, communication is key, and consistent communication is critical. To date, the statements on the novel coronavirus from health officials, and from President Trump himself, had centered on containment and not prevention of community spread through mitigation. This was the wrong approach. This focus on containment would lead the American public to believe that the virus was primarily outside the United States and that, if it crossed our borders, it could be stopped immediately. Comparing the new virus to SARS and MERS would also feed the American people the false expectation that it would be limited in spread, as was the case with those viruses. Meanwhile, comparing it to ordinary seasonal flu would give the public the sense that not only was this coronavirus fairly harmless, but that it could be treated like the flu—which is diagnosed by its symptoms and which doesn't deeply impact families and communities. It trivialized the virus and its threat.

I communicated to Matt that we needed to break this chain linking the novel coronavirus to SARS and the seasonal flu and reprioritize testing, full mitigation, mask wearing, improved hygiene, and more social isolation. To that end, a coordinated, concise, and carefully worded series of communications would be necessary to get the messaging right both within the White House and, more crucially, to the American people. The private sector had taught me the importance of message segmentation based on demographics. Communication must constantly evolve based on continual feedback from the community you are addressing.

I was giving Matt a lot of ideas, and I was glad to be able to help him, and the country, but I had one lingering regret. I texted Yen: "Every time I turn something down, I feel like I am making yours and your children's lives more difficult, as Matt has to work more. But I am trying to support behind the scenes."

She responded: "That's funny, because it's not true. Don't feel bad. Our lives will continue to be difficult until he finds a new job. He thinks you should take over Azar, Fauci, and Redfield's jobs, because you're such a better leader than they are. He has been underwhelmed thus far."

Though I would come to see in Matt's comment the sense of foreboding it carried, I dismissed it in that moment, taking the statement as hyperbole. Yen was a sympathetic spouse, expressing Matt's frustration, the depth of his worry, and the inaction he was seeing out of the Department of Health and Human Services, which includes both the NIH and the CDC.

Though Matt was unnerved by what he was seeing from the federal public health officials, I trusted Bob and Tony. This trust was reinforced at my meeting on January 31 with the African Diplomatic Corps. Everything Drs. Fauci and Redfield said about their approach made sense based on the information available to me at that point. While President Trump had casually dismissed the coronavirus's potential threat to the country, there were good people with great minds and effective strategies at work on it in the United States. Two of them had just shared further evidence of that: at the meeting with the African Diplomatic Corps, Tony presented on the work being done to accelerate therapeutic treatments and vaccines to combat the potential pandemic, and Bob talked about the work the CDC was doing to test and confirm cases. Neither of them spoke specifically of asymptomatic silent spread or of the role testing should play in the response, but I didn't read too much into this omission. And I didn't have time to speak to either of them specifically about testing, as they had time only to run into the meeting, make their presentations, and then run out. Later, it would become clear they were working on the China travel ban that day.

As the first month of 2020 came to a close, I believed that those with the most acute vision with regard to the virus would prevail and that the United States was in good hands. I could sleep well knowing that the full force of what the nation had at its disposal—the public health agencies, researchers, laboratories, and medical professionals—was in place. I wanted to be certain that I could say the same for Africa. For the foreseeable future, I would ally my interests with where my duties called me. As far as I could assess, my focus on Africa was not misplaced. The threat to an already vulnerable region was far greater than it was to the United States, with its expansive, and expensive, medical industrial complex.

Still, it was hard to shake that trained sense of worry that came from experience. I found myself thinking back to a call I had taken two days earlier, on January 29, from my deputy, Dr. Angeli Achrekar. Right away, something in her voice quickened my pulse. She had called me from beneath a stairwell at the airport in Addis Ababa, Ethiopia. She confirmed what I'd suspected: A tide of Lunar New Year celebrants from China inundated the lounges and departure areas.

The levy that had kept SARS largely isolated to Asia almost two decades ago hadn't just broken; it didn't even exist anymore.

CHAPTER 3

That Others Might Live

Like many people, I sometimes give in to the temptation to anthropo-morphize a virus. It can become an "opponent," a "villain," the "en-emy" in a war. The media will often sensationalize it, call its spread "remorseless," refer to it as "evil." But, of course, a virus doesn't have a mind or a morality. A virus simply does what it has evolved to do: reproduce and thus go on existing. Viruses uniquely need a host—actually, a host *cell*—to reproduce. Viruses lack all the machinery to self-replicate, so they need to invade and take over our cells. A virus must find new hosts to invade each and every day. The viruses that are most accomplished at this move through the air invisibly. Viruses with the highest transmission potential are those that infect their hosts but don't immediately kill them. Instead, they produce minimal symptoms, allowing the host to live long enough for the virus to fully replicate and be shed, through secretions or droplets, or to remain suspended, invisible, in the air, silently present, awaiting a new host. Successful vi-ruses produce a chain of transmission, spreading from person to person, minute by minute and day by day. With cases on the rise, this novel coronavirus was clearly doing its job. It was less clear, though, if we were doing ours.

As January ended, a flurry of events made me realize that officials in the United States were starting to take the novel coronavirus, and the evident disease it was causing, more seriously. Most notable was the official announcement that the White House Coronavirus Task Force had been created and would be headed by Alex Azar. Still, many of the efforts the task force appeared to be focusing on—various travel restrictions, symptomatic screening, and voluntary quarantine—belonged under the same heading as before: containment. As the month drew to a close, the primary belief that we could prevent the virus from coming ashore in the United States was still very much alive, still replicating in the minds of those who should have known that containment had already failed.

I assumed that Bob and Tony, both of whom were also task force members, were involved in these travel restriction and quarantine orders. The screening of incoming passengers would now take place at twenty major airports around the country. Also by the end of January, the CDC had issued a Level 3 travel alert, its highest, recommending that all travelers avoid nonessential trips to China. This was an important move, but it came too late.

Cases rose rapidly everywhere in early February, so much faster than with SARS. Many people like round numbers and likely paid attention to the fact that, on February 5, Wuhan's health officials announced that they had reached more than ten thousand "confirmed" cases. Given China's spotty record on honest reporting and the likelihood that its consideration of silent spread was minimal, the real number was, of course, much higher. My PEPFAR data team at the State Department and I talked about whether it was actually one hundred thousand or more. Looking back, I see our number was almost certainly closer to reality than theirs.

While I continued to believe the worldwide numbers showed that silent spread was fueling this outbreak, hard evidence of this was still missing. Without aggressive testing, no country would know how widespread the virus was already. As much as I believed we were headed toward a global pandemic, a silent invasion of the virus across the globe, I lacked the kind of evidence I could share to get countries, including the United States, to take aggressive proactive measures. So, I kept observing, focused on ensuring that Africa would be alerted, and waiting for data that was more defensible.

I wanted to prove that silent spread was at work in this outbreak. The cruise ship the *Diamond Princess* would provide the evidence. On February 7, I learned that the liner had experienced a viral outbreak and was currently at port in Japan. Some passengers were reported to have pneumonia-like symptoms. Japanese medical authorities investigated thirty-one suspected cases of the novel coronavirus and confirmed the presence of the virus in ten passengers. On February 3, Japanese authorities promptly issued orders that the ship's passengers be held on board in isolation for fourteen days.

In medical science, we often create experiments and computer simulations to replicate real-world conditions and outcomes. Right away, I saw this tragic situation for what it was: If no one were allowed off, the cruise ship would become a kind of human petri dish, essentially a closed environment in which the virus could reproduce, spreading (or not) from person to person within the confines of the vessel. I am sure that, in this moment, many experts believed that locking down the *Diamond Princess* would contain the virus. Unfortunately for those aboard her, they were now subjects in an ad hoc laboratory experiment— except, tragically, the stakes were much higher.

An eighty-year-old was among the original passengers. He had been aboard briefly and then disembarked back in Hong Kong. Ten days earlier, he had been in mainland China but far from Wuhan. Later, he would tell officials that he had a mild cough for several days. He was mildly symptomatic but throughout the course of illness, which was ten days to two weeks, he was actively spreading the virus. On February 1, five days after he disembarked the cruise ship in Hong Kong, he became sicker, went to the hospital, was tested, and became a confirmed case of novel coronavirus infection. He likely thought that what was hospitalizing and killing people was something severe and wasn't at all like what he was experiencing.

On February 1, Japanese health ministry officials had informed the ship's owners that an infected passenger had been aboard the *Diamond Princess*. Either in concert with the ministry officials or on their own, management determined that with so few symptomatic individuals present the virus presented a low risk to the health and safety of their ship's passengers and employees. They repeated the mistake of not accounting for the asymptomatic individuals already spreading the unseen virus through the air.

However, the ship was placed in quarantine and would not be allowed to dock, and her passengers were prevented from disembarking. The captain informed the passengers and crew about the one passenger who had contracted the virus. He also informed them that they would be quarenteed for the next fourteen days. He asked anyone who had gotten sick while on board to report to the medical facility. There, they would have their temperature taken and answer some symptom-based general health questions.

That same night, Japanese health officials went door-to-door on the ship to identify those they believed needed to be tested for the virus, basing their decisions on the passengers' either being symptomatic or sharing a room with someone who was. (Even this early in the crisis, Japanese research scientists had developed a test for the disease based on finding traces of its RNA in the nose.) They took 253 swabs from individuals based on the incomplete criteria they'd established.

Surprisingly, normal social functions and interactions among passengers and the crew were allowed to continue throughout the next day. The younger crew members also circulated among the passengers.

Unfortunately, if there was silent asymptomatic and presymptomatic spread, many of them would already have been exposed. Later reports indicated that the captain's announcement had the opposite of its intended effect: passengers congregated, talking, no doubt, about the situation they had found themselves in. Some may have worn masks; others did not.

The Japanese health officials believed they had moved very quickly. The cruise ship's management team believed they were enacting all the right protocols. Their initial, symptom-based screening had discovered only 31 people, among more than 2,600 passengers, who were believed to be infected with the novel coronavirus. The health officials viewed this as a low-risk-of-further-transmission scenario and believed the virus was now contained. They were, in their estimation, being properly judicious—protecting those on board with proactive isolation and those on shore by restricting who could leave the ship. Still, it was dismaying for me to see: If the *Diamond Princess* case was any indication, public health institutions around the globe, likely taking the lead from the WHO and the CDC, remained firmly focused on detecting the presence of the virus through symptoms, not testing.

Simultaneous to the news reports about the *Diamond Princess*, her passengers took to social media to describe what was going on aboard the cruise ship. Passengers were told to monitor their own temperature.

If it increased to above normal, they were to contact the medical services personnel on board the ship. Some passengers were allowed on deck and were required to wear masks and to keep six feet away from one another. These protocols were in line with best practices, but to what degree other passengers stayed in their own rooms and stringently followed the guidelines was difficult to assess.

Meanwhile, with all the focus on the passengers, I wondered what was happening with the crew. On cruise ships, the crew is usually younger than the passengers, and they usually live in small, shared spaces while on board. The crew needed to eat and drink, and they would have to remove their masks to do so. I also believed the younger crew was more likely to have mild or asymptomatic disease and were likely silently passing the virus to one another and, potentially, to the passengers. In my mind, the *Diamond Princess* was a microcosm of what could happen in the wider world: the aged and vulnerable being protected and taking measures, while the young were asked to monitor themselves for symptoms. This created the false impression not only that if you didn't have symptoms, you didn't have the virus, but that younger people seemed unaffected by the viral spread—a fallacy that would help the virus spread widely as time went on. Additionally, with no emphasis on testing everyone on board every day, the virus would likely live out its biological imperative largely unimpeded.

I believed the *Diamond Princess* could still chart a successful path forward. Going back to the 2002–2003 SARS observations and my general understanding of the willingness of Asian populations to act upon government health regulations, including wearing masks, I felt that the *Diamond Princess* could fare better if those on board complied with the precautionary measures. If the Japanese officials on board achieved a high degree of compliance for masking and social distancing, if they made the silent spread visible by testing all crew and passengers daily and isolating the newly infected, they just might successfully mitigate against ongoing spread.

And so, the *Diamond Princess* became my evidence base. Data from its passengers could prove two things: that this was a highly transmissible virus due to its silent spread and that preventative measures, if enacted and complied with, could slow the spread. I started to track the ship's case numbers as closely as those from around the globe. If I was right about Wuhan and its enormous silent spread transmission, I surmised that we'd witness the same scenario on the *Diamond Princess*. Unless mask and testing mitigations were followed rigidly, cases would dramatically increase.

Amid the unfolding drama aboard the *Diamond Princess*, one development made talking and writing about this novel coronavirus a bit easier. On February 11, the WHO stated that scientists had more properly identified the characteristics of the virus. It was now called SARS-CoV-2, and the disease it produced was known as Covid-19. I once heard that to name a thing is to exert some power over it. I hoped that would be true in this case, but unfortunately, naming it merely gave our tormentor a new identity.

As I made final preparations for my PEPFAR conference in South Africa, which would commence on February 17, the confirmed positive cases on the *Diamond Princess* rose. Even during my twenty-four hours en route to Johannesburg, February 14–15, the cases of Covid-19 rose: 10 . . . 61 . . . 135 . . . 174 . . . 218 . . . 285 . . . Once I arrived in Johannesburg, I continued to monitor the situation. The numbers continued to ascend from the first day of my scheduled three weeks there: . . . 355 . . . 454 . . . 542 . . . 621 . . . 634 . . . 691 . . .

The numbers were emphatic and clear. And these were just the *visible* cases: the Japanese continued to restrict testing primarily to symptomatic passengers and those in close contact with them, while the crew most likely had significant asymptomatic infection. Seeing spread like this in spite of the containment measures in place meant the virus must have been circulating widely, and spreading asymptomatically, before the quarantine was put in place on February 4. To those who would listen, silent spread was announcing its presence.

From thousands of miles away, the *Diamond Princess* was the wake-up call I was sure would spur the CDC and other public health agencies to further action. We were all now able to see the results of this floating laboratory experiment writ large. The documented spread was intense, going from 1 to 691 confirmed positives in only three weeks—and those were just the people with symptoms. If they had been testing more widely, among asymptomatic people, the real number could be two to three times greater: 1,200 to 1,800 infections. Despite the measures the Japanese health ministry had put in place, this explosive growth was clear evidence of silent spread.

Containment was not working—not aboard the *Diamond Princess* and not in other parts of the world.

In the United States, this wasn't the story the Trump administration was telling. The *Diamond Princess* cases failed to shift the rhetorical focus on containment, and neither the president's nor Health and Human Services' actions were proportionate to the threat the cruise ship's living experiment had exposed. While the president remained focused on the United States' low case count, HHS had been preoccupied with limiting travel. They only screened by symptoms the thousands of Americans and permanent residents, and the one hundred thousand American cruise ship passengers returning to the United States from around the globe. Instead of tamping down fear, as the president's remarks seemed designed to do, the administration and federal public health officials should have been warning us about what the *Diamond Princess* had shown. Instead, their language stayed much as it had been.

Containment. Containment. Containment.

Unfortunately, rhetoric would harden into inaction.

I hadn't spoken to anyone on the White House Coronavirus Task Force since my January 31 African ambassador meeting with Bob and Tony. I tried to silence the alarm bells going off in my head by reminding myself that the confidence I had in their abilities wasn't misplaced. I felt sure that they almost certainly had access to data I didn't, data that had perhaps lessened their level of concern. But I struggled to reconcile the concern I felt with the lack of public action, especially in light of the demonstrable threat of silent spread. Where was the testing? Where was the warning that this virus could be spread by those without symptoms, so Americans could be on guard? Where was the full-throated preparation we at PEPFAR were doing to ensure the safety of those in Africa?

In the coming weeks, it would become even clearer that the *Diamond Princess* was a warning to us all, a show of force by a virus demonstrating just what it was capable of. Unfortunately, warnings work only if people are willing to act on them.

AT LEAST THE JAPANESE had a reliable diagnostic test. The same could not be said for us.

When Bob Redfield spoke at my African Diplomatic Corps meeting on the last day of January, all seemed to be moving along nicely toward the development of a diagnostic tool that could be rolled out quickly and in large numbers. Over the first half of February, though, it became clear that neither was true.

On February 8, we learned that contaminated testing materials had caused the tests the federal agencies had created to result in false positives. This production issue was resolved, but the delay was costly. We should have been on the brink of having millions of test kits manufactured and shipped to the states. Now, rather than ensuring that those kits were deployed at the first sign of an outbreak, the CDC had to validate a new test. This general disregard for and mismanagement of testing, as compared with the rapid focus on vaccines at the NIH, would continue to haunt us throughout the spring and into the summer.

We were no longer days behind; we were weeks behind. Early in a pandemic, days and sometimes even hours can be crucial. You are attempting to slow the spread before the curve becomes exponential. Basically, losing one week early on feels more like losing four weeks. Being four to six weeks behind at this initial stage meant that, by the time the federal agencies had caught up to where they were before the testing debacle, they were actually sixteen to twenty-four weeks behind the progress of the viral spread.

Though the overall number of confirmed cases in the United States was still relatively small when news of the faulty tests first broke in early February, the failure significantly delayed the states in their ability to test. While even under ideal circumstances it would have been difficult to get the state public health agencies to test widely at this early juncture, their ability to use tests merely in so-called surveillance mode (a common practice to see what viruses are circulating in a population) was severely compromised. At precisely the moment when we needed to be testing widely to see where and how the virus was spreading, we were flying almost totally blind. Tests should have been broadly available; instead, they became a precious commodity.

While the failure of the tests was deeply disturbing, I was also concerned about some of the decision making at the CDC. When the tests were designed back in January, they were built mainly for use and analysis by our country's Public Health Laboratories and their often unique equipment. This made the kind of rapid, widespread testing of symptomatic and asymptomatic people we needed quite hard, if not impossible. Each state has at least one Public Health Laboratory. As a part of the global health security focus, many of these critical laboratories were expanded significantly with CDC pandemic preparedness funding prior to Covid-19. These labs are more like research facilities than diagnostic labs that patients would access. In fact, the general public is probably unaware of the function or location of these sites. The CDC uses the tests performed and analyzed at these labs to answer specific, and often isolated public health questions. One of their most common basic functions is to track the strains of seasonal flu circulating in a state to see how they compare to that year's flu vaccine. From the thousands of symptomatic flu cases in a state, a Public Health Laboratory may test a few hundred collected viral samples over the course of flu season. Often, the results of the test aren't returned to the hospital, clinic, or patient.

These CDC Public Health Labs are an important but niche research asset. Because they were designed for research or limited surveillance, they are not high-throughput (i.e., fast-processing, high-volume) clinical or commercial labs. Worse, most have been significantly understaffed for decades. As a result of all this, the Public Health Labs were quite unsuited for testing on the scale needed for this current crisis. And yet, the CDC had created a test that could be processed only at these low-throughput facilities, with their somewhat slower, older, more labor-intensive, lower-tech platforms and equipment. In comparison to higher-throughput, commercial labs, the CDC versions could process fifty to one hundred samples every four hours, far less than the five hundred to one thousand samples every four hours processed at the clinical and private diagnostic labs. If you think of them like software, the operating system at a Public Health Laboratory would be version 1.0 when the vast majority of clinical and hospital laboratories would be version 10.5.

Almost no hospital, clinic, emergency room, urgent care clinic, or commercial lab uses the unique platforms the Public Health Laboratories do. The vast majority of the laboratory capability in the United States lies with private hospitals, clinics, and commercial labs. So, designing a test that relied primarily on the Public Health Labs meant that the bulk of U.S. processing equipment was sitting idle. With the CDC limited to about 0.1 percent of the country's laboratory capacity, all those high-throughput systems, the 99.9 percent of our national laboratory capacity, were sidelined when they were needed most.

The only reasonable explanation for this oversight was that the CDC had never envisioned using its test on the scale necessary to test widely and identify silent spread from the asymptomatic and presymptomatic cases. If you test only the symptomatic, you don't have to account for the three- to tenfold increase that silent spread accounted for. In the case of SARS-CoV-2, you might be able to do significant and impactful research somewhere down the line using these facilities, but you wouldn't be able to test proactively to identify where the virus was in the present moment. In other words, you could look into the past, but you couldn't effectively view what was happening in the present enough to shape the immediate future.

A more tactical approach to testing development and production in an emergent crisis would have been to work with all the assets at Health and Human Services and the commercial manufacturers to create a test format that could be processed by university research, public, and private labs. This would have ensured the rapid development of tests for all our preexisting platforms across the United States. From where I was sitting, on the outside, I didn't know whether this was because of some failure on the part of either the CDC or another of the departments within HHS, like the Office of the Assistant Secretary for Preparedness and Response (ASPR). The bottom line was that when it came to testing, cooperation with the private sector wasn't happening.

The testing bottleneck that hampered our earliest response to the pandemic and continued for far too long had its roots in this early decision not to engage the private sector. That choice reflected the CDC's belief that the virus it was then facing was like one it had encountered in the past. The United States had never needed to test on this scale before, and the CDC believed that it wouldn't be necessary in this crisis, either. Unfortunately, all the test kits in the world are useless if you don't have enough capacity to process them.

The problems didn't end there. In declaring a public health emergency on January 31, the United States had initiated what are called "emergency use authorizations." These allow the Food and Drug Administration (FDA) to take emergency measures to protect the nation's public health in the face of a chemical, biological, radiological, or nuclear threat. Infectious diseases fall under that umbrella of possible dangers. The point of the emergency use authorization is to help make medical countermeasures (vaccines, drugs, biological therapeutic products, decontamination systems, and a host of other devices and equipment) readily available more quickly to serve the public's needs. The system is also designed to speed the usual labyrinthine, but necessary, FDA approval process for vaccines and medicines. Still, just how much more streamlined the other approvals would be under emergency use was difficult to gauge. Having any one agency managing the approval process for as many measures as might be needed in a pandemic would put tremendous pressure on a small number of staff. This would be particularly true when it came to reviewing the new testing kits.

An FDA team worked diligently to evaluate and approve the CDC's initial test. This was the good news. However, central to its approval was a statement that declared the tests approved for use with *symptomatic* people. This one statement created an immediate and terrible restriction. So, not only had the initial CDC test experienced false positives and been developed in a format that would limit its processing primarily to the Public Health Labs, but it was approved solely for use on those showing symptoms. The test was expressly not approved to do the most important thing we needed it to do: diagnose and isolate the asymptomatic, presymptomatic, and very mildly symptomatic people who were spreading the virus unknowingly.

Screening and testing only those with demonstrable symptoms would account for, at best, between only 25 and 50 percent of infected people. You can't stop outbreaks by detecting fewer than half the cases. If you have incomplete data, then you might draw incorrect conclusions.

As it turned out, that's just what happened.

FROM SOUTH AFRICA, I tried to sift through the implications of the testing debacle, but something else, aside from the immediate logistical setbacks of the failed tests, troubled me, something larger and more ominous: bias.

The failure to course-correct on the silent spread following the *Diamond Princess* disaster, combined with the design of tests solely for the Public Health Laboratories, demonstrated that those in charge weren't looking at this situation through the correct lens. The CDC obviously expected the new coronavirus to behave like seasonal or pandemic flu, and they assumed they'd be able to spot it circulating in the public in the same way they spot the flu every year: by relying almost exclusively on symptoms. This was a huge miscalculation.

Historically, the CDC's flu-testing effort is focused more on identifying the particular strain of flu that's active in a community, rather than preventing its spread. Millions of people get the flu each year, millions have flu-like symptoms, but very few people are actually tested for the flu—the vast majority are diagnosed on the basis of their symptoms alone, with doctors reporting cases through "presumptive diagnosis" rather than "definitive diagnosis," which requires actual testing. Also, by and large, the CDC's flu surveillance system, built with ease of burden in mind, works very well, we think—for the flu. (Ease of burden means not taxing too greatly the limited public health labs we have with processing too many tests.) We expect that within a relatively small degree of variance, around twenty-five to thirty-five thousand people per year will die from the flu, and year after year, this has generally proven consistent—but this is a modeled estimate, not a definitive number.

Of course, we all prepare for the worst-case scenario of an influenza virus outbreak. In 2005 and 2006, while I was at the CDC, the White House Homeland Security Council outlined the "National Strategy for Pandemic Influenza" and the "National Strategy for Pandemic Influenza Implementation Plan." In respiratory pandemic circles, leaders often talked of the risk of widespread avian flu or other H1N1 or H5N1 infections. We were always on high alert for these zoonotic viruses. This was evident in 2009, when the swine flu spread across the United States. We learned critical lessons from the 2009 flu execution, but those lessons were very much focused on vaccine distribution and employing the full clinical capacity of the U.S. pharmaceutical industry, including pharmacies.

Part of the CDC's past success at combating the flu had been attributable to the establishment of the Pandemic Influenza Implementation Plan, an annual desktop exercise. It simulates a flu outbreak to see how the plan would help mobilize a response to a viral outbreak of the H1N1 or H1N5 variety. The belief was that by being fully prepared for the flu—in particular, the H1N1 and avian flu strains—the HHS/CDC preparedness plans would, according to the CDC's website, "enable HHS to respond more effectively to other emerging infectious diseases as well."

That was a worthwhile goal, to be sure. But my long history in the military had taught me that, while plans are great, you need to be flexible as the situation on the ground unfolds and not become locked into "the plan." In the military, you are taught to allow your response to evolve, and evolve rapidly, based on your frontline information. In the case of SARS-CoV-2, those in charge had to observe what this unique virus was actually doing and be agile enough in their thinking and actions to change strategy early and aggressively.

You also have to learn the nature of the opposing force and how it operates. Simply put, the CDC was prepared for the type of viruses with which it had demonstrated past success: H1N1 and H1N5. But SARS-CoV-2 was a different kind of virus. The CDC believed its plan would be effective enough to deal with other emerging viruses as well as it did influenza viruses, but that expectation didn't account for institutional inflexibility and the variability of a different strain of virus. As a result, our nation's top public health agency would cling too long to the wrong response model, one based on a different type of virus, while not accounting fully for every type of spread, not just symptomatic. We can't implement a response based only on past experience or on a model that worked for a different disease.

Later, many people would be led to believe that the Trump administration threw out the Obama pandemic plan, causing a delayed and ineffective response. This is not true. The pandemic plan is and has been continually updated from administration to administration through the CDC. Covid-19 merely revealed the holes in that plan. The perceptions and assumptions that created those holes were present in previous iterations of the plan throughout multiple administrations, both Democratic and Republican.

Experience can be a help or a hindrance. In the early 1980s, with AIDS, we simply didn't know what was making people die, and we felt helpless as an unidentified virus silently spread. After being caught flat-footed by HIV's asymptomatic spread, which took us years and thousands of lives to understand, I found my guard going up whenever a new virus seemed to exhibit even the faintest tendency toward asymptomatic behavior. Obviously, I knew that this novel coronavirus and HIV transmitted in very different ways, but when evaluating the data, I kept an open mind regarding the former's possible silent spread.

This past experience with HIV, along with my limited evidence from other countries and from the *Diamond Princess*, told me that this was transmitting differently. But those two things—past experience and existing evidence—would not be enough to move opinions at the CDC. Based on what my experience working at the CDC had taught me about its intransigence, I knew that I would need more. No one, no organization, wants to be wrong, but to show that an organization as large and important as the CDC has been so wrong requires a lot more proof. And at that point, I couldn't prove conclusively to anyone, even myself, what I fully suspected was true.

Given my position outside the domestically focused federal agencies and the White House, all I could do was share the information I had with someone I trusted. I saw the gaps in the flu model approach and communicated my concerns with Matt Pottinger early and often.

He got it. He advocated for a different, more urgent approach.

Though I had armed him with sufficient rationales he could present to others, the administration remained stuck in the flu groove. They stuck to the game plan without realizing that the rules this virus played by were dramatically different. Matt sensed that because he wasn't a medical doctor nothing he said would carry the authoritative weight of the truth I was relaying through him.

DESPITE MY GROWING DOUBTS, I held out hope that the CDC would somehow be able to right the testing ship, that it would see the errors in its approach thus far and move to correct them. This was easy to imagine because I remained on the outside at this point, with no idea what was happening behind the scenes to address my concerns.

From his perch on the inside, Matt was less optimistic. As early as February 13, the day before I left for South Africa, Yen and I exchanged texts. Matt had told her that there was a lack of leadership and direction in the CDC and the White House Coronavirus Task Force. Both had failed to assess the significant lessons from the *Diamond Princess*. Instead, their primary concern about the situation was that a small group of Americans was on board the ship. The State Department had a plane at the ready to fly to Yokohama to bring them back home. I agreed that this was an absolute must-do and immediately I laid out a simple plan of action: getting them onto buses masked and socially isolated, driving them to the airfield, repeating masking and distancing on the plane, testing them repeatedly. Yen wrote that Matt had said a plan similar to mine was offered up, but Bob Redfield and Alex Azar had wanted another night to think about it. Yen then wrote exactly what I was thinking: unless they were evacuated quickly, everyone on board that ship was going to get infected.

In their hesitation to immediately protect the Americans on board, I sensed a larger pattern at work, one of limited action and too-cautious overthinking. I was all about now-now-now action; they were all about "wait for irrefutable data." I also heard faint echoes of Matt's comments about a lack of decisive, proactive leadership. It's understandable—no one wants to make a mistake. But in this case, waiting to be absolutely right would result in broader community spread and more fatalities, not fewer. With pandemics, acting rapidly and aggressively based on limited data, pushing the envelope and being prepared to fail spectacularly, can actually save lives.

These potential errors of omission were compounded as I looked at the messaging coming out of the White House. Throughout February, I became increasingly concerned about the Trump administration and how it was—or, in most cases, wasn't—communicating with the public. One of the most essential components of any public health initiative is clear, effective, consistent messaging, stating over and over what you know and don't know. Yet, any optimism I had about the consistency in their messaging had been quickly dashed against the rocks by weeks of deflections, diminishments, and reversals.

At a campaign rally on February 10, days after the ill-fated CDC tests were shipped more widely, President Trump said, "Looks like by April, you know, in theory, when it gets a little warmer, it miraculously goes away. I hope that's true. But we're doing great in our country." The next day, he doubled down on his overly optimistic assessment: "In our country, we only have, basically, twelve cases, and most of those are people recovering and some cases fully recovered. So, it's actually less."

A government should not base its public health policies and emergency measures on miracles and hope. No one preventive measure is ever 100 percent effective, not even vaccines. You must have overlapping measures, duplications and redundancy. Whether it is containment, vaccines, or miracle cures, believing too much in any one thing—putting all your eggs in one basket—is dangerous. It gives the public a sense of invulnerability, of absolute protection, allowing them to take risks when you need them to take precautions. It's why we layer protections in cars—crumple zones, airbags, seatbelts—each element adding some overlap and redundancy, resulting in more lives saved. This is what true mitigation looks like.

The president's confidence in containment and control of this virus led many to underestimate its potential impact. The American people were the very ones who needed to take this virus seriously, and were the ones who would eventually need to comply with whatever miti-gation measures were recommended or mandated. His overly rosy as-sessments only clouded the main issue: Containment alone would not work here. We couldn't prevent the virus from breaching the invisible geographic borders. It was already on planes, ships, cars, buses, and subways, moving around the globe at speed. The virus had not been contained anywhere—not on the *Diamond Princess*, nor in Wuhan—so there was no reason to believe the United States would prove to be the exception. But this logic didn't stop the administration from suggest-ing that we had little to worry about. On February 2, President Trump told Fox News's Sean Hannity, "We pretty much shut it down coming in from China." As late as February 24, the president tweeted, "The Coronavirus is very much under control in the USA." On both those oc-casions and others, he fundamentally misrepresented the urgency of the situation and contradicted what we were already learning from other countries.

Another problem was that, at nearly every turn, someone in a posi-tion of authority was characterizing the risk to average Americans as "low." I agreed that the risk for very serious illness might have been low, but the risk of getting infected was not. This virus was proving to be highly transmissible. The curves I was tracking across the globe clearly told that story. This virus was more contagious than SARS-CoV-1 and the seasonal flu. Other evidence existed to demonstrate this.

After all, the risk to the average passenger on the *Diamond Princess* certainly wasn't low, nearly one-third of passengers showed symptoms in less than three weeks. And if the virus was being transmitted widely, the chances of someone with a weakened or aged immune system con-tracting it were greatly increased. More exposure meant more infec-tions, which meant a greater frequency of serious illness and death.

I wanted to believe that, behind these optimistic assessments from the president, an aggressive and coordinated mitigation response was being planned and set in motion by HHS, that the White House and the federal agencies were working full-out, and not basing their approach on hope. I believed they weren't publicly contradicting the president, perhaps believing that it was important not to create panic, but were working feverishly behind the scenes.

But then I saw Tony and Bob repeating that the risk to Americans was low.

On February 8, Tony said that the chances of contracting the virus were "minuscule." On February 29, he said, "Right now, at this moment, there is no need to change anything you're doing on a day-to-day basis." I now believe that Bob and Tony's words had spoken to the limited data they had access to from the CDC. This lack of comprehensive domestic data, and the reliance on projections from this very limited data, haunted the pandemic then and over the months and years to follow. If you weren't testing enough, then you were dependent on the CDC's interpretation of the data and on the flu model, and that could lead to vastly underestimating the extent of the silent invasion.

Maybe I just had a wider lens. I had access to more unreported, real-time global data, but maybe they had data in the United States that I did not. I trusted them, and I felt reassured every day with them on the task force. In the meantime, all I could see was how high the risk to Africa was, and I continued to concentrate my efforts there.

In hindsight, I know now that the overly optimistic language being used by the president and agency leadership was rooted in the lack of definitive data and at least partially rooted in a debate I'd experienced many times in my career at the CDC and with other medical professionals. When you have a virus that produces a spectrum of disease ranging from no acute symptoms in some people to deaths in others, you need to carefully explain in real terms what this means at the community level. When you have inadequate data to make a definitive statement, you say so, even if it's not politically expedient to do so. You inform the public that we don't know really how widespread the virus is because we don't have the data systems in place to fully track it. Too often, scientists believe they need final, perfect data; or that scientific concepts are too complex for laypeople to understand. So, rather than finding the right communication strategy to convey difficult ideas, they oversimplify. But oversimplifying the message can be problematic.

Equating Covid-19 to seasonal flu (an illness people knew) implied that we knew more about it than we did. And not making Americans aware of the potential depth and breadth of the threat caused some to dismiss the virus as nothing more than a bad cold or the flu. Later, when we tried to convince them of the unique seriousness of the virus and of their need to protect vulnerable people in their families and community, they became confused, still believing the earlier oversimplified statements.

I don't believe this level of simplification is ever needed. I believe that it's the role of public health officials to explain complex ideas without reducing them to irrelevance. Not being completely transparent early on about what is known, what is not known but suspected, and what needs to be studied will only create later confusion. During my time on the task force, I heard repeatedly "The public will never understand the nuance. So, we are just going to say X, and not include the Y part." But when the Y part is critical, and it later appears, people question everything you said before and, consequently, everything you tell them afterward. Talking down to people is never effective. Raising their awareness through effective communication is. Arming the public with every detail, telling them what we are studying, prepares all of us to learn together.

Months later, in September 2020, it was revealed in interviews Bob Woodward conducted with the president in March that Trump's optimistic public comments contradicted his very real concern about the dangers the virus posed. If the president had just gone public and said these things to the American people it would have gone a long way toward conveying the seriousness of the situation then and now—no scientific nuance required.

This is a dangerous virus.

It transmits easily from person to person.

We need to protect the vulnerable, the elderly in the community and in nursing homes.

You'll want to protect your family, your friends, and your local community.

The virus may not be in your area yet, but it likely will be at some point.

To keep it from doing great harm, please do these things: practice good hygiene, socially distance, wear a mask when indoors, and if you can't keep safely apart, limit your interactions with others.

And most important, remember this: You may not feel ill, but you can still be infected. By getting tested regularly, you can prevent harming your family members, friends, and coworkers and still go about life fairly normally.

We will keep you informed, day by day, of everything we are learning.

*Even if we don't have the final study or the final answer, we will let you
know what we are doing.*

*We will let you know what we think will work while we get the data to
ensure that it does.*

We will learn together. Day by day. Together.

I was stunned when I learned in September of the president's remarks
to Woodward, and I remain unable to square the former president's comments from that March interview with what he was saying publicly prior
to that. In a White House in which people were unable to keep anything
secret, I never heard anyone say that the president had expressed such
sentiments to them regarding his understanding of the level of seriousness the virus presented. He never said anything like that to me, Bob,
or Tony; nor to any of the agency leads of the various divisions within
HHS; nor, to my knowledge, to any of his closest advisors. Word of such
remarks would certainly have leaked, but it didn't. Among the various
mysteries surrounding this pandemic, President Trump's comments to
Woodward remain shrouded in what I can only think of as ego. Perhaps
the president wanted Bob Woodward to know that, at a time when only
a select few had the "privilege" or the depth of insight or the ability to
see into the future, he had seen the truth of the situation.

Among those select few, working in deep background, was Irum
Zaidi, my PEPFAR chief epidemiologist and data person. Irum assisted
us in setting up databases around the globe so that we could bring ac-
countability and transparency to the sixty-thousand-plus clinical sites
PEPFAR supports in nearly fifty-five countries. Through her work,
Irum knew another "data person," who had access to figures about the
novel coronavirus from around the world and very specific data from
China. This individual was taking a great risk in passing it along to
Irum, and his courage serves as an example for all of us.

According to the data shared with us, we learned that the virus was
preying specifically on the elderly and laying waste especially to people
with comorbidities (diabetes, lung disease, and the like). This was in
line with what the world would eventually learn about the risks of
severe Covid-19, but Irum and I understood these risks in mid- to late
February. We passed along the need for greater vigilance and testing
to our African PEPFAR people. Dr. John Nkengasong, the head of the
African CDC, spoke to all those assembled in Johannesburg, his infor-

mation striking the perfect balance among preparedness, prevention, and not panicking. Africa needed to be ready.

I had to believe the same diligent preparation was occurring in the United States. The president might be presenting one scenario, but Bob and Tony, the agencies, and the scientists had to be responding to a different one, right? I stood astride a fulcrum of hope and fear, optimism and frustration.

I shared my fears and frustrations about the U.S. response privately with Matt, Yen, and others. Publicly, I stayed in my lane—Africa, where HIV/AIDS continued to rage. I wasn't going to step outside my chain of command (the State Department and my global mandate). The wheels in government grind slowly; I understood that. Besides, domestic policy was outside my area of responsibility.

That said, I cared deeply about the United States and what Covid-19 might do to the American public and my family scattered around the country. I had no qualms about continuing to assist Matt from the shadows, out of the spotlight. He was inside; I was outside. And I'd never been one to pull rank and publicly criticize my colleagues' actions if I wasn't there to see what they were seeing. That wasn't my leadership style, and I knew that encroaching and criticizing seldom worked.

When I sent my insights to Matt, he would decide if they were useful and whether to share them with the White House Coronavirus Task Force. Fearing blowback for stepping outside my area of responsibility, I asked him not to use my name when discussing the opinions and data I was providing. If it meant influencing the White House's approach, I was fine with my ideas being seen as his.

ON FEBRUARY 23, A Sunday, I was working all day, and so was Matt. He reached out to me again, this time by phone. There was urgency in his voice. He was especially concerned that at a recent task force meeting, Alex Azar and acting White House chief of staff Mick Mulvaney had made the decision to ask Congress for, as Matt later put it, "a tragicomically tiny budget supplemental" for therapeutics, vaccines, PPE, and testing. The ask was for less than a billion dollars, a figure that didn't even account for the money needed to develop vaccines. It was not terribly surprising: the figure was proportional to the low level of threat they believed the virus posed.

Matt wasn't calling just to gripe. He asked, again, that I come work at the White House. Only, this time, he was no longer looking for me to be a voice, a spokesperson. The situation at the White House had become critical. He needed me to step in and take on a more comprehensive role. Without a background in public health and medicine, Matt was just another nonexpert in the room competing to be heard.

I told him again that I couldn't leave Africa and PEPFAR with critical planning hanging in the breeze. The next day, Matt phoned again. (He was certainly persistent.) He believed that the situation had deteriorated even more, both within the White House itself and globally.

"I've sent some more suggestions to Yen," I told him. "Can't we keep doing things this way?"

But Matt, to his credit, continued to press me; the status quo was untenable. The White House was vetting other names for a task force response coordinator role. Matt wanted to know if he could include mine. I told him I needed a night to think about it.

I spent most of the night awake, considering my options. I went over in my mind all the reasons I'd given before about not wanting to work in the White House. But this time was different. When Matt made me his first offer, back in November, it seemed like the want-to-need ratio was 85/15. When he came back to me in 2020, at the end of January, I'd adjusted this to 80/20 want-to-need. Now, though, I saw it as 25/75. The balance had clearly shifted.

The virus was spreading silently. Without extensive testing capacity, the United States would be completely unprepared. The administration had already done and said things that would undermine Americans' faith in their public health care system's guidance. Without that faith, how effective could any ask for behavioral changes be?

But Matt's urgency represented another degree of concern: the unknown. If he was this concerned, what else was happening? What else would happen? With one of the highest security clearances, Matt had access to all kinds of information that I did not.

I knew if I said yes and got the job, I would likely stay until the crisis was over or the Trump administration ended—whichever came first. Was I prepared to make that commitment? I texted Yen; I trusted her opinion. She knew my background, and she knew me. She also knew the difficulties in the White House. Amid all the back-and-forth, she wrote me a short, simple text: "Remember, your country needs you," followed by a winking emoji.

Had that message come from anyone else, it might not have tipped the scales. But, somehow, that simple statement of "need" weighed just enough to confirm my decision. Yen was a mother; she knew about doing what was needed, about setting aside your own wants and desires to help others. Her message echoed what I'd been saying to myself. For so many years, my decisions had been guided by this principle: Go where and do what is most needed.

On February 26, Matt called me expressing greater worry. He told me that every moment I delayed making my decision, I could potentially be costing American lives. Having served in the military for twenty-nine years, I knew that when a senior leader asks you to take on a mission for the American people, you commit to the mission. And you stay on mission 24/7 until it is completed or you are removed. That's what soldiers do. That's what anyone called to serve does. As simple as Yen and Matt's message was—that I was needed—it contained multitudes.

The next day, I phoned Matt to let him know he could add my name to the shortlist.

If they decided they needed me, I was in.

CHAPTER 4

Where Is the Response?

Flying thirty thousand feet above the Atlantic with no view of the earth below me, I was struck by an analogy. That I could envision what was below me but couldn't see it was pretty much how I'd experienced the Trump administration's response to the outbreak thus far. All this would change soon enough. In less than a day, on Monday, March 2, I'd be on the ground, walking into my new role as White House coronavirus response coordinator. Then I'd be able to see if perception aligned with reality.

I had my reservations about working in what looked like, from the outside, a dysfunctional White House. Before agreeing to come on board, I'd shared my thoughts with only a few people, primarily my PEPFAR deputy Angeli Achrekar, Irum Zaidi, and Yen and Matt Pottinger. In fact, Yen and I had a bit of a laugh when she asked me what my husband thought of my taking on a new role. I'd told her that, given that I was still in South Africa and he was in the United States, I hadn't yet told him (not to mention my adult daughters) about the possible White House move. I was that concerned about information being leaked. Who knew who was monitoring our communications? I'd committed, and I was going, but in those long hours on the flight home, I found myself reassessing just what I was actually entering into.

From my perch at the State Department, I had seen three years of

high turnover and continual news reporting of chaos in the White
House. A pattern emerged: talented and respected people went in and
then abruptly left. Some of them wound up testifying at congressional
hearings. More concerning, I saw too many political casualties among
retired military members serving in the White House. H. R. McMaster,
John Kelly, James Mattis—none had survived that undisciplined environment, not one. I didn't know these men personally, but I'd served
under men like them. Worldwide, they had overseen high-stakes life-
or-death situations where events went sideways, where rules of engagement had to be funneled through Department of Defense lawyers, and
where different agendas, military and political, conflicted. If those experienced, never-give-up-on-the-mission leaders hadn't survived, what
chance did I have?

From the outside, this administration seemed too disorganized, too
negative, too chaotic, too disrespectful, too vindictive, too scattered,
and too mean. Because I wasn't a political person, I had only two connections, besides Matt, to the people in the White House. The place was
filled with business-oriented people who were tight with one another
and who had made millions, perhaps billions, in their careers. Prior administrations, and the country, had become increasingly partisan and
distrustful of civil servants. Add to that volatile mix a looming presidential election.

I wanted to do what was best for my country, but I also had to consider what would happen to me. Accepting this position would end
my federal civil service career. That was a precondition of taking the
job that I forced myself to accept, especially given the polarization of
the political parties and, particularly, the broad negative view of the
Trump administration. That end was a tough pill to swallow. I'd had a
great career, one that I was justifiably proud of. I was helping to control
the HIV pandemic by using granular data and by driving the policy reforms necessary to establish equity, remove barriers, and ensure access
to services for everyone, independent of tribe, race, gender, sexual ori-
entation, wealth, political party affiliation, age, or geography. This was
all a dream no one had previously thought possible, but it was made
real through many people constantly pushing. My colleagues viewed
me as a hard-driving force focused on those who needed our help the
most, addressing both the social and medical reasons for vulnerability.

Of course, there was also my selection itself. I had to consider the possibility that I was a "gendered" choice, the lone female medical doc-tor on a team that had six male doctors already on board. I also worried that ulterior motives had been at work in the selection process. Would I, as a multidecade civil servant, become a victim of the civil servant/politician divide? Were people like me and Tony, who had an even lon-ger career in civil service than I did, being positioned as scapegoats in case things went sideways? Those with long-term Trump connections could insulate themselves from the president's hot-tempered wrath. (Not to mention that the roster had already changed, with Secretary Azar out as head of the task force and Vice President Mike Pence in. What this foretold was still to be determined.) By having us around, the "politicals," when things went awry, could say they'd gotten bad advice from the "public health experts." They could easily blame the civil servants and characterize us as members of the "deep state." These were realistic possibilities.

From past experience, I was clear-eyed about what I might face in working with this administration. In 2019, I faced off against President John Magufuli of Tanzania. The previous president, Jakaya Kikwete, had been very supportive of our comprehensive response to the HIV pandemic, but when President Magufuli came in, he replaced the min-ister of health and enacted a series of regressive policies that, if allowed to stand, would have severely limited our ability to help not just gay men, but the highest-risk group, ten- to twenty-four-year-old girls. We had worked hard for many years and had made significant progress. President Magufuli staunchly opposed our programs for the most vul-nerable and reversed laws that required young women to be at least eighteen years of age before being allowed to marry. Magufuli seldom seemed to have the best interest of his most vulnerable Tanzanians at heart. Still, over time, we'd been able to claw back all but one of the losses we suffered under his administration. After that experience, I wasn't going into the White House task force afraid of hard work or dealing with those who didn't prioritize best public health practices.

I had dozens of reasons not to do this job, and only one reason to do it: Matt seemed certain I was the missing piece. He knew I had worked on RNA viruses like SARS-CoV-2, from the laboratory bench to the community, developing tests, therapeutics, and vaccines. He knew I had practical on-the-ground experience in what was needed to combat a pandemic. He knew I had convinced other presidents and prime ministers around the globe to enact politically unpopular policies crucial to saving their constituents' lives. He also knew that, being ex-military, I was mission-oriented and that service mattered deeply to me. He also knew my focus on data and my relentless drive to use it for decision making to drive policy change.

More than all that, I think he believed I was resilient enough not to get distracted by the constant political swirl around 1600 Pennsylvania Avenue and beyond. He thought I could survive in this White House. I wasn't so sure. I was confident in my abilities, but less sure of how they would be received in a potentially hostile environment.

For the rest of the flight, I focused on the mission. This began with considering my first interactions with members of the task force and what elements of the response to prioritize most. I had to better understand who and what I would be coordinating. The task force had been meeting since the end of January, and like any new kid showing up after the start of the semester, I was already behind on some of the social aspects—how the various personalities in the room were meshing and what was happening behind the scenes at the agency level. The task force members must already have established a working rhythm and relationships.

To that point, Matt had offered only somewhat vague impressions about Tony and Bob and his general sense that the task force's lack of urgency was problematic. I've always trusted my ability to read a room, but the stakes here were very high. I was confident that I could present a plan of action for how to proceed, but I certainly wasn't going to be able to come in and dictate terms to anyone. Any whiff of overbearing leadership would be detected instantly, and I'd likely find myself backed into a corner. Forcing myself into a defensive position would serve neither mine nor the country's interests.

As for what we were up against, as the jet carried me across my Rubicon, I reviewed all available past U.S. pandemic preparedness plans. Fundamentally, crisis management problems already existed. The pandemic preparedness plans could clarify which agency under HHS was the lead. Was it the CDC or the Office of the Assistant Secretary for Preparedness and Response? This was the first time the United States had faced a substantial pandemic with both ASPR (which was formed in 2008) and the CDC (established in 1946) in place. Having competing factions within the same department could foster execution and accountability, but would the two agencies play well together? If not, how was that failure already impacting the situation?

Not having a clear sense of who was leading on what front could cripple coordinated preparedness efforts. Similar confusion over agency oversight existed when it came to developing vaccines and therapeutics (Biomedical Advanced Research and Development Authority [BARDA] vs. the NIH). Knowing who was responsible for what aspect of a mission was essential to successful planning and execution. So was agreeing on a core set of principles to guide the response. The lack of urgency and task confusion (as we'd seen already with the testing debacle) had put us behind, but the lack of consensus on this virus's silent spread as a guiding principle could prove crippling.

I wrote a list of action items and charted a way forward. Gaps existed, but from thirty thousand feet in the air, my perspective was distorted. I needed to see how large those gaps really were. I'd devote my first week, then, to on-the-ground reconnaissance.

Testing and communication were two of the largest gaps to be filled immediately. Scaling production to meet the needs of the entire United States required aggressively enlisting the private sector. One of my priorities that first week would be meeting with commercial test developers to close that capacity gap with a rapid increase in testing production. I assumed that some version of this conversation with the commercial test developers had already taken place with someone on the task force or at the White House, but I wanted to make sure this partnership with the private sector was being prioritized.

To communicate more accurately, we had to replace two of the administration's persistent damaging messages with correct ones: Covid-19 was not the flu, and the risk to Americans was not low. The risk of death may have been lower depending on age and underlying conditions, but the risk of infection was not. Also, the healthy and young were unknowingly spreading the virus. That the CDC was relying on its flu model approach behind the scenes was bad enough; but the White House's rhetorically comparing the risk of this virus to the seasonal flu was far worse. Most Americans heard this claim, and once that low-risk flag was planted, it was going to be hard to remove it. I was determined to do so.

My initial reconnaissance mission would also prioritize understanding the administration's views on masking. Matt had told me that there was little to no support within the task force for their use. I couldn't tell precisely from where in the White House the resistance was coming or what form it was taking. Certainly, the administration's public statements weren't making clear what its stance on masks was. If the administration, following the CDC's lead, believed that this was a flu-like event, then any type of intervention that didn't fit a flu scenario would become a harder sell. We had to make that sale, but it would be like getting a car buyer to opt for the extended warranty before buying the actual car. Communications, testing, and masking were important, but first things first—we had to sell the car and get awareness of asymptomatic spread driving the response.

With the assistance of my PEPFAR colleagues and others, I'd gathered, to that point, international data. The task force would be basing its approach on domestic data, and I wouldn't know until I got to the White House precisely what that looked like. Further complicating things, I had no idea how quickly I'd be able to get domestic data in order to make clear the charts and graphs I would use in lobbying for action. I always made my own graphics and PowerPoints to convey critical concepts visually, using PowerPoint for its graphs, not text. I had always found that a graphic could tell a complex story it would take hundreds of words to convey.

For all the logistical issues that testing would present, it would be far more challenging to change entrenched positions in the medical professionals at the CDC and elsewhere in the administration about what this virus actually was: a silent spreader. I couldn't begin to win over the politicians to the idea of silent spread, or of the drastic actions we would need to take to curb it, if I didn't even have the doctors on my side. My sense, my hope, was that the doctors would see what I was seeing and what it revealed about silent spread. Even if, publicly, they weren't in a position to act on it as aggressively as I felt they should have, at the very least, they had to recognize its role. Maybe they hadn't seen it as soon as I had, but by now, they must have seen it. Either that or they had access to numbers they believed would prove their case, numbers that, from my vantage point on the outside, I hadn't yet seen.

Still, in the back of my mind, I continued to hear Matt's voice—his sense of urgency, how hard he'd worked to get me on board. The foreboding behind his words was palpable, and the sooner I was behind the curtain, the sooner I could begin to understand firsthand what was motivating his concern. I desperately wanted to arrive at the White House and learn that Matt's fears were misplaced, that the machinery behind the scenes was functioning, albeit slowly, much as it should have been, or that, perhaps most optimistically, I was simply wrong about the silent spread.

In fact, at no time in my life had I ever wanted to be wrong about something as I wanted to be now.

I LANDED IN DC at around noon on March 1. That evening, as I prepared for the next day, I realized I had no idea how one was supposed to dress for this job. I spent considerable time thinking about the impression I wanted to make on the president, the vice president, and the other members of the task force the next day. Though I was ridiculously jet-lagged from my all-night flight, I managed to think clearly. I picked a black dress with a black jacket emblazoned with buttons. It looked very much like a military uniform, and that was the image I wanted to project. With that outfit and my highest heels (which made me more than six feet tall), I wanted to send a serious and commanding signal: I might be the only female physician on the task force—indeed, I might be one of only a few women in the room—and the odds might be stacked against me, but I would look like I was in command-and-control mode.

That first day, I was greeted by two representatives of the vice president—his chief of staff, Marc Short, and an aide named Olivia Troye—before meeting with Vice President Pence. Going in, I anticipated feeling a bit of kinship toward him. He was in a somewhat similar situation to mine, having been appointed to head the task force just seventy-two hours earlier. The vice president and I had met very briefly once, when I asked him to speak at an AIDS Day event in 2018. But for me, he was still essentially only an outline. Once in his presence, I quickly filled in the details. I looked around his extraordinarily organized office. Photos of different family members surrounded him on the north and west sides of the room. Biographies of leaders filled the bookshelves on the east side of the room, while the south side contained a working fireplace. The room reinforced my previous impression of him as the prototypical reserved, solid, down-to-earth, family-oriented midwestern man.

We exchanged brief greetings and very few pleasantries before I got down to the business at hand. As the head of the task force, the vice president, above all others, I believed, had the president's ear. Sticking to my plan, I addressed the major points immediately: the president had to understand that we weren't facing the flu and that the risk to Americans was not low. Again and again, I emphasized with the vice president, and with many others that first week, that any comparisons of Covid-19 to the seasonal flu, any references to its case mortality rate in relation to the flu's, any belief that detecting SARS-CoV-2's presence solely through symptoms in an individual or in a cluster would be a real blow to the safety of the American people. Testing was key.

I was pleased to hear Vice President Pence set the tone in that first meeting with me, and then later with everyone else, by saying bluntly and unashamedly "I don't understand that" whenever he didn't fully grasp some medical or technical concept. In an environment where bluster and stubborn know-it-all-ness too often prevailed, Vice President Pence admitted that he was no expert and that much of the material I was presenting to him about virology and epidemiology was foreign to him. He also impressed me by trying to contextualize my points: "Why is that important? Explain to me why you think this way." From the outset, I greatly appreciated this kind of honesty and directness. Even better, throughout the eleven months we worked together, he kept this spirit of inquiry and active curiosity alive in himself and encouraged it in others. He was willing to listen, to learn, and to act. Because I was willing to take the time to consider all questions, and had come in prepared with a plan, he listened to me and seemed to trust me. I received no signal that he was resistant in any way to my presence or to my specific recommendations. He grasped my sense of urgency and committed his staff to translating the priorities I listed into immediate actions.

I didn't have a team at this point; it was just me in the White House. Although Matt and NSC staff were supposed to find me office space that first day, it was the Office of the Vice President that found me a place. Within hours, they had moved Olivia Troye, who now worked directly for Marc Short, into that same space, four feet from me. Later, I'd learn that Olivia had been part of the National Security Council, detailed from DHS, but she was now an aide to the vice president's White House Coronavirus Task Force.

At the time, I didn't think much about what this shift from NSC quarters to the vice president's locale meant. Later on, I'd learn that the move was not about space allocation, but was politically motivated. It was the opening salvo by Marc Short and his team to "protect the vice president and the president," to scrutinize my possible impact. Olivia Troye would be responsible for monitoring my every move and reporting back to Marc Short.

As I bounced from one meeting to the next that first week, I tried to get a feel for the dynamics at play. In getting to see the personalities in action, I gathered that my fears about the vice president's and my dual entrance into the task force, combined with Alex Azar's perceived demotion, would complicate matters further. As head of HHS and the task force, Azar had been in power. Now on board as the coordinator, I was unsure where I fit in or, critically, where Azar thought I did.

From the outset, Alex Azar's hostile body language and clipped verbal responses to me revealed his dissatisfaction with the roster shuffling. As he spoke, he often leaned back with his arms folded—a defensive posture. People on the defensive, or people who feel they have been slighted, can become vindictive and don't always act logically or stay focused on the task at hand. We all needed Alex and his leadership, and I needed to find a way to make things work. From using data collection and analysis to structure the public health response (CDC), to developing therapeutics and vaccines (NIH and the private sector), to having these therapies and vaccines approved (FDA)—the Department of Health and Human Services and its agencies were central to the Covid-19 response and to protecting Americans. As it turned out, one of the reservations I'd expressed to Matt from the beginning, my relative lack of experience in navigating unfamiliar domestic and highly territorial waters, was going to be an enormous challenge. I knew global; I didn't know domestic.

While operating in the shadows I had intuited from the ambiguously written pandemic response that a struggle within HHS was occurring. The task force meetings I attended that first week revealed that conflict was playing out in real time. In a large bureaucracy, interagency squabbles can, obviously, be very counterproductive. This particular border war was between two agencies geographically distant from each other: the CDC in Atlanta, headed by Bob Redfield, and ASPR in DC, headed by Dr. Bob Kadlec. The gap to be bridged was between the CDC technocrats and the ASPR bureaucrats; it was between who would plan the response and who would execute those plans. That HHS had to be in charge was undeniable. The question was: Would it be the CDC or ASPR that took primacy in ensuring a comprehensive response and continual course correction in real time as the pandemic evolved?

Geography matters, and ASPR personnel and leadership were in the same building where HHS and its secretary resided. From the start of the pandemic response, this meant that ASPR was in the best position for face-to-face conversations between HHS and the task force simply by our walking down the hall. Meanwhile, the CDC's team was mostly on a conference line. I considered all CDC personnel as essential front-line workers whose efforts would be enhanced by working on-site in Washington, DC, and in all states. In times of crisis, when data moves quickly and responses need to be fast and agile, physical presence can be critical. As so many of us have discovered as a result of living our work lives digitally during the pandemic, the dynamics of in-office and remote work environments are not the same. Although there was some rotating of CDC personnel in DC and a handful of them on C Street, 99 percent of that agency was in Atlanta and, like many white-collar Americans, working from home. This lack of continuity would hamper their efforts.

Worse, I would soon realize that the most senior CDC technical people overseeing the pandemic were not in the room at HHS. Instead, they were manning their own Emergency Operations Center in Atlanta. The CDC senior technical people were the ones we needed available long term in Washington. They could be responsible for explaining to the bureaucrats the more difficult-to-grasp, technical, scientific details of the current crisis. Their absence there made a hard situation even harder.

In those initial days, because I didn't know all the players, I relied heavily on people I knew I could trust—mainly Matt Pottinger, Tony Fauci, and Bob Redfield. When I had questions or concerns or important information to relay, I went to those I knew best, Bob and Tony. Conversely, I felt certain that others on the task force had their reservations about me. Could they trust this outsider? No, they didn't know me. They weren't thrilled that I had come along, they didn't believe they needed coordination, and they potentially believed that my very presence had knocked their agency down a peg or two. Pecking orders and territorial boundaries mattered.

Vice President Pence appeared to be one of the few people who was fine with my being there, but his staff, accustomed to protecting their boss's self-interest, didn't appear to trust me. Marc Short, the VP's chief of staff, headed the list of those whose antennae seemed to be up. At that early stage, I didn't know Short nor understand his motivation, but I knew he was political and that he was there to closely monitor the political aspects of the response. I couldn't tell if he was protecting the vice president or, ultimately, the president. In the end, it didn't matter. The men's interests were intertwined. Vice President Pence was my direct report, and whatever role Short played with the vice president, it didn't interfere with me having access to Pence, and we worked closely and well together.

I SPENT MOST OF my first week observing where things actually stood relative to where I thought they should be. I asked questions. I listened. I didn't want to come in pushing from the get-go. I wanted to learn where we were on crucial fronts like testing and masks, but more important, I wanted to discern how the task force was (or was not) building the role of silent spread into their thinking.

At one meeting early that first week, Bob was presenting to us on the state of the epidemic. He handed out an Excel spreadsheet printed on a single double-sided page. It summarized case data and deaths by state—just those generalized numbers, which came from the states themselves. This thirty-thousand-foot view was not what I had expected or what was needed. I was hoping that the task force would at least have had more precise data by county and zip code by now, something that would show us exactly where the virus was in the states, counties, and municipalities, so we could plot from those coordinates and focus attention and resources in real time. I was looking for numbers that would allow us, at worst, to project where the virus was likely to emerge next or, at best, through testing, to spot the early onset of clusters in precise locales.

Instead, what we had been handed was without interpretation, analysis, or projection. It simply showed where we were twenty-four hours before. To me, numbers alone, without interpretation or projection, don't mean anything. All this data would show us was where the sickest of the sick were now. Without more granular data, like test positivity and new hospitalization rates, we couldn't see which hospitals would need support, where the hardest-hit communities were located, and who was most susceptible to severe illness. What we had wasn't enough to project into the future or prevent future community spread elsewhere. Numbers that don't contribute to a solution mean nothing.

While I didn't say anything to Bob at this point, I knew we needed more precise data points than just case totals. There had to be secret data somewhere in the CDC's Emergency Operations Center, data coming in from the states that was being used to model the next move. This single piece of double-sided paper couldn't be the extent of what the federal government knew and believed about cases of Covid-19.

After that meeting with Bob, I quietly started to inquire about getting the raw data, shaking the trees a bit to find out who had access to it. I asked Bob Redfield and others where the data was that I could really sink my teeth into. Where was the demographic data on the infected—broken down by race, ethnicity, underlying conditions, age? These figures wouldn't just identify the infected; they would identify how this virus operated, an essential component to formulating a pandemic response. Knowing whom a virus infected led to understanding how it behaved.

From the onset of the outbreak, other countries, like Italy, despite being in the midst of their own Covid-19 crises, had been continually providing comprehensive hospitalization and fatality data, aggregated by age and gender, in real time to Irum and me. That is the way global support is supposed to work, and that is how Europe had responded to this crisis: with the specific demographic information necessary for us to understand how this virus was acting in ways that were different from previous viruses. Changing the narrative so that we all agreed that this was not the flu required my knowing (by race/ethnicity, age, co-morbidities, and so on) who was getting seriously ill, where, and when.

I felt tension tightening my neck and shoulders. I had to unclench my jaw. How was this really happening? How in the world could we not have clear data at this point? What the—? It was March. Bob and the CDC had heard about this two months ago, and this was all we had? A static, partial thirty-thousand-foot awareness. I pressed the flats of my hands into my eyes and shook my head. I had expected something very different, but now I could see not only that the data-reporting structures that had taken years to build in Africa weren't present in the United States, but that we had days to get the same job done here. I thought of those photos of the Chinese building hospitals in ten days.

That kind of construction was easy compared to what we now had to undertake.

Had the United States really been flying this blind for this long? We had known about this virus for weeks. We had all seen the overrun hospitals at ground zero in China. How was it possible that we had so little understanding of what was unfolding on our own shores? Following the science actually requires getting the data to develop the evidence base that creates the "science"—we didn't have that critical data.

My notes from March 4 say it all: "The White House Task Force has no unified source of truth to understand the scope, scale, and spread of Covid-19 in near/real time. The CDC has an incomplete subset of information (CDC case data) among multiple fragmented efforts on various legacy (on-premise hardware) infrastructure. A significant wealth of critical information exists distributed across the state/local, public and private entities and requires White House direction to centralize."

To cut through this immunologist-speak: we were dangerously behind the eight-ball on this one. The CDC had precious little systematic data, and what it did have had not been curated efficiently or effectively. It existed on outdated hardware, using software not meant for this kind of pandemic. Things were so bad in fact that, in 2020, the exchange of information within some states was often transmitted by fax. Hand-entered data was then passed along to the CDC electronically or by fax. Clearly, data collection at the CDC was inadequate.

One of the gaping holes in the U.S. pandemic preparedness plan was not having a capable data collection system. Data is everything in a pandemic. Data shows your gaps; it shows where communities have an effective response; it lays bare the truth, where things are deteriorating and where they are better. It allows you to stay laser-focused and develop evidence-based policies. Without comprehensive data, you won't have a comprehensive response. Without data, you don't know what is working and what isn't. You can't see who needs help and who doesn't. And you can't manage what you don't measure.

I knew that hospital data existed in deep detail elsewhere. Once the coding of illnesses became how health care providers were paid, they had developed or purchased highly sophisticated data systems. Whether it was in doctors' and nurses' notes or in barcoded entries for medicine and supplies dispensed, in tests conducted or results confirmed, real-time data did exist.

Despite the Health Insurance Portability and Accountability Act (HIPAA), which ensured the confidentiality of these records, we had been extracting this data globally for years. With the patients' personal information removed, the remaining data allowed us to see what was happening without breaching patient privacy. In all the years of pandemic preparedness, certainly we had the ability to protect patient information while getting real-time data around specific symptoms and laboratory results. Where was it? Why didn't the CDC have access to all of it from every hospital and clinic and not just from a select few? I didn't give up looking.

In the meantime, I continued to monitor the more demographically specific global data that Irum was collecting through her international contacts. I also reached out to the HHS Office of Global Affairs to help me get data from Europe. This wasn't going to be sufficient to convince those in the White House, but it would help me build the case. Most notably, I thought it could move Bob and Tony toward my belief in silent spread. In my mind, the biggest red flag was the older age of those admitted to hospitals with serious disease. Given that everyone infected did not go on to become seriously ill, hundreds of people would not be showing up to hospitals seriously ill, as they currently were, if there weren't community spread. There had to be a lot of asymptomatic younger people and people in their forties and fifties with much milder infections driving the spread to these more vulnerable individuals.

Before one task force meeting that week, I turned to Bob and Tony. The task force meetings often started five to ten minutes late, and we had arrived early, so we used the extra time to confer. Huddled together, the three of us discussed where we were on silent spread.

"Our people tell me that it's not greater than twenty percent of total cases," Bob said.

"Based on what?" I asked.

"The *Diamond Princess*, for one."

"But that data isn't accurate," I said, incredulous. "Those numbers were based on primarily the symptomatic cases and the roommates for the symptomatic cases. I'm not telling you anything you don't know, Bob, but the older passengers were mostly symptomatic. They didn't account for the younger, asymptomatic cases, the crew members."

"Those are the numbers we saw. We can't project based on what we didn't see."

This was a real sticking point. We knew how many people were on board that ship. That was the denominator. But we didn't know the full numerator—the total number of cases—because not everyone on board had been tested. You can't derive an accurate percentage that way. If you have a bag of ten apples and bite into one that's rotten, you can't say that 90 percent of them are bad based on that one sample. Your estimate will, more than likely, be wrong. How would we produce an accurate projection? By sampling all ten of the apples, of course. But if your belief is based only on appearance and not on core empirical evidence, then your percentage estimate will be close to useless.

And then there was the surface sampling on the *Diamond Princess*. It was bad enough that not everyone on board her was tested every day, but after passengers disembarked, surfaces on the ship were swabbed for the virus. Of course, the PCR test showed SARS-CoV-2 everywhere. RNA fragments can survive on surfaces where the full-length infectious virus may not. What was left behind, therefore, couldn't infect people. Experts should have understood this immediately, or investigated it quickly. By doing that, they would have eliminated the possibility that surface contamination was contributing to spread. Inadvertently this swabbing led to the "surface transmission" theory of Covid-19 spread, producing the run on disinfecting wipes across the globe, the religious quarantining of packages, and the habitual wiping down of all groceries for weeks. And it was another distraction from the true source of continued spread: the infected experiencing no symptoms and transmitting the virus to others.

In this case, though, it wasn't solely the CDC that had underestimated the silent spread figure. They were basing their number on what their Japanese colleagues had calculated. The CDC's rank-and-file scientists and researchers accepted the Japanese numbers and used them in their own silent spread projection. I saw the initial calculation as flawed; therefore, any projection based on it would be equally flawed.

I felt that silent spread could account for as much as 50 percent of total cases; the CDC believed it was closer to 17 percent. This lower figure would make silent spread far less of a concern for them. Their estimate would have had me sleeping better at night—only, I didn't agree with it.

As it turned out, Bob and Tony didn't agree with the CDC's number, either. While they weren't willing to go quite as high as my 50 percent figure, their thinking was closer to 25–30 percent—still high, but not quite as dire as I saw things. Also, I believed the spread was distinctly age-related, with younger adults more likely to be asymptomatic. Though I was somewhat disappointed that Bob and Tony weren't seeing the situation as I was, at least their number supported my belief that this new disease was far more asymptomatic than the flu. I wouldn't have to push them as far as I needed to push the CDC. Silent spread demanded dramatic action outside the task force, and I would need both Tony and Bob fully behind me to build support for any action needed to address it.

Despite disagreeing on the numbers, Tony did totally support aggressive action based on Italy's data. So, until I gained access to (or, in the end, had to create) the full range of data we needed domestically, I relied on Italy's data to make my case. Tony knew a lot about the Italian public health care system and its researchers. Both were part of one of the most sophisticated and advanced health care systems on the European continent. Tony had worked with a number of prominent Italian doctors and researchers over the years. I curated all the global data and presented all of it—from South Korea, Japan, Spain, and France. But to make the most important and specific points I needed to, I chose the Italian data because it offered the best opportunity to get Tony farther down the path toward accepting my conclusion regarding silent spread.

From the Italy data in particular, I was able to see that we were about ten to fourteen days behind where the Italians were. And at that moment in Italy, things looked grim. By early March, the virus had already spread to all regions of the country, and by March 9, all sixty million of its residents would be in lockdown. An Italy-like surge in cases was coming to the United States. We needed to prepare now for what Italy was experiencing, even though it seemed likely we were already too late.

WHATEVER SUSPICIONS THE VICE president's staff had about me, they helped when I asked for it. One of my requests within the first two hours of my arrival was that I be able to meet with representatives from the private sector's laboratory test developers and heads of commercial labs, to scale up testing and processing immediately with their high-volume, high-throughput systems. It still wasn't clear to me whether any such meeting had already happened, but better to be redundant than let something slip through the cracks. By Wednesday of that first week, I was sitting down with private developers.

I'd been envisioning this meeting ever since I recognized how flawed the first tests were and how misguided solely relying on the Public Health Laboratories had been, and it didn't take long to confirm my fears about private-sector testing: this encounter, incredibly, was the first time they were meeting with a federal representative about testing. Sometimes the federal government keeps the private sector at arm's length and engages with it sparingly; some in the public health field are inherently distrustful of the private sector's for-profit motivation. I'd had a very different experience in Africa and Asia with the pandemic response to HIV and TB/HIV. Approaching the private sector strategically had helped us address the gaps we were unable to fill ourselves. When fully supported and allowed to operate without political constraint, the public and private sectors can move mountains together.

In our conversation that day, it became clear that by using every available testing platform, from Roche, Abbott, ThermoFisher, Hologic, Cepheid, and Becton-Dickinson, we could increase testing capability by a factor of one hundred, and then one thousand, within weeks. I trusted these developers because they were the backbone of PEPFAR's viral load monitoring and point-of-care HIV and malaria testing across sub-Saharan Africa. Using their tests and equipment, we had moved very quickly from presumptive diagnosis and treatment in rural areas to full, definitive diagnosis of each and every patient. If these companies could create high-throughput nucleic acid tests and point-of-care tests for Covid-19—simple tests that could be used in communities here and overseas—we'd diagnose people on the spot, and not in days. How to deliver and then process those tests was still a sticking point—a typical distribution and collection problem, but on a massive scale—but the developers' representatives felt it was achievable. Likewise, these companies could resolve another major problem with testing: production. They could manufacture as many tests as we needed. The only problem was that manufacturing—indeed, all these things—would take time we didn't have. And then there was the need for swabs to take samples from the nose, transport media to move the samples to the lab, and then test tubes, pipettes, and PPE for the laboratory workers.

This was the essence of the problems we faced. The testing manufacturers offered solutions, but this meeting should have happened weeks ago, and it might not have happened at all if I hadn't asked for it. We should have already had in place the tests I was requesting. If, four weeks earlier, when the first CDC tests failed, the White House, the CDC, or ASPR had sat down with the testing developers, we would already have had hundreds of thousands of daily tests by this point. We could have seen the silent spread. We could have been more proactive. We could have warned the American people.

Instead, with testing availability limited and processing so slow, we were still seeing only a small fraction of the cases, the sickest and the hospitalized. We were still weeks away from conducting a significant number of tests. Losing those early weeks of full proactive preparation because of the CDC's failed tests had been troubling, but learning that we'd also lost the four weeks since then was devastating. We were weeks behind in seeing the silent invasion that I believed was already occurring.

In many ways, this was close to a worst-case scenario. In later conversations, I asked one of the manufacturers critical to the production of high-throughput and point-of-care tests if anyone in the administration or from the CDC had called them in January and February 2020. Their reply was that, in fact, it had been the other way around: the manufacturers were the ones calling the CDC to understand what it was doing and how they could help. The manufacturers were held at arm's length by the CDC, ASPR, and HHS. They were told their assistance wasn't needed. At a congressional hearing on March 13, Bob Redfield was asked who was in charge of testing. He said the CDC's responsibility was to ensure state Public Health Laboratories had access to the tests. When pressed on how much funding was needed to bring public health infrastructure up to speed, he wouldn't commit.

It was difficult for me to understand why the CDC wasn't aware of who was in charge of this important aspect of the response. Testing was baked into both the prevention and the control of any disease. Overseeing state Public Health Laboratories access wasn't enough. Reluctance to private-sector involvement compounded the problems the agency was creating.

When I eventually asked Bob Kadlec of ASPR and Steve Redd, who was at the CDC before joining me in an advisory role, why there had been no prior conversations with the private sector on testing, they essentially pointed fingers at each other and their respective agencies, saying it was the other guy's job. While institutional ambiguity was obviously a problem in other areas of the response, the lack of follow-up likely stemmed from the CDC's remaining committed to the flu model, which relied not on testing but on symptomatic diagnosis. Also, not securing private industry assistance was part of a larger pattern of CDC behavior. For years, those in charge at the highest levels believed that as the premier public health agency in the country, the CDC didn't need much outside help, occasionally engaging a few university scientists and researchers. For the most part it believed that it could manage and lead on its own.

On March 14, Alex Azar assigned Admiral Brett Giroir, MD, to lead the comprehensive testing effort—that is, to be the "testing czar." Brett immediately focused on all aspects of testing, but he was behind and could only react as one crisis after another complicated his efforts. The CDC/ASPR conflict had resulted in an absence of stockpiles of swabs, pipettes, and transport media (solutions that keep viral samples alive until they can be tested). Although I didn't know her well at the time, task force member Seema Verma, the administrator of the Centers for Medicare and Medicaid Services (CMS), would be instrumental in ensuring that testing was prioritized and readily available to the public. She did this by getting CMS to pay for the tests and to pay enough for the tests to drive expanded use of them.

The private sector also needed to be engaged with mask production. Within the first forty-eight hours of my arrival, I found out we didn't have a significant stockpile of the N95 masks and other PPE we needed for health system personnel. This was a stunning and disturbing realization, given how vital N95 masks were for keeping frontline health care workers safe. In January and February, U.S. manufacturers had shipped millions of dollars' worth of masks and PPE to China. The federal government had encouraged them to do so. Because the leadership believed that the virus could be contained, they didn't proactively increase the supply of vital equipment. That should have been one of the top priorities in January and February. So was developing a needs-based distribution system for those supplies, which also didn't happen until after I came on board and rectified an inequity I saw. Fifty percent of the needed stockpile had already been distributed to states based on their populations rather than their need—another flaw in the pandemic preparedness plan. This meant that we distributed based on equality (the same amount for every state), not equity.

Equity means getting the right things in the right amounts to the places where they are most needed. Many of the states that received those supplies never saw significant Covid-19 cases in the first March-through-May surge. Triaging supplies is always a problem in pandemics, and this costly mistake substantially depleted the stockpile without adequately addressing need, worsening an already critical supply chain issue. Those precious supplies needed to go to the ten largest metropolitan areas seeing significant hospitalizations. Reducing health care worker exposure to risk in a timely manner would have gone a long way toward reducing stress on the system and on those on the front lines, reducing some of the trauma those workers experienced. Health care workers should never have had to reuse, ration, or only use partial PPE. We should have never had the surgeon general urging us to not purchase masks (and creating all kinds of mask-related messaging problems down the line) as a result of that early failure to see the clear signs that indicated to me the enormity of what we faced.

I didn't know if anyone had even gone to the states to assess their level of preparedness. Surely, some states had their own plans in place and may not have needed the same level of federal intervention. Would public health officials in those states make it clear that they actually didn't need what the feds had to offer? No. Everyone took what ASPR and the Federal Emergency Management Agency (FEMA) sent them.

To help address the N95 shortage, on March 5, I boarded Air Force Two with the vice president and the FDA's Stephen Hahn to meet with PPE maker 3M, in Minneapolis. Steve was new to federal service in the way I was new to the task force. He had come to the federal government to lead the FDA only in December 2019. Smart, dedicated, articulate, and passionate, he didn't take himself too seriously and was easy to talk to. We immediately connected.

With 3M, we learned that the United States' shifting of protective mask manufacturing and production overseas over the past two decades had had unintended consequences. Almost all the surgical masks, hospital gowns, and gloves used in the United States came solely from overseas, mostly from China and Malaysia. We did not have any real domestic production except of N95s, but we could ramp up domestic production of these only so much. We didn't make any of the essential medicines or critical PPE in this country anymore. This needs to change. The United States needs a secure and independent supply chain.

Shockingly, most of the N95 masks (those that filter out 95 percent of airborne particles) are produced for and used by construction workers. We make thirty-five million of them in the United States each month, but only four million go to health care workers. Those made for construction workers don't have the fluid barrier on the outside layer that is meant to prevent a large volume of blood or other body fluids from penetrating the mask. Otherwise, they are identical to medical-grade N95s in all their components, including, crucially, the ability to filter out the same size aerosols and droplets responsible for this viral outbreak.

If the FDA approved the use of these construction-grade N95s by health care workers, we would have nearly ten times more masks available immediately for those on the front lines. On top of everything else, most of the supplies of medical-grade masks, gowns, and gloves ASPR had ordered wouldn't be delivered until June. The FDA's Steve Hahn understood this and immediately took action, and overnight, under FDA emergency use authorization, the newly approved construction masks increased availability by a factor of ten. This would help, but we still needed more.

Looking back now, I believe this shortage of critical PPE (especially N95 and surgical masks) in February, at the very beginning of the crisis, had pushed the task force to think of masks as a scarce resource that had to be hoarded, instead of a public health measure that could slow the spread. This projected scarcity put masks in the wrong category—they were reserved for health workers only. And because N95 and KN95 masks (the latter certified in China) became precious, it was harder to advocate for broad adoption of them. Meanwhile, those responsible for the shortage weren't going to press for the use of masks of any kind, given that it would only expose to the public the depth and breadth of the issue and the lack of proactive preparedness by the leadership. Now, with masks in short supply, anyone in the White House or the federal agencies who was innately opposed to masking had yet another rhetorical crutch to prop up their position. Frontline health care workers needed them, the thinking went, not the general public. The former was true; the latter was not. The government's own long-term failings, fourteen years of underfunding the active stockpiling of needed pandemic supplies, was secretly being used to justify a suboptimal health care approach.

To be clear, recommending that masks be reserved for health care workers did not solve the supply shortages that had health care facilities rationing masks and other protective gear. The lack of masks and PPE for workers treating infected and high-risk patients created a secondary health crisis among those essential medical personnel. Compounding one mistake with another dug us deeper into a hole.

We could have climbed out of that hole if the CDC and other researchers had immediately started to test the effectiveness of other types of masks—cloth ones or less robust surgical masks. People in Asian countries had used cloth masks to prevent viral spread. But because the CDC had not studied other types of masks, it wouldn't formally recognize them as an effective mitigation tool. N95 and KN95 masks are named for the percentage of particles they block. They are the most effective masks readily available, able to screen out the smallest of particles, such as aerosolized ones. But surgical masks can be nearly as effective against respiratory droplets, which are larger than aerosolized viral particles. Their level of effectiveness varies, but they are still effective. The same is true for cloth masks, including ones made in the home.

The minute ASPR and the CDC knew that N95 mask supplies were limited, they should have sought alternatives and tested those, using simple lab experiments. Those studies could have been done in days. (Eventually, in early fall, the mask material tests were done, but by then, too much damage had been done, too many minds hardened.) In the absence of definitive homegrown studies of mask efficacy or mask materials, the CDC wasn't going to commit to recommending masks one way or the other.

Scarcity, plus this lack of verification of effectiveness, shaped much of the ambiguous initial messaging on masks. Had mask tests been done in February, we would have carried a consistent message about the initial effectiveness of cloth masks from early in March. (The more infectious variants that followed over a year later required an upgrade to KN95 and N95 masks, which by then were plentiful.) Instead, the messaging was mixed and confusing—as evident in the tweet that task force member and U.S. surgeon general Jerome Adams tweeted on the day I joined the task force: "Seriously people. STOP BUYING MASKS!

They are NOT effective in preventing the general public from catching #Coronavirus but if health care providers can't get them to care for sick patients, it puts them and our communities at risk!" Adams was used to working in the hospital trenches. He wanted to make sure his frontline colleagues were protected. Just weeks later, he corrected his position and even showed Americans how to make their own masks.

Still, not only was there no evidence to support his "NOT effective" claim, but every piece of anecdotal evidence we had at that point—from SARS in 2003 to Asia during early Covid-19 in 2020—showed us that masks of all kinds effectively slowed the spread of the virus. However, unless this narrative evidence were "validated" by researchers and approved by the CDC, the task force couldn't fully refute the surgeon general's position. Even in the absence of definitive studies, the CDC could have said what was clear to the naked eye: *Throughout Asia, people have adopted the wearing of cloth masks in public, and the case numbers there are low and their outbreaks more quickly controlled. Based on this, we think cloth masks could be effective here in the United States. We recommend you make use of cloth masks in public spaces while we work with our scientists to study their full effectiveness in the laboratory.*

The American people would have understood clear messaging that laid out what we knew, what we didn't know, and what we were recommending until we got the data we needed.

But the CDC didn't do this. More frustratingly, it didn't even say, "Wear a mask. It might save your life." Those at the CDC held themselves back. They didn't value learning from others around the globe. They didn't make commonsense recommendations while they waited to obtain the laboratory data so critical to their process.

Without a cultural memory of masks having been effective in the past, and with no governmental efforts to work around this, the public was, initially and very understandably, confused. To their credit, many people cut through the noise and wore masks regularly. Many more would have if we'd been clearer.

The CDC prides itself on being a preeminent medical and scientific authority. Historically, its adherence to scientific rigor has suited it well, but if a level of rigor isn't adaptable to the moment, it can become an Achilles' heel. In the midst of an enormous health care crisis, you can't wait until a definitive, tightly controlled, randomized, replicable, peer-reviewed study has been done to make a recommendation that will likely save lives while costing none. Telling people to wear masks early on had little downside. Even so, the CDC couldn't bring itself to make a less-than-precise statement on mask efficacy, just as it couldn't bring itself to rely on data and studies from other countries.

The push and pull in research are often between rigor and speed. But during a pandemic, when days and sometimes hours matter, when time truly is of the essence, you can't worship at the altar of rigor at the expense of speed. Yet this was exactly what the CDC did.

For many weeks and months to come, I fretted over how many lives could have been saved if the CDC had trusted the public to understand that it would take time to get the science right, but that, in the meantime, masks would do no harm and could potentially do a great deal of good in preventing people from transmitting or contracting the virus. Were masks 100 percent certain to work? No. Would cases go down to zero if masks were worn? No. But were masks doing something to slow the spread? I thought so, and clearly China, South Korea, Singapore, Thailand, and Vietnam agreed, because they were all aggressively promoting mask wearing, and their case numbers fell.

But this simply wasn't enough for the CDC, because the United States had no historical data to rely on to validate masks' effectiveness. Anecdotal evidence wasn't going to be enough for them to make a recommendation, and what anecdotal evidence there was didn't come from experiences in this country. An outbreak of this severity had never occurred in the United States during the CDC's lifetime. Collectively, the CDC, other agencies, and the public had no historical memory of the 1918 Spanish flu outbreak. The belief seemed to be that this early twentieth-century tragedy was *then* and this was *now*. Science and medicine had advanced to such a degree that we didn't need to rely on past practices or what had worked elsewhere. We could apply a modern, fully American response and evolve it based on data derived from cases on our own shores. Sometimes, common sense butts heads with scientific certainty, and that was the case here.

Relying nearly exclusively on data native to the United States was wrongheaded, and this problem was made worse by the CDC's lack of a fully integrated data collection system that gathered data from all hospitals and clinics. Because some in the CDC used the most rigorous research standards for evidence-based recommendations and analyses, the lag time between data collection and analysis was too great to support proactive responses.

The CDC never wanted to use so-called dirty data—that is, unvalidated numbers, even those from reliable sources in real time. The figures might have come from hospitals or clinics that weren't perfect about every granular detail. Though the basic information needed to see the big picture and formulate a response might have been there, data entry personnel, the CDC worried, might not have gotten every element right. The CDC also wouldn't use data from sources they believed hadn't met their highly exacting standards before. This is best practice when conducting a clinical trial or producing a scientific paper for publication, but in a pandemic crisis, "dirty data" has its uses in showing a general trend. It can be reflective of what is happening in the general population, not just within a select subset of individuals. Population data from all sources is frequently better than so-called clinical data, which is more narrowly focused. Indeed, sometimes, highly controlled data collection like that exhibited by the CDC, as compared to looking at the entire community in real time and real life, can result in significant bias.

Many times, at PEPFAR, data on our efforts in Africa was shared with the domestic HIV response team at the CDC. Our hope was that those responsible for the domestic program on prevention could learn from our work abroad. But the CDC scientists wouldn't verify it because they themselves hadn't done the work to create the program or collect and analyze the data. Instead, they took the time, often years, to gather and report the U.S.-based data on which to base recommendations. Wasting time and effort in a public health crisis like HIV/AIDS, which had been well managed, was inadvisable. Doing the same in an emergent and, likely, long-term public health care crisis was unconscionable.

Bob and I would talk about this throughout the spring of 2020. He recalled what it had been like for him to come to the CDC in 2018. He had a passion to change the course of the opiate epidemic in the United States, which CDC teams had been working on for years, but when he asked for the current data, he was given two-year-old statistics.

"'Where are the current figures?'" he told me he asked them. "'We're still validating it. It's still under review.' Two years after the fact and not reflective of what the present circumstances were."

"That's crazy," I replied.

"That's the reality," Bob said. "So, do you really think I can push my people on Covid-19 data and get anything close to real-time analyses and recommendations? We're not built for that, as much as I would like us to be."

In a crisis, you need to get all the data you can to integrate it into actions and improve and validate your response over time. You need to improve the actual data and data collection rather than merely improving the modeling of the data to fill in the gaps where data was not collected.

I could have, and should have, done better myself to push the importance of masks, testing, and data collection. I saw all these issues, many of them predating my arrival, but I was uncertain how to change the dialogue. I knew it would take years to reverse the institutional current flowing against me. Still, institutional intransigence was no excuse. I should have found a way to push harder. I did push for tests to be performed on the effectiveness of various mask materials, but I was never able to overcome the collective resistance to doing so, and in the end, the testing came from studies in Japan published in October 2020. This took far too much time and energy than it should have. I believed that masks were effective as a mitigation tool, but I didn't yet have the evidence needed to convince the skeptics, and I couldn't see any way around these obstacles other than pushing federal agencies to test non-N95 alternatives, which I did. Even so, I should have found other ways over, around, and through. This was the ethos that drove my mission.

After that trip to 3M in Minneapolis, the vice president, Steve Hahn, and I continued on to Seattle to get an on-the-ground reality check on what was happening in the first city knowingly dealing with Covid-19. The schedule was packed with meetings with state and local officials, public health leaders, and frontline health care workers. We met with Washington governor Jay Inslee and various U.S. representatives for the state. We also toured the State Emergency Operations Center and then did a press prebrief, followed by a press statement.

The meetings were critical for outlining what I had already been hearing. The state was already short on tests, testing supplies, and all forms of PPE. Just a few weeks in, and they were short on everything. We had become a country of just-in-time, expecting packages of supplies to arrive on our doorstep or to our hospital supply room or clinic overnight, and if not overnight, then within forty-eight to seventy-two hours. Critically, even in those early days, Washington State's hospitals and clinics were short-staffed. Caring for Covid-19 patients was time intensive and complicated. We went through the nursing home data. The state's nursing homes were ground zero for our most vulnerable Americans—elderly and often with comorbidities, everything that, as we had learned from Italy, Spain, South Korea, Japan, and China, put them at the highest risk.

The Washington State Emergency Operations Center was filled with dedicated personnel working around the clock—coordinating among FEMA, HHS, the federal agencies, and local officials to determine implementation and equipment requirements. Governor Inslee made some political statements during the press conferences, but the vice president didn't take the bait. He understood the pressures the governor was under. He knew governors didn't have deep expertise in public health, let alone with the deadly virus they were now confronting. It was scary and unnerving. For me, seeing the stress in the eyes of Washington State's public health leader haunted me over the next many weeks as what began in Washington began to play out elsewhere in the country. I stayed in touch with the Washington State health leader over the next several weeks to get his insights and take them to the task force.

Flying back home that day, I thought of something that had happened earlier that morning. In the task force call prior to our visit to Seattle, Joe Grogan, a senior advisor in the Executive Office of the President, had expressed his concerns about what was transpiring in that city with regard to the virus. He had been worried about the severity of the pandemic from the beginning, but on that morning, his political comment was representative of what I would later see. Joe understood that we were at the beginning of a massive increase in infections. As he put it, we would see a "massive surge in new cases reflecting community spread." He wanted to make it clear that, while in Seattle, we shouldn't imply that things were okay. This pandemic was serious.

I agreed with him. I didn't intend to say anything of the kind, and I certainly didn't make any such misleading statements during our visit. But I didn't agree with his rationale. Joe was afraid that if, after we left Seattle, the cases there increased, "We will own the new increase in cases reported." He didn't want it to appear that we, as representatives of the White House, were in any way accountable for the rise. He didn't want White House fingerprints anywhere around Seattle's outbreak.

During that phone call, he had recommended that we announce new measures while in Seattle. He would have preferred that this be done with Governor Inslee's support, but "if we have to jam him, we have to jam him," he said. He concluded by reiterating, "We will own it."

In that moment, I think Joe was clearly verbalizing what some were already saying behind the scenes in the West Wing—that is, this was not what he personally believed, but something others believed. Joe was an ally who understood how serious the crisis was and would become. He wanted me to hear what others were thinking, and he gave voice to those thoughts.

The message from those others seemed to be: The more we did, the more we would "own" the response—all the good and all the bad. I should have paid more attention in the moment to what he was trying to tell me. He'd issued a warning. Some roaming the halls of the West Wing believed that the less we did, the less we would be accountable for whatever was about to happen.

Too late. I believed we already owned it. Now I just had to make them realize it.

CHAPTER 5

Shaking the Flu-Like Message and Model

Sunday, March 8, marked the start of daylight savings time. We sprang ahead, and I felt the loss of that hour more profoundly than ever before. Still working on jet lag and a series of late nights, I had woken up at 2 a.m. to do my usual review of the worldwide numbers that had come in overnight. In the United States, the seven-day average had gone from seventeen cases at the beginning of the week to sixty-three by Saturday. Slowly, the numbers had crept up linearly day after day, even without the testing we needed. We were "seeing" only the very sick. This slow linear growth was the prelude to the leap I had been envisioning and fearing for us: the exponential phase of the surge, signaled by a doubling of the daily cases we had witnessed in Italy, Spain, France, and Germany.

These numbers were only one concern among many. Normally an optimistic person, I wouldn't have taken the job if I didn't believe I could effect some positive change. But sitting there in the dark, the room cast in shadows by the glow of my computer screen, I felt the doubt and fear brought on from my first full week finally creep out of the shadows and into the light. They had unmasked themselves; I smiled regretfully at the thought.

Joe Grogan's statements during the Seattle trip about our "owning it" weighed heavily. Acknowledging the need to be held accountable and fearing the consequences of being accountable are two very different things. Suspecting that Joe had voiced the prevalent attitude throughout the White House, one that Matt had also identified, was disconcerting. Up to that point, no one else had flown these colors so blatantly. To counter this desire to evade responsibility, in my response to him then and at every opportunity moving forward, I refocused the message.

The federal government *was* responsible. Only it could perform certain functions, one of the most critical being comprehensive data collection in real time across all the states and territories. Only the federal government possessed the enormous resources required to accomplish all facets of the response. We were accountable, and we needed to act.

To counter the mixed messaging reaching the American public when we had fewer than one hundred cases, I asked the vice president to bring the media's medical correspondents into the White House that first week so I could articulate for them my initial core messages:

This is not the seasonal flu.

This disease, in specific age groups, is deadlier than flu. In others, its presence is invisible.

We cannot completely contain this outbreak. It is out there. It is spreading.

We need your help in communicating this message effectively.

In particular, CNN's Sanjay Gupta was a key component of my strategy. From the onset of the outbreak, while I was in South Africa, I had watched as Dr. Gupta served as the senior CNN medical correspondent and the network's chief Covid-19 correspondent. He specifically spoke about a mild disease—another way to describe silent spread. I saw this as a sign that he got it. As a doctor himself, he could see what I was seeing. He could serve as a very good outside-government spokesperson, echoing my message that family members and others they were in close contact with could unknowingly bring the virus home, resulting in a catastrophic and deadly event. It would be important messaging going forward, but it could do nothing to slow the spread that had already been set in motion.

The truth was: we were behind on everything. I'd been in the White House a week, and in that time, I had not heard a single piece of news that placed us on track with where we should have been—save for a single meeting the president had convened about vaccine development that put the time line at twelve to eighteen *months*. We also spoke with the pharmaceutical companies creating therapeutics. These would be available within months. I felt that the vaccine estimate was fairly conservative, one that we could almost certainly beat with the right approach and constant attention to the details of each phase of the clinical trial enrollments. The president was actively engaged with the vaccine companies, repeatedly pushing them for more faster. He had also attended the testing meeting with the commercial developers and urged them for faster and more. The vice president continued to prod them during our weekly calls.

Best case, though: a vaccine was a long way away, too far in time to take any consolation from that one meeting. The concern I'd heard in Matt Pottinger's voice was now running through my own mind. There was no way around the fact that the response to date had been an interrelated series of failures, starting with the unfounded belief that the virus was being, and would remain, contained.

Seattle, and the on-the-ground realities I had heard and seen there, had driven home the hard truth of the consequences of these failures. Insufficient supplies of PPE, severely limited testing, the imposition of a death sentence on many nursing home patients as the virus was silently brought to them unknowingly by the staff—all this amounted to a specter of death hanging over the coming weeks. I was suddenly brought right back to those dark days at Walter Reed dealing with AIDS. What was happening in Seattle would soon be happening all across the country, in larger cities that were just as unprepared—New York, LA, Chicago, Houston. When the relentless wave hit in a little more than a week, every shortage, every misstep, every failure of vision, would be laid bare. And there would be no way to change course without drastic, seismic action that would be as unpopular as it was essential to saving lives.

The question I wrestled with was how to change minds within the DC bubble, how to point out the groupthink that was shaping the faulty response. I needed those in the White House to switch strategies on everything, from masking to testing to silent spread. But more than anything, I needed to figure out how to make them accept and prepare for the unacceptable: a shutdown to flatten the curve, to prevent every hospital in the United States from being overwhelmed. We were weeks into silent community spread. Containment measures had inevitably failed, so, the next logical step was to protect hospitals in states with community spread. In the other states, we could actually prevent the silent invasion and the initial community spread that eventually led to the vulnerable.

I SPENT THE BETTER part of that Sunday formulating my plan. While I had prepared for a lot of eventualities and had sketched out many plans before arriving at the White House a week earlier, I had not accounted for the dire scenario we were now in. Had this outbreak occurred under any other administration, I would have gone into work the next day and stated the best course of action immediately, no matter how painful or extreme it sounded—because any other president would have wanted to know just how bad things were going to get and what could be done to prevent the worst case.

Except, this wasn't any other president or any other White House. This was President Trump and the Trump White House. I was standing on constantly shifting sand, among political players I didn't know and a president who apparently liked his news served good and upbeat, or not at all. The right approach here would be as important as the right pandemic response plan—perhaps even more so.

I had always prided myself on being a hard-charging, get-shit-done public servant. Complacency doesn't suit me; I've always been a doer. And I'm not always patient. Recognizing my own tendencies and flaws was nearly as important now as assessing the current situation. Unlike the CDC, which had approached the novel virus with a standardized response, I had to modify my approach to meet the situation I had found on the ground.

I'd expected some level of resistance coming into the job. But what I encountered wasn't just different thinking, but a different attitude toward action. I believe in proactive intervention, but the response to date had been stuck in reactive mode. I was always in prepare-now mode. I'd have a Plan A, B, and C ready in case Plan A didn't work. But the Trump administration and the government agencies had been in a wait-and-see mode. When they acted, they made mistakes, and they didn't like being seen making mistakes. This made them gun-shy.

For this reason, charging hard wouldn't move either the politicians or the doctors. That things had been allowed to get to this point spoke volumes about how deep the dysfunction ran. Because negative patterns of behavior had already been established, they would be that much harder to reverse. I wasn't starting at zero; I was starting at negative twenty. I was the one who was going to have to adapt, not them.

I would do this, I decided, by several means. The first was to continue the same global numbers drumbeat. I had to be selective, however. I would emphasize Europe, given that many in this White House and on the task force understood Europe better than they did Asia.

That Sunday, March 8, as I was wrapping my head around how to make myself heard, Italy announced, with a screenwriter's timing, the shutdown of its northern provinces. Its strict measures affected sixteen million people living there, more than 25 percent of its population. The spread there was placing significant strain on hospitals; Italy's sophisticated health system was faltering.

National government decrees banned movement into and out of areas in the north except for emergencies. The message was clear: stay home to stay safe. They meant it, and they would enforce it. Festivals, football (soccer) matches, church services, and other large-scale events were canceled. Italians needed passes just to go to the grocery store. (Researchers, by tracking cell phone signals, showed that mobility dropped by 90 percent.) Those who violated the decrees were subject to fines and imprisonment.

Because the Italian situation was rapidly deteriorating, I pushed the rest of the task force to acknowledge that, as Italy went, so went the United States. I believed that we were already into the same viral spread Northern Italy was experiencing. Our hospitalization rates were rising, but with testing still mired, case rates painted an inaccurate picture. Our projections put us where Italy had been ten to fourteen days earlier. They were now in lockdown, where we likely should have been and would need to be in a week's time.

At this point, I wasn't about to use the words *lockdown* or *shutdown*. If I had uttered either of those in early March, after being at the White House only one week, the political, nonmedical members of the task force would have dismissed me as too alarmist, too doom-and-gloom, too reliant on feelings and not facts. They would have campaigned to lock me down and shut me up.

Little did I know.

I had been the only woman in the room before and was familiar with the slippery slope of being pegged as "hysterical" and "overreacting."

I'd experienced having every point I made followed by statements like "You know how she is. It's not that bad." This response had been my constant companion throughout my career—as a woman, I was primarily dubbed too aggressive, too pushy, or too direct. I had been in the room when this was said about other women, and I knew the same was said of me when I was not present.

Spooking anyone with extreme scenarios wasn't going to work in those early days. I couldn't use Italy's actions as a model for our own, but I could use them as a canary in the coal mine to demonstrate how bad things could, would, get. The virus's current crushing impact in Europe had its roots in the early steps those countries had taken, many of which were very similar to our own. Their leaders' efforts were largely ineffectual because they missed the early silent spread. As a consequence, the virus was in full, exponential growth across Europe. I needed to use these European facts to get the White House to act quickly, as Italy had now acted.

Still, it would be a tough sell. During the conversations I participated in that first week, I saw that "American exceptionalism" was a very real problem, one that was making the situation harder. People in the White House thought the United States was special, that our health care system was special, that the biological rules this virus played by didn't apply to us. This misguided sense of infallibility would prove deadly. Every task force member needed to understand that our hospitalization numbers and deaths wouldn't be better than those coming out of Europe. They would be worse. This would be difficult for them to accept.

To counter this, the second front of my plan was to use the time allotted to Tony, Bob, and me at our task force meetings over the next week as strategically as possible. I would build toward an end-of-the-week pitch to this administration for the most aggressive mitigation steps possible. Knowing it would never fly, I wouldn't seek a "shutdown" in the Italian sense, but the closest possible thing I could get from this administration.

Unfortunately, this would mean not taking the full measures necessary at that point. I had to accept this reality. Some would think they were too drastic; some would say it was up to the states; some would just ignore the issue. Still, anything short of a similar, flattening-the-curve-type shutdown would leave us exposed to the fully overrun hospitals and fatalities seen in Italy. We needed to act, but the worst possible thing I could do at this stage would be to push too hard and end up with our doing nothing. We were so far into community spread that we now had to flatten the curve. By enacting any mitigation strategies, we would be acknowledging that the continuous rise in cases was inevitable. Spreading that increase out over a longer span of time would keep the rate of rise lower, so cases wouldn't overwhelm health services, supplies, and health care providers. This wasn't an optimal mitigation strategy, not even close, but it was the best outcome we could hope for at this moment, with where we were heading, and with whom I was dealing.

What I would pitch to the vice president at the end of the week would function like a circuit breaker. It would be a set of measures designed to slow the spread and buy us time. I wouldn't use the alarmist language that could set off the economic team's and the president's tripwires. I didn't yet know how my proposal would make its way to President Trump—whether the vice president would bring it to him or I would. Ultimately, it didn't matter who the messenger was; all I cared about was the time line.

In a week, we had to sell our version of a shutdown. Together, Bob, Tony, and I needed to take a step-by-step approach to lead everyone to the eventual conclusion that this makeshift shutdown was the only option to protect every metropolitan area in the United States from that initial surge. If we moved too quickly to get them to that point, we'd lose the narrative and our audience. Regardless of how I dressed up the language, shutdown was a terrible outcome for any White House, but in this one, it would be anathema. The resistance to it would be overwhelming.

That's why the third, and perhaps most crucial, part of my approach was to try, once again, to get more of the medical establishment to acknowledge the greater degree of silent spread. The fundamental disagreement over the role of silent spread was the common denominator for all the problems that had led us to this moment. I had to find a way to change minds on this—starting with the rank and file at the CDC. Thought leaders in Atlanta had continued to discount the role of silent spread, and the cascade effect of the CDC's "flu-like" characterization, along with its influence on states' responses, could do great harm. And unfortunately, Bob could only do so much to influence the CDC rank and file.

The CDC is populated with many talented and gifted professionals, many of whom also happen to hold liberal political points of view.
As a political appointee, Bob was stained by his association with a
Republican/conservative administration. The rank and file would question everything that came from Bob and this White House. Whether
it was because of my background in the military (once considered a
bastion of conservatism) or because I was now part of the White House
response team, I suspected that my views would be examined skepti-
cally through these two political lenses, traditional conservatism and
Trumpism, as well. Fair or not, scientifically objective or not, this was
the reality.

To begin to confront this, the day before, on Saturday, March 7, I'd
spoken to Tom Frieden, the CDC's director under President Obama from
2009 to 2017. I wanted another CDC ally, and Tom was well respected
within CDC circles specifically and in the field of public health generally. Though Tom was no longer the head of the CDC, he still had
some influence—considerable influence, I thought—over the opinion
leaders there and those in positions of power within the agency. I was
direct with Tom: I needed the CDC, most notably its most senior officials
Anne Schuchat, Nancy Messonnier, and Dan Jernigan, on board the
silent spread train. I also needed the CDC's talent and reach. Critically, I
needed the agency to evolve beyond temperature checks and symptoms
to see the silent spread and to move to proactive testing as a pillar of the
public response.

Tom wasn't quite there yet on the magnitude of the silent spread.
On our call, he mentioned that same 17 percent figure derived from
the CDC's interpretation of the *Diamond Princess*. I wasn't deterred. I
believed we could bridge that gap. I sent him the analyses Irum and I
had done of the data. He promised he'd review them. I knew he'd look
at the data without bias, and I was confident that this would get him
to my side eventually. I couldn't say the same for the folks at the CDC,
given how entrenched their position was, but Tom could prove to be
a large weight on the scale, tipping those flu model hard-liners in my
direction. Getting Tony and Bob was key, but getting Tom was arguably
more important, given that the climb up the medical mountain would
be insurmountable without more support from inside the CDC.

To do this—to tell this story for Tony, Bob, and now Tom—I needed
numbers, all the numbers. I needed access to whatever data Bob and

Tony and the CDC had. I needed to understand the full breadth of what data was being used to make decisions. Based on what I'd seen that first week, the data was far less than I had hoped for and much less than I needed. But I was still convinced there had to be something more granular, something that could tell me with greater specificity who was getting sick, how old they were, and precisely what areas of the country they lived in. Digging around for this data became one of my main priorities for the week ahead, so I could marry the U.S. data with the European data to show clearly where we were and where we were headed. Once I had the data, I could become more aggressive with my lobbying efforts, because by then, the math would be incontrovertible. *Patience.* I wrote the word at the top of a report I'd been reading. I didn't like approaching problems too slowly, too stealthily, but I had to do what was needed.

As I shut down my computer the Sunday night prior to the task force meeting the next day, I reminded myself that this administration had spent the last three years railing against the "deep state," accusing professional civil servants like me of supposedly undermining President Trump's leadership. Weeks of inaction had gotten us here, but once again, I reminded myself that I couldn't risk falling farther behind by rushing too hastily. At this early stage, marching in with bold pronouncements was only going to get me fired or, worse, ignored.

CHAPTER 6

No Numbers, No Forecast, No Plan

I knew the week ahead would present opportunities to put all the different elements of my plan in place, but I started the second week with one clear focus above all others: to use the best available data to demonstrate the way to move forward. To do this, I needed help.

I already had a clear sense of whom I needed for my small but essential data team: people whose tirelessness was a testament to their character. At first, I thought I would be able to find them on my own, but then Irum called me after she'd arrived from South Africa. She was coming voluntarily. She knew I would need support, and she was offering it. I needed her aboard immediately, and she was.

By Friday of the first week, I'd requested two others from the White House: Daniel Gastfriend from the Office of Management and Budget (OMB) and digital services expert Amy Gleason. Daniel had studied public administration and international development at the Harvard Kennedy School. He had also worked in South Africa, Uganda, and India. I hoped that having someone from within the White House Office of Management and Budget would help offset the "deep state" perceptions. Daniel was also young, very talented, and a tireless advocate for public health responses, possessing a wealth of knowledge about health security.

Amy Gleason had worked for years in the medical data field before joining the White House in the United States Digital Service team in 2018. Her previous experience with OMB and the White House would help in her management of services—and my management of perceptions.

Others I had worked with for years from the CDC volunteered to help me. Chuck Vitek was a CDC epidemiologist I knew and trusted. He was board certified in infectious diseases and had a strong public health background, having worked in the U.S. Public Health Service's Commissioned Corps. Behind the scenes initially and out front later, Dr. Sean Cavanaugh was invaluable. With prior CDC and Department of Defense experience, he understood what frontline work was all about and the need to do it now and not wait. Bob Redfield said he'd be willing to release Steve Redd from his job in the CDC's Emergency Operations Center in Atlanta. Steve served as an important conduit to CDC staff in Atlanta, aiding us greatly in aligning our mission with theirs.

I was glad to have CDCers in-house to interface with the agency's data people. Steve was senior, smart, and a critical influencer within the CDC. If I could convince him and Chuck of the depth and breadth of the silent spread, both would be helpful in convincing senior CDCers that their flu model of tracking symptoms was wrong for this virus. We were fortunate in that Steve knew the individuals as well as the group-think that often characterized the CDC's positions and policies.

They knew what I was up against and how the negative perception of working in this White House might affect their careers, but they volunteered anyway. I will never forget all of them or the commitment and sacrifice they displayed in taking on something so hard and never ending.

We set up a command suite in a large conference room in the Eisenhower Executive Office Building (EEOB). Though the team comprised just five people on-site plus Sean, they were all outside any political bubble. I could trust them to look at the numbers and provide an unvarnished analysis free from a hidden or political agenda. There would be no groupthink within my inner circle.

Doubt existed about the extent of the silent spread. In our first week working together, we'd wind up in spirited debate. The working group demonstrated what I had hoped to find when I first came into the task force: people working from the best available information, using their diverse knowledge, skills, and experience to arrive at the best approach to the problem at hand.

Right away, I started pressing Steve Redd on the CDC's data streams, and he explained that the CDC didn't have the demographic data I was looking for. Worse, the data it did have would never help paint an accurate picture of this pandemic outbreak. When it came to an outbreak like Covid-19, the CDC reporting system was riddled with problems. Hospitals weren't required to report their data, and the voluntary data-reporting facilities that existed weren't distributed evenly around the country. Models based on this geographically biased data would, therefore, always produce inaccurate projections. Reports the CDC did receive included symptoms but no confirmed diagnoses of an illness. These symptoms were used to cluster illnesses into less specific categories, like "influenza-like illness." Without knowing for sure what was causing the symptoms, we faced a whole new kind of ambiguity in the data sets. Was it flu or Covid-19, or was it both? Without a laboratory test, it was impossible to separate the two.

The CDC also had no laboratory reporting to it outside its own Public Health Laboratories. There was no way, therefore, to see the total number of tests performed, the total number of positive tests. Test positivity rates were a way to understand whether we were finding most of the infections and the only way to define the level of asymptomatic spread.

There was also no comprehensive hospital reporting of even symptomatic respiratory diseases. The CDC should have demanded that 100 percent of U.S. hospitals report all their Covid-19 cases and pushed for definitive laboratory confirmation of all cases.

Instead, the CDC relied on another reporting system—the National Hospital Security Network for its hospital figures. The NHSN was used primarily to track hospital-acquired infections and community-acquired infections that required hospitalization. It was also used to produce projections about the depth and spread of flu outbreaks. Unfortunately, this voluntary CDC-funded system was in place in only 40 percent of hospitals across the country. It could never provide us with an accurate view of what was happening in the remaining hospitals, only in modeled assessment. Worse, these figures often took weeks to get to the CDC. With only 40 percent of the delayed hospital data available to project what was happening across the country, we'd have a less accurate geographic assessment of the depth and spread of the outbreak.

It was also only a window into those sick enough to be hospitalized. In parallel to utilizing global health security funding they were tracking a small subset of United States emergency rooms symptom-based illnesses including influenza-like illness. Again, only those sick enough to come to emergency rooms. It wouldn't help us account for silent spread, and any projections based on this biased and delayed data when used to create a predictive model would be inaccurate.

Also, the NHSN was especially problematic for use with Covid-19 specifically. If, in early March 2020, two people with "flu-like" symptoms went to a reporting hospital, they would both have been recorded in the CDC database as experiencing "flu-like" illness. If those patients weren't tested for Covid-19, then they would remain in that flu-like illness category until tests determined whether they had seasonal flu or Covid-19. Even worse than this incomplete and slow accounting of hospital cases, relying on symptoms as a diagnosis meant the system once again offered no visibility into how Covid could be spreading asymptomatically. The only time the system could be trusted to identify Covid-19 disease by symptoms was when the seasonal influenza and influenza-like illness (ILI) declined and one could clearly see the second peak in the spring of 2020 due to Covid-like illness (CLI). The public health labs could test samples for influenza (H1 and N1) strains, but their platforms weren't easily adaptable to test samples of suspected SARS-CoV-2 infections. It took some time before those lab tests were available to identify definitive positive cases, further delaying the CDC's ability to see and to communicate what was really going on.

Along with these delays, because this system was used mainly for infection control and analysis, it didn't track the hospital PPE utilization or need. Consequently, there was no centralized built-in alarm to alert us when protection for frontline health care workers was running short. We needed to move to a proactive system. We couldn't wait for the phone calls from hospitals notifying us that they were totally out and in crisis. I was told by CDC officials that tracking hospital supplies was not their mandate. They wouldn't add the supply element module to their system. Their refusal would require us to create yet another reporting system for hospitals. Now they would need to submit data three times. First for clinical care based on codes, next to provide data to the NHSN, and now data to the federal teams to support supplies. This resulted in them triplicating their efforts when they needed to simplify their response due to the coming crush of patients. Without a federal level reporting system, requests would come in from the states and local officials in a far less organized manner than what was needed.

If 100 percent of hospitals, emergency rooms, urgent care clinics, and pediatric offices had been part of the CDC's early warning system as well as definitive laboratory diagnosis of ILI, Covid-19 would have been spotted far sooner on our shores, and we could have reacted much more quickly. Without a doubt, there are Americans, particularly in New York City, who went to the doctor, hospital, or ER with Covid-19 in February 2020 and were sent home with a symptomatic diagnosis of the flu, without a flu test being performed. If they had been tested for flu (or for all viral respiratory diseases), a red flag would have gone up when the results came back negative. Medical personnel likely would have realized that this person was sick with the same mysterious respiratory disease reported in China, and that was not the flu. That's how new viruses typically get discovered: doctors run normal laboratory tests and, through negative results, learn that it's none of the usual infectious suspects. Through negative tests for the flu, this mysterious illness would likely have led local doctors and, eventually, the CDC to connect the dots to what was happening in China in February 2020.

Instead of those early Covid-19 patients being caught in our early warning system, no red flag was raised, and those cases began to spread in the United States. For weeks the U.S. would stay stalled in the "containment" phase rather than aggressively mitigating. The Plan A that had long been established and agreed upon was based on this symptomatic detection. There was no Plan B then, and to this day there still is no Plan B. We would work to resolve some of these issues, but at this critical juncture, while I was trying to formulate a case for shutdown by any other name, the existing CDC data wouldn't be of much help to me.

When I first went to Africa twenty-two years ago, no comprehensive laboratory or health data system existed. It took several years to build one for HIV, and we added additional disease models over time, while also including laboratory information systems. These databases and programs integrated data from various sources, allowing us to account for different aspects of the infections and disease. This comprehensive data system and population-based surveys allowed us to track the quality of care and the outcomes and impact of the program. These systems were the backbone of our collective ability to begin to control the HIV pandemic in Africa. From 2014 on, we did all this efficiently and accurately, while also adapting the health data system flexibly, to meet our needs and the specifics of a disease. Shockingly, this wasn't the case with the CDC's system in 2020, and it still isn't today.

If we could design and use these essential tools in resource-limited settings like Africa, why hadn't it been done in the United States? My assumption had been that the CDC had a best-in-class system to gather baseline data from all the hospitals and nursing homes in the country. If you have a new infectious agent you're worried about, a health data system is where it would show up first. But the CDC didn't have data collection like this in place.

The United States needs a Plan B, one in which all community-acquired infectious diseases are reported and definitive laboratory diagnostics performed. We need to move into the twenty-first century, using our amazing technical capacity linked with existing coded electronic medical records to create a national database on current community infectious diseases, so we can spot the new ones as soon as they emerge. This integrated data system would also ensure widespread testing capacity and better antivirals for treatment. Not only will that help us begin interventions sooner, but it will give pharmaceutical and biotech firms an earlier start on developing viral specific therapeutics.

After I learned about the CDC's data collection issues—the delays in reporting as well as the vastly incomplete data which was missing age, race, and ethnicity information—the CDC's reluctance to accept my ideas around silent spread began to make more sense. While I had been poring over the more complete international data from Italy and other countries, data whose implications were obvious to the trained eye, the CDC had been focusing on more limited, domestic data. It wasn't seeing the asymptomatic spread because its early warning system was designed to show the exact opposite: growth in the number of people with significant symptoms, and only those who sought treatment in a clinic or hospital—with luck, one of the 40 percent the CDC was monitoring. This was why it was so hard for CDC researchers and officials to see the role asymptomatic cases were playing in fueling that growth—the CDC was operating in a bubble it had itself created. It was almost as if it was working to formulate a response to a completely different virus than I was. At this point, in the second week of March 2020, Johns Hopkins was doing a better job of aggregating count and location data across the globe than the CDC was.

Making matters worse, the CDC had put this symptom-based, incomplete flu model data through the sophisticated predictive model it had developed to show how serious the novel coronavirus outbreak would become in the United States. It then used those results to determine what was happening in the country. This compounded the agency's initial errors by making predictions based on the poor visibility of the original data.

Yet the CDC believed steadfastly in the accuracy of its predictive models. If it had been the National Oceanic and Atmospheric Administration doing the same thing to create a weather forecast, outdoor plans might have been disrupted. But when it is the premier public health agency in the United States looking at an emerging pandemic, the consequences are far greater. No one individual dropped the ball. The system had been designed this way, and its problems were deeply entangled with how the CDC had related to the states for years. While it might sound overly ambitious to imagine every hospital in the United States reporting data to the CDC every day, we made it happen by June 2020. So, it wasn't that hard to envision every one of those six thousand hospitals reporting detailed data—if only the CDC had set up the infrastructure for this. The same held true for nursing homes: there was no mandatory reporting out of nursing homes, and yet there were only fifteen thousand of them in the country. Seema Verma took this on and made it happen by June 2020, proving that the problems with mandatory reporting weren't intractable. What it took was finding the will to do it.

This lack of will was the root issue. The CDC has never required anyone, neither the states nor hospitals, to report data outside a specific list of required reportable diseases, and that reporting is often done weeks to months after the cases are diagnosed. Every year, the CDC disburses an enormous amount of taxpayer money in the form of block grants and cooperative agreements to all fifty states, local governments, nonprofits, educational institutions, and for-profit groups, but it has been historically unwilling to attach strings to those federal dollars in the form of mandated data reporting and accountability for the dollars spent. Work plans are often submitted and reviewed, but there was no required reporting of outcomes or impact. Had the rates of diabetes, hypertension, and obesity declined? Were the dollars linked to the reporting of respiratory diseases (including viral ones like Covid-19) and to improving their laboratory diagnosis in real time? Without mandatory reporting, we lacked the data to answer these questions.

Similarly, despite the federal funding going to the states, CDC personnel believed they had to be invited into individual states during an outbreak or other public health emergency. So, unless a state wants them there, the CDC lacks the power to do anything directly. Basically, data reporting became something of a states' rights issue. Knowing they would need permission to come into these states, the CDC didn't want to create any ill will through mandatory data reporting that might contribute to the states' refusing them entry. The compromise that had been reached with the states came in the form of "voluntary" reporting, and the CDC lived in fear of having the states turn off this voluntary data spigot. At the CDC, the mantra appeared to be: Tread lightly, don't press too hard, and be grateful for what little data you have access to, no matter the delay or incomplete nature of the data.

This jurisdictional tension would play out in many ways during this crisis, but at the start, it led to the CDC's not pushing the states too hard for detailed data and relying heavily on its preexisting voluntary reporting systems, including those where the states used paper and fax machines to transmit data. Any remedy to this system would be tantamount to making what had been voluntary, mandatory, and getting to mandatory reporting would take changing legislation and guidelines. But these actions were too time consuming and too complex to take on in the immediate. (Eventually, after a great deal of lobbying on my part, I was able to get mandatory reporting of all laboratory PCR test results, both positive and negative, from the states to the CDC. This required getting the White House to weigh in and make sure that this change was included in the Coronavirus Aid, Relief, and Economic Security pandemic-relief bill known as CARES, passed weeks into the pandemic, after the situation had already grown dire.)

Before these systems were up and running, both Roche and Abbott, due to the great work by the private sector in developing and shipping tests by the end of March, were sending me their nightly equipment reports from their platforms, so I could see hospital and commercial laboratories' test positivity rates. In many labs, that figure was between

30 and 50 percent. This gave us some insight into how bad the next few weeks would be. Night after night, month after month, this data arrived from laboratory after laboratory. Having all this data allowed us to triangulate it with other figures and use the findings to guide our actions. This contribution from these labs demonstrates the value of mandatory data reporting. We got there, but if we'd had that more complete data set earlier, we'd have had a greater sense of urgency sooner, urgency that could have been translated into more proactive measures being implemented sooner.

In the end, the flu bias in the CDC's data system and the agency's jurisdictional limitations exposed, once again, the rigidity in its thinking—only, now, instead of on masks and tests, it was on data collection. To work around this systemic inflexibility would require more than just a tweak. We would have to significantly modify the models used to produce accurate projections, something that would require between weeks and months of work. More time lost meant more lives lost. This was the reality we faced in early March 2020.

As is true for so much of this story, the failure here does not belong to any one institution or person. It's bigger than that, broader than that. It defies easy explanation or reflexive answers. It's easy to lay the blame at the feet of the Trump administration, saying things like *Well, this all happened because the Trump administration gutted the CDC.* But it's crucial to understand that the CDC had been fully funded for global health security and domestic health security since 2009. It had received millions upon millions of dollars in global health security funds every year for precisely the situation we were now in. They made decisions that resulted in the absence of a comprehensive national clinical and laboratory data system. No one else is accountable for that.

In other words, this specific issue was not a Trump administration problem; it was a CDC problem. No one in any presidential administration had instructed the CDC to design their data system this way. This was a plan the CDC had devised and funded. It was a twentieth-century solution for a twenty-first-century world.

It's convenient to solely blame President Trump, and while he certainly deserves his share of the blame and then some for his response, blaming only his failures and letting the CDC and other HHS agencies off the hook for poor institutional choices over the years would be a recipe for repeating the same mistakes in the future. From masks to tests to data collection to flu bias—the CDC made several very real and very consequential decisions that shaped the early months of the pandemic and are still shaping events today. To this day, the data is incomplete. It is collected and analyzed too slowly. Small convenience studies are done rather than comprehensive country-wide analyses of everything from testing to sequencing to the impact of differential mitigation. As I write this, there remains a lack of emphasis on active testing, and there is still no sense of urgency to find the silent community invasion that occurs days to weeks before symptoms. This is bigger than who the president is or which political party is in charge. In many ways, anti-Trump sentiment has prevented people from seeing the full spectrum of the breakdown at the CDC in the pandemic's early months and that continues today and still needs to be addressed.

Ultimately, the CDC's greatest sin was not recognizing and acknowledging that it had sinned. Instead of fixing the issues, it hardened its belief that its approach was the best because they were the best.

Simply put, I, along with many others, saw that the CDC's numbers were incomplete. By plugging its flawed data into their models, the agency's researchers were never going to produce an accurate forecast. Even much later, when more data was available, they clung to their initial projections. As difficult as it is for me to believe and to write this, it's true that one of our nation's finest medical/scientific institutions, one of the world's leading public health entities, believed that the projections it had derived from its model of cases and hospitalizations were more accurate than the real, comprehensive data itself.

This really stunned me. I pushed back on this notion constantly, but they were immovable.

This resulted in consequences in that first year that carried into 2021. Each and every surge through 2020 and 2021 was visible early, with rising test positivity in the younger age groups appearing days to weeks prior to hospitalizations increasing. Early and expanded testing of community college students along with sequencing would have allowed us to "see" the virus in the community early and see the new variants much earlier. But each time, the alert came too late, sometimes because of incomplete or late data—or, maybe because each time, we believed it would be different. Reacting late each time and not testing broadly enough cost lives then and now. Later, when we were behind in finding new variants, we had to rely on other countries, where the data was available. The same was true for vaccines and the durability of vaccine protection: we had to rely on the well-collected data from other countries to make decisions in this country, the United States, the country with the most advanced IT and data collection capacity, which we could have tapped into from the private sector.

When it comes to bad data producing bad projections to guide actions, it's hard to determine the specific long-term consequences. What I can say is this: real-time data allows you to understand the present and predict the future with a fair amount of accuracy. This is especially true when tracking relative rates of increase and not just an absolute number. The confusion around these two concepts—how fast an increase is occurring versus what the overall tally is—continues to distort perceptions and delay actions. If a wildfire is approaching, knowing how fast it is moving and not how many acres it has consumed so far will better guide your immediate action. The rate of rise determines the level of urgency more than an accumulated total does. The first of these helps predict what is to come; the latter characterizes what has already happened. Too often, the CDC relied on this "rearview mirror" approach instead of using data to determine a future course of action and make corrections.

The more information you have, the better able you are to show people, with a high degree of certainty, where they're going. Incomplete data opens us up to inaccurate interpretations, mischaracterizations, and politicization.

Perhaps it's not surprising, then, that all three of these happened next.

ON WEDNESDAY, MARCH 11, we gathered in the Oval Office for a face-to-face meeting with the president. The only item on the agenda was issuing another travel ban—this one to and from Continental Europe.

I wasn't looking forward to this meeting. Two days earlier, the task force had moved in favor of the European travel ban over the strong objections of Trump's economic team. Now we came together to debate it all over again—only, this time, we were playing to an audience of one. I immediately thought the ban made sense, but to convince others, I had to appeal to their primary interest: the economy. President Trump believed that his policies had been the prime mover of the economic engine, creating the "best economy in the history of the United States." It was going to be difficult to engage him on public health without acknowledging the economic impact. Consequently, I had to tie the two interests together, and how I united those strands would be critical.

To that point, my interactions with the president had been very brief and limited mostly to the formal meetings. Once in the Oval Office with the entire task force and various members of the president's team of advisors, I wasn't sure what to expect. I looked around the room. Everyone seemed comfortable. I wasn't. All the reservations I had had about accepting this job were seated, with a few exceptions, around the room. I didn't feel I was surrounded by enemies; I felt surrounded mostly by unknowns.

One known was this: this president wouldn't have much patience for lengthy explanations or rationales. I would have to be on point and concise. I couldn't do anything that would reveal my true intention—to use the travel ban as one brick in the construction of a larger wall of protective measures we needed to enact very soon.

The vice president opened by stating that the task force had agreed that this European travel advisory was necessary. The president took this in and then immediately, and perhaps predictably, turned to Steve Mnuchin and asked him for his thoughts.

I had come to see Secretary Mnuchin, Trump's secretary of the treasury, and Larry Kudlow, his director of the National Economic Council, as the frontline players on the administration's economic team. In addition there were the most important backbenchers, Marc Short, Stephen Miller, and the communications team at the Office of the Vice President, who, prior to my arrival, had agreed that the imposition of flight restrictions on China was warranted; now they didn't see this to be the case with Europe.

Their reason? Secretary Mnuchin spoke for those opposed to the European travel ban. It would have a much greater impact on the U.S. economy and—left unsaid but clear—on the stock markets than the China ban had. He listed the reasons: With many more multinational businesses with offices in the United States, Europe, and the United Kingdom, many more people flew back and forth from Continental Europe. This could further cripple the airlines. I also heard talk about the effect on the gross national product numbers. Not just people, but cargo flew on those passenger planes. What was this going to do to the world's supply chains? What would that mean for retailers and others along the way?

When Secretary Mnuchin was done speaking, the president turned to Bob, Tony, and me. Bob spoke first and in slightly more detail, elaborating on the points the vice president had made. He kept his focus on public health matters, making the case for why the travel ban was the right course to protect people. I made the same argument and noted that European cases were higher than China's had been when the ban on visitors from there was imposed. I had a somewhat simplistic, but nonetheless viable, argument to counter that of the economic team: I always believed that the cost to human lives with this illness would be far greater than other costs down the road. If we limited the number of cases, if we mitigated against the virus in the short term, we could stabilize our hospitals sooner, save more lives, and immediately reduce these other, longer-term economic effects.

Outwardly, I was calm, but inside, my emotions were churning. I had that new-kid feeling again. Not only was I aware of how important this travel ban was to my plan to help save American lives, but I was feeling the pressure of having to disagree with one of the president's most trusted cabinet members. Steven Mnuchin was articulate, logical, and brilliant. Over the last three years, he had worked side by side with the White House to improve the economy. I wasn't lacking the courage of my convictions or my expertise. Still, in a high-stakes situation such as this one, *who* was saying something mattered more than *what* was being said. Mnuchin didn't have a more valid argument than mine, but he had established credibility with the president that I lacked.

Mnuchin cut me off several times to rebut some of my points, but I kept my composure and carried on. I got the sense that he wasn't used to anyone challenging his perspective. I felt satisfied that I had clearly communicated the China-then, Italy-now formula for decision making.

Bob and Tony backed up my additional points, which was key. We needed to be as consistent in our position as the economic team was united in theirs. The president then consulted with his security people, including Ken Cuccinelli of Homeland Security, who spoke of the logistical matters inherent in such a ban and pointed out that we would need considerably more personnel to enact the proposed measures.

At that point, the president looked around the room and then at his watch. "I don't see anything that looks like consensus. Go into the other room. Figure this out and get to a point you can all agree on."

An aide got his attention; it was clear we'd been dismissed.

I was dismayed but not surprised. On the one hand, I saw this as standard operating procedure, a kind of "tell me what I want to hear so I can do what I want to do and have justification for it based on your recommendation" scenario. On the other, I saw it as an approach that might work to my advantage.

Given how entrenched both sides were, true consensus was unlikely. Instead, I figured that we'd get to something more like capitulation. Some would concede, going along with the majority point of view simply because they knew that was what was being asked of them. Others would feel like they didn't have much skin in the game in terms of what would be asked of the departments they represented. Some truly were conflicted and could be swayed.

I looked at it this way: If their concern over the economic impacts of the European travel ban was this high, how were they going to respond to the more extreme messages I planned to deliver in a few days? I really believed that what I was advocating for—my impending gambit for big-picture mitigation measures that would have us enacting more Italy-like measures—was a higher ground worth dying on. The travel ban was among the first steps up that steep hill. Retreat here, and it would make the upcoming struggle even longer and more difficult. I had the stamina to hold that ground, and Bob and Tony were with me.

After we reconvened in the Cabinet Room, I reviewed with them the most up-to-date data I had collected from Europe, South Korea, and China. Predictably, the number of domestic cases and fatalities was rising, but more important, in the two days since I'd presented on Monday, the cases were now (even with the extremely limited data reporting we had to work with) rising faster. Cases were essentially doubling each day, exhibiting the characteristics of exponential growth, a rate of increase that could be explained only by the underlying vast silent community spread.

Whether because of the dark picture my words painted or a lack of understanding of what I was arguing, in those early task force meetings, I saw a lot of distanced looks when I talked numbers and mentioned things like linear-versus-exponential growth.

Once I'd presented the data, the economics team pushed back again. They said the U.S. case counts were low and would stay low. They firmly believed the economic risks to the American public were far greater than the risks to their health. What I was presenting was theoretical, they said, based on assumptions I was making from Europe. America was different: We could do what everyone else had failed to do. We were better.

The discussion took on the form of a debate team meet: *I see your incidence and fatality rates, and I raise you the GDP.* America's wealth and its collective faith, the economics team seemed to believe, would protect us. (I was grateful to my high school and the debate team training I received there for helping me sharpen my discourse and responses.)

Seeing this line in the sand drawn so starkly between the economy and public health so early in my tenure was as concerning as it was illuminating. Seeing who was on each side of that line and what their position was, I found beneficial. In theory, weighing the cost in human lives versus economic costs, and the downward spiral that poverty and a loss of livelihoods might bring, was germane when crafting a pandemic plan. But when it came to the crisis we were all now facing, I saw the economic consequences as a "tomorrow issue," while human life was a "today issue."

Steve Mnuchin spoke again. Much as I had done, he reiterated his most salient points. What he was forecasting could be potentially catastrophic to the economy as a whole. He wrapped up fairly quickly, letting the authority he brought to the room do a lot of the talking for him. It was as if he were saying, *Whom are you going to trust? Someone who's helped navigate this country into such a strong economic position, or her?*

When Steve was done, I took a moment.

As a woman, I was used to certain adjectives being used to describe my aggressive pushing for lifesaving policies—most not complimentary. Behavior in men that was described as "focused" or "driven," became "bitchy" when I exhibited it. *Why is that bitch so aggressive and so direct?* Never: *Why is she direct and authoritative regarding actions to save human lives?* Again, a man in my current position would have been viewed as a visionary, "willing to take the hard line to save lives." They would have been applauded. I was always disparaged. I had experienced this time and again over the decades—and to be honest, it hadn't improved that much anywhere in the world.

As evenly as I could, I continued: "I showed you my data supporting my position. I showed you how this could overwhelm our hospitals." I looked around the room. "I understand. I've painted a disturbing picture. But it's a realistic one, built using the best data available, to make projections based on what's happening in the real world right now."

I paused, before adding, "Where is *your* data supporting your dire economic predictions?"

No one interrupted me. No one countered my last point. But no one looked very happy or relieved. An uncomfortable silence lingered; a line had been drawn. At that moment, I couldn't see how deep and broad that line would become. Those early months would come to be defined by the medical professionals on one side of the divide and the economics team on the other. What on the surface looked like agreement was, for some in the room, a call to action.

That moment marked the last time the White House senior advisors or components of the economics team would show up to a meeting like this without data. From that day, they would begin, and continue, to bring forward "alternative" data, analyses, and projections. This was not Secretary Mnuchin or Larry Kudlow, but others on the economic teams. Much like the CDC, they'd find or create models to back up their previously drawn conclusions. Like the CDC, they were working backward from a premise. I was looking forward, using data to clarify what the future held.

I'd asked where their data was. I wouldn't ever get it, but I would eventually receive a lot of analyses of the data *I* had presented—much of it distorted or inaccurate, but it had the *appearance* of being factual. These distortions would become part of their standard operating procedure: Find some way, any way, to use figures, no matter how inaccurate, to support their conclusion. Interpreting data would become the new battlefield. While making the right policy decisions mattered deeply to me, using the right data to arrive at the right conclusions mattered just as much. In my mind, the two were inextricable. I operated not on gut checks, but fact-checks.

At that meeting in the Cabinet Room, I watched as clusters of advisors from both sides formed huddles and then broke up, heads inclining toward and then away from one another. I don't know if it was immediately after this meeting or later that Tony and I earned the nickname "Dr. Doom." Left unanswered was how my doom regarding human lives was any different from theirs predicting dire economic consequences.

Ultimately, on the question of the European travel ban, we didn't vote by secret ballot, we didn't do a show of hands, we didn't reveal our position by voice acclimation. We shuffled papers, clicked pens, and scooted back chairs. We were heading back into the Oval. We'd reached "consensus": we were recommending the ban. I did a quick review as I made my way back to the Oval Office: Derek Kan, of OMB and the Council of Economic Advisers (CEA), was very much on the economic supremacy side of the argument, but I sensed that Mnuchin and Kudlow, despite the clear economic impacts, were listening to the health arguments. (They did this throughout the remainder of the pandemic response. I came to see them as distinctly different in this regard from the rest of the CEA members.) I had a good feeling that two other key influencers on the president's team seemed to get the "seriousness" argument. Marc Short had the vice president's ear, but Jared Kushner and Hope Hicks had that most elusive of things: the president's attention.

Knowing who could best champion specific elements of the Covid-19 response would be key to navigating the White House. Even in those first few days, it was clear that the president was constantly soliciting opinions from his senior advisors, but also from his outside influencers and from random people who crossed his path. I was not, and would not, be present for the majority of these interactions, and as such, I needed to ensure that someone in the room was armed not just with the data, but with the correct interpretation of the data.

Jared was one of the key people I needed to sway. I hadn't spent much time with him yet, but he was present in the Oval Office meeting and the one in the Cabinet Room. Based on his brief interactions with me, I suspected that the Italy scenario as a predictor of the United States' future had registered with him on some level. I watched him from across the large table in the Cabinet Room, when I and others were talking. He was tracking the conversation and seemed concerned when I talked about the frightening scenario on the horizon.

As we filed back into the Oval Office again, Matt Pottinger waved me over, pointing to the seat he had placed immediately in front of the Resolute Desk. I took the seat, and he retreated to one of the yellow couches. He wanted me there, as visible as possible to the president, as a reminder that there were two sides to the argument and that public health, symbolically and practically, needed to be front and center.

In the end, there wasn't much reason for me to be placed there now, because with very little fanfare, the vice president promptly informed the president that we'd reached consensus. The European travel ban would be announced.

The president nodded, and we all left the room.

GETTING THE TRAVEL BAN through was a crucial first test of my data-driven approach. That it worked would, I hoped, make the end-of-the-week pitch for our version of a flattening-the-curve-to-protect-hospitals "shutdown" easier. But as the travel ban debate had shown, we were going to have to overcome resistance, primarily from a specific wing of the economic team, especially the Council of Economic Advisers. If I read the room right, both Secretary Mnuchin and Larry Kudlow were concerned about the potential of SARS-CoV-2 wreaking economic and public health havoc across areas of the country. The travel ban would impact a relatively small portion of the population, along with, primarily, the airline industry and associated businesses. It was unclear how the economic team would respond to measures affecting a far greater portion of the population and the economy. For this, it wouldn't be enough to point out the seriousness of the pandemic and the need for action; we would have to develop a specific set of steps to recommend.

On Monday and Tuesday, while sorting through the CDC data issues, we worked simultaneously to develop the flatten-the-curve guidance I hoped to present to the vice president at week's end. Getting buy-in on the simple mitigation measures every American could take was just the first step leading to longer and more aggressive interventions. We had to make these palatable to the administration by avoiding the obvious appearance of a full Italian lockdown. At the same time, we needed the measures to be effective at slowing the spread, which meant matching as closely as possible what Italy had done—a tall order. We were playing a game of chess in which the success of each move was predicated on the one before it. We needed to ensure the virus didn't spread across the country. We had neither the hospital personnel, ventilators, nor the supply capacity to endure that type of surge across the country in metro after metro. With our response so far behind the virus, with no therapeutics to save the sick, we would have to do something drastic.

By Tuesday, March 10, we had initial drafts of the guidance and easy-to-comprehend slides to share with the nonmedical members of the task force. Ultimately, based on the feedback we received, we'd either share them as is or revise them before I met with the vice president in advance of presenting them to the full task force on Sunday. For the rest of the week in our task force meetings, I slipped in oblique references to future recommendations. For example, I'd present a graph that read, "Encourage your employee to . . ." followed by suggestions about how to keep the workplace safe, ratcheting up the urgency of my language incrementally. Bit by bit, I moved the pieces on the board in advance of delivering the full flatten-the-curve message. I wasn't making slashing attacks with bishops or rooks. I was subtly moving my pawns, not wanting to put anyone in a defensive posture too soon.

Knowing the objections to the flatten-the-curve measures that were sure to come from the economics side, we decided to take these on directly, highlighting "preemptive" and "low-cost" interventions. By emphasizing "preemptive," we'd send an important signal: some states still had time to prevent community spread. The "low-cost" interventions would be social distancing and enhanced hygiene vigilance. The first cost nothing; the second would involve the purchase of sanitizing and disinfecting agents. If the cost to the public was low, and if the behaviors were ones they typically engaged in, the public would be more likely to respond positively. The sooner they adopted and stuck with these measures, the sooner businesses could fully reopen. It would be a win for the "consumer" and a win for business owners.

To that end, we also made the difficult and significant decision not to include a specific reference to wearing masks. Without the full support of the CDC on nonmedical-grade masks, and with the administration's ongoing resistance to their use, I couldn't afford to have the entire mitigation program meet with strong opposition because of one provision. I was okay with this omission because, if the public followed the rest of the guidance, they would be staying at home more, reducing viral transmission opportunities. In retrospect, in focusing on the needs of the many, this safer-at-home approach put essential workers at greater risk. Either when delivering to homes or still encountering customers in stores, they faced a greater risk of exposure and they weren't protected. This still haunts me. Four weeks later we were able to put the importance of masking in the opening guidelines. But for that four weeks, we put Americans at substantial risk.

I organized the presentation around three key areas: work, school, and family. These didn't necessarily represent all of my priorities. Opening with "work," I felt, would appeal to the economic team. Getting them started off on the right foot would make the next two steps easier. We'd highlight how we could keep the economy running while ensuring the safety of those who got the pistons pumping. If business owners acted proactively and got positive results, then nothing would feel punitive.

School guidance was second on the list, for reasons both rhetorical and practical. I wasn't sure how many in the room had school-age children or grandchildren, but it was likely many. Keeping students, and the teachers for whom school was a workplace, safe was essential. This would lessen the impact on parents, allow them to continue to work, and maintain that cherished economic momentum. The medical side of the task force had to do triage here as well. Stabilize the situation first so that, downstream, we wouldn't find ourselves in a more devastating crisis resulting in enormous loss of life. Fortunately, the European data on the virus up to this point had shown that children and young adults did not seem to become as seriously ill as older adults.

The last of the three, and the one I personally prioritized, was family. Because data had shown that seniors and those with underlying medical conditions were particularly vulnerable to being more severely affected by the disease, and were dying from it at a greater rate than young people, I used more direct language in our guidance. I sensed that American seniors would recognize their vulnerability and would be more willing to mitigate and protect themselves than younger people. I was very concerned about multigenerational households, like the one I lived in. Millions more of them existed and housed more people than care facilities. Within such households, the interplay between the unknowingly infected and highly contagious and those most susceptible to severe disease and poor outcomes produced the greatest likelihood of the most dangerous consequence of silent spread: hospitalization and death.

While I was writing the mitigation guidelines to present to the task force, my daughter and I continued to enact the practices we'd established weeks ago in the home we shared with my parents, who were ninety-one and ninety-five years old, and my grandchildren, both toddlers. My daughter also had someone coming in to provide child care. The weekend I returned from Africa, everything had changed. Given my fears about asymptomatic spread and the potential for my grandchildren (the two-year-old was in preschool) to bring the virus home and infect their grandparents, the entire household went into full lockdown. No one in, and no one out. Both sets of grandparents stopped their visits with the grandchildren. My daughter canceled child care, and the two-year-old stopped going to preschool. Without any explicit instructions, my daughter researched how to support the family with food and supplies and set up the most comprehensive supply chain to the house I had ever witnessed, all while still working full time from home.

Toilet paper, paper towels, and disinfectants were delivered to the house; we were dependent on her supplies throughout the pandemic. The same alert went out to my brother, his children, and other family members. We had vulnerable adults at risk in every household.

The guidelines that came out of the task force were meant to address another audience: the CDC. To break free from the constraints the CDC had placed on itself with its myopic flu model response, which had minimized the notion of silent spread, we couldn't just call them out. Despite our differences in approach, we needed them. To that end, many of the guidelines we incorporated were based on CDC best practices. Undercutting the CDC at this stage wouldn't have done anyone any good. The agency was sticking to its guns, and so was I. Yes, silent asymptomatic spread was a significant contributor to the outbreak, and all the advice we were giving to the American public spoke to that point, but without mentioning the concept by name.

While the guidance was written to make it hard for anyone—the CDC or the economics wing of the task force—to object, we still had to anticipate future greater resistance. I was already thinking this: Because we were dealing with a large number of political figures on the task force (essentially, everyone except Tony and me), in the White House, and in the administration throughout the many agencies, we knew we couldn't call any measure a "regulation." The Trump White House had spent the last three years eliminating so many of them in so many areas where it believed regulations interfered with economic progress. While rules and regulations were absolutely what the present crisis called for, and would help force the states to adopt these measures, the presence of the word *regulation* in the guidance would immediately have ended all our shutdown efforts. The guidance would never have made it past the White House gatekeepers. We therefore constantly emphasized that we were making "recommendations," not establishing rules. This principle would guide our approach and our messaging within the task force, with the administration, and with the public. At this stage, we used the word *encourage* in the heading for each of the three main areas of focus.

The key was to establish a serious tone and let the states use the guidelines to justify more aggressive action. The White House would "encourage," but the states could "recommend" or, if needed, "mandate." In short, we were handing governors and their public health officials a template, a state-level permission slip they could use to enact a specific response that was appropriate for the people under their jurisdiction. The fact that the guidelines would be coming from a Republican White House gave political cover to any Republican governors skeptical of federal overreach and would lead to most states' implementing clear regulations themselves.

With the benefit of almost two years' worth of behavioral data, it's easy to look back now and read a certain amount of naïveté in our belief that people would prioritize the health of others over their own personal liberty. Of course, we anticipated that some people would resist, but we strongly believed that most people would want to keep others in their community safe by doing the right thing en masse. Early data from Europe showed widespread compliance, with interaction with mobility decreased across the continent by 75–95 percent. The tricky thing was finding agreement over precisely what the "right thing" was. Unfortunately, we knew that, at some point, we would have to account for how politicized and divided the country had become. At that early stage, though, we had to focus our energies on how politicized the White House bubble was.

MY NEXT MOVE IN the ongoing chess game was to quietly convene an ad hoc group at my home. On Saturday, March 14, Joe Grogan, Steve Redd, Tony Fauci, and Tom Frieden joined me. Olivia Troye showed up unannounced. At the time, I didn't think much about this, but clearly, she had come under the direction of Marc Short to see what I was doing.

The objective of this meeting was twofold: The first was to refine and finalize the presentation I'd be making the next day at the task force meeting. The second was to position ourselves to block any of the president's escape routes. If all our pieces were aligned properly, he'd have no choice but to agree to our flatten-the-curve measures. A few of our moves would be dictated by the schedule. On Sunday, I would meet with the vice president before the afternoon's task force meeting. In that private session, I hoped to get him on board with the recommendations we were drafting. With Vice President Pence in position, the president's options would be even more limited. Going against the man he himself had put in charge of the task force wouldn't look good, and the media were sure to get wind of how things played out. If I could get all those pieces in place, we would get to step one of a "circuit breaker" to flatten the curve, prevent the virus from spreading across the United States, and save as many lives as possible.

I was fairly certain the vice president would go along with what the task force recommended, but the president was always going to be a wild card. For this reason, I had invited task force member Joe Grogan to join us at my home. A longtime member of the administration and an important conduit to the president, Joe was a former member of OMB and the current director of the Domestic Policy Council, a key group in determining whether our recommendations could be enacted. He was therefore a critical lynchpin in getting my strategic plan approved. If Bob, Tony, and I could present a unified phalanx of support for the measures, Joe and, to a lesser extent, Olivia Troye, would take this positive show of strength to the Oval Office, to Marc Short, and the vice president. I'd seen the president's desire for consensus before. Demonstrating that we had it would go a long way toward convincing him to approve our recommendations.

I was also hoping to leverage Joe's prior "We will own it" statement to my advantage. I counted on him telling the vice president and president that it was better to act now than endure the dire consequences of inaction, when the blame would rest squarely on the president and his advisors. Joe would be the key to these discussions.

The trick was getting the medical side in agreement on silent spread; I planned to have Tom Frieden help bring the CDC along. Like me, the CDC wanted to do everything to stop the virus, but the agency needed to align with us on aggressive testing and silent spread. Those in my camp had nuanced sensibilities regarding data and modeling; Tony's preferences were very much on my mind. At that point, the president trusted Tony, and to put it bluntly, Tony didn't put much faith in predictive models because they were too dependent on the initial assumptions made to create them. He was right. Baseline assumptions and, consequently, what they predict can be too open to subjective perception. Tony absolutely believed in data, but he had a strong preference for quality-controlled data from well-designed clinical trials. With no such trials in place, Tony had little faith in projections beyond what the data was saying about the present moment. As a scientist, Tony very much wants to see substantiated, validated facts first, so I planned on leaning hard once again on the quality of the data I had from Italy.

During the course of the week, I'd continued to be in touch with Tom. If I could get him to agree that these shutdown-like measures were necessary, other CDCers would fall in line. While he still wasn't fully committed to backing my 50 percent asymptomatic figure, he understood that the nation, with the administration leading the way, needed to take immediate action. What was most important was flattening the curve. With this virus, we were way past proactive mitigation, as was Europe. At this point, getting the doctors, including Tom and Tony, to be in complete agreement with me about asymptomatic spread was slightly less of a priority. As with masks, I knew I could return to that issue as soon as I got their buy-in on our recommendations. For now, I was focused on marshaling support from the public health side for this de facto shutdown.

The setting for this meeting, my home, was far less formal than the conference rooms at the EEOB or the White House. Though the tone was different, informal and interactive, the intent was similar: to exchange information with Tom and Tony and make sure I included their insights in the final guidelines. As with the travel ban discussions, I was prepared with data and graphics. I went through our guidance recommendations, beginning with the workplace category.

I expected some pushback from Joe, given that this was the most sensitive area for the administration. There were a few chuckles at the bullet point that read "Consider adjusting or postponing large meetings or gatherings." At one point, I saw Joe nod his head and heard him say, "Seems very commonsense."

"That's the point," I said. "Commonsense. Actionable. Letting every American know what part they should play."

As I expected, Joe asked Tony, the White House medical advisor, to share his perspective. Essentially, Tony said the projections we'd based on the European data were likely to be true: We were past the point of containment. We had to limit people's exposure to the virus and do exactly what flattening the curve is intended to do.

One of the key decisions made at the meeting was on the fifteen-day time frame during which we'd ask all Americans to do their part. Instead of just asking people to be vigilant during the flu season and to cough and sneeze into their elbows instead of their hands, we'd ask for other behavioral changes more in line with silent spread mitigation. To arrive at "fifteen days," we had relied on the CDC's estimates of this virus's full transmission cycle (from inhalation of droplets or airborne particles to viral shedding to infectivity), a maximum of fourteen days. This was the justification for the exposure and quarantine times being used around the globe. Fifteen days was the minimum required to have any impact. I left the rest unstated: that this was just a starting point.

With the presentation finalized and the meeting ended, I tried to get as much rest as I could before my final presentation to the vice president. I had shown him an earlier version and planned to present the final one to him prior to the task force meeting the next day, Sunday.

When I met with Vice President Pence before the task force meeting I brought along all the final graphics I wanted him to present to the president. As in chess, it's helpful to know your opponent's tendencies, and the president liked visuals. I also factored in how Vice President Pence was likely to react. A former governor, he was predisposed to believe that it should be up to the states to take whatever measures they deemed necessary. They should decide whether to seek the support of the federal government. That's where my "recommendations" and not "regulations" move came into play.

Knowing the vice president's immense respect for governors and their leadership independent of political party, I had slightly altered my approach in my presentation to him. I knew that, when communicating our guidelines to the president, the vice president also would have to walk that fine line between nudging and pushing, making it seem as if our great idea had come from someone other than the task force. The psychology involved in this was maddeningly delicate.

At the conclusion of our meeting, I felt Vice President Pence understood the gravity of the situation and the urgency of what we were suggesting. Acting now was critical. The vice president and others needed to believe this was consistent with federalism and that the governors would "own" the final decision making. I believed in that moment that we had obscured the larger truth—that the White House was essentially recommending a flattening-the-curve "shutdown" to states with significant viral spread and extreme reductions in mobility and gatherings in other states to prevent surges there. After all, once the White House put out any serious recommendations, the governors and the American people would see it as open acknowledgment that what we were facing was serious and needed to be addressed immediately.

I wanted immediate confirmation that our plan was a go, but I wouldn't be part of whatever meeting was held to inform the president. It was as if I had set up the chess board with a checkmate in three moves only to have another player edge me out of my chair to execute the final move, presidential approval.

I spent an anxious few hours trying to distract myself while the vice president met with the president—until I received an email from Jared Kushner that we were to proceed as planned. The vice president and Jared had gone to President Trump and gotten the green light. Somehow, it felt anticlimactic, but I told myself not to look a gift horse in the mouth: the White House would post "The President's Coronavirus Guidelines for America" on its website, and the following day, President Trump would announce the campaign, which was eventually given the name "15 Days to Slow the Spread."

It didn't take long for the cost of the disconnect between the task force and the CDC over silent spread to be laid bare. Late in the evening on Sunday, March 15, I received an email from Olivia Troye. She forwarded a string of emails in which a member of the administration expressed concern that the CDC had independently issued directives regarding mass indoor gatherings: Their recommendation was that they be limited to no more than fifty people, while the recommendation in our presentation had capped the figure at ten people.

This was exactly what I didn't need—neither the mistiming of the message nor the message itself. Ultimately, cross-household gatherings needed to be stopped entirely, households isolated from other households to prevent further spread. Limiting gatherings to ten had been a first step and was consistent with my spoonful-of-sugar approach. Starting at fifty, as the CDC was advising, would require many more moves to get to zero. All week in the task force meetings, with the CDC's Bob Redfield present, we'd talked about that need and that number: ten. I'd even shared with Bob the proposed bullet points containing that figure. He hadn't raised any objections, either in our meetings or privately with me, to using ten as the upper limit. This fifty figure must have emanated from the CDC rank and file, and Bob was probably caught off guard by it, too.

Of all things to suddenly drop through the roof, this discrepancy in numbers was the last thing I expected. By using fifty as the upper limit for gatherings, the CDC was giving those in the White House who already thought I was overreacting the exact ammunition they needed: *See, even the CDC doesn't see the virus as that serious a threat.*

And that's exactly what happened. Both behind the scenes and in direct emails, Marc Short of the Vice President's Office pointed out the fifty-versus-ten conflict. He wondered why the White House plan was not aligned with the CDC recommendations. He also asked if there was a scientific basis for fifty versus ten, or even five? He was trying to flip the script on my travel ban message to the economics team: *Show me the numbers.* For people like Marc, the discrepancy was about how serious this viral outbreak was, and he was implying that I was exaggerating the seriousness by a factor of five. Was everything I was saying, every figure I was citing, similarly off by that factor? What if it were off by more?

The real problem with this fifty-versus-ten distinction, for me, was that it revealed that the CDC simply didn't believe to the degree that I did that SARS-CoV-2 was being spread through the air silently and undetected from symptomless individuals. The numbers really did matter. As the years since have confirmed, in times of active viral community spread, as many as fifty people gathered together indoors (unmasked at this point, of course) was way too high a number. It increased the chances of someone among that number being infected exponentially. I had settled on ten knowing that even that was too many, but I figured that ten would at least be palatable for most Americans—high enough to allow for most gatherings of immediate family but not enough for large dinner parties and, critically, large weddings, birthday parties, and other mass social events.

The number mattered also because of this: When Americans heard "50," I knew they would round up to 100. With the CDC guidance, large weddings, funerals, and birthday parties would proceed unchanged. If they were told "250," they would hear "500." Similarly, if I pushed for zero (which was actually what I wanted and what was required), this would have been interpreted as a "lockdown"—the perception we were all working so hard to avoid.

In response to the White House query about the "correct" number, Tony couldn't commit. Privately, he wrote to tell me that he didn't want to publicly contradict the CDC, but that he also believed fifty was too high. As he told me, he couldn't give Marc Short an answer about the scientific basis for any number the CDC put out; that wasn't the task force's call to make. He wondered whether a call to Bob or Anne Schuchat at the CDC was in order.

This wasn't a case of passing the buck; it was a function of Tony's political sensibilities. He didn't want to undercut Bob's or the CDC's already shaky authority. As scientists in a task force outnumbered by politicians, economists, and bureaucrats, we needed to present a more or less unified front—otherwise, a strategy of divide and conquer could have come into play. The CDC had erred in coming out with its announcement, but this fifty-versus-ten dispute was merely a microcosm of the fundamental disagreement among scientists over the nature of the virus's transmission. We could agree neither on silent spread nor on how large a spread people could put out for friends and family. Surprising as it might seem in hindsight, this was a rare case of the White House taking the more cautious road than the CDC.

We were so consumed by the work we were putting in to develop the guidelines that became part of the 15 Days to Slow the Spread campaign that it was only after the president's press conference on March 16 that something occurred to me: When we'd argued in favor of the European travel ban, we'd engaged in lengthy debate, with President Trump sending us away to achieve consensus the way a judge might instruct a hung jury to keep at its deliberations.

In contrast, for as much effort as we had put in and as much anxiety as I'd felt over 15 Days to Slow the Spread, we'd received little to no pushback that I could discern from the CEA or the other economics-minded factions in the White House. I thought again of Joe Grogan's comments about "owning" this. It was possible that the president and his advisors saw the fifteen days as more of a cover-your-ass move than an effective means of combating what I and others saw as the most serious public health crisis of our lifetimes. The administration might have been worried about owning the potential failures, but its members, at that time, didn't seem overly concerned about owning the responsibility for reducing the scope of the crisis.

I wondered if I'd oversold the notion that the guidelines would be low-cost, low-impact. Had we not made it clear enough that this was a serious matter? With whom had the president consulted before agreeing to announce the guidelines? If there was no infighting or debate, did that mean there had been complete agreement—or was it an indication of indifference? Did the president know that the practical effect of the guidelines would be that governors would begin to shut down their states?

Indeed, following President Trump's announcement of the guidelines—almost on cue—the recommendations served as the basis for governors to mandate the flattening-the-curve shutdowns. The White House had handed down guidance, and the governors took that ball and ran with it. Based on the weekly governors' call and my other interactions with them, I could tell they'd been looking for the White House to take the lead in letting the American public know how serious the situation was. With the White House's "this is serious" message, governors now had "permission" to mount a proportionate response and, one by one, other states followed suit.

California was first, doing so on March 18. New York followed on March 20. Illinois, which had declared its own state of emergency on March 9, issued shelter-in-place orders on March 21. Louisiana did so on the twenty-second. In relatively short order by the end of March and the first week of April, there were few holdouts. The circuit-breaking, flattening-the-curve shutdown had begun.

CHAPTER 7

Turning Fifteen into Thirty

No sooner had we convinced the Trump administration to implement our version of a two-week shutdown than I was trying to figure out how to extend it. Fifteen Days to Slow the Spread was a start, but I knew it would be just that. I didn't have the numbers in front of me yet to make the case for extending it longer, but I had two weeks to get them. However hard it had been to get the fifteen-day shutdown approved, getting another one would be more difficult by many orders of magnitude.

In the meantime, I waited for the blowback, for someone from the economic team to call me into the principal's office or confront me at a task force meeting. None of this happened. Maybe it was because the stock market had stopped its plummet; I didn't know then, and I still don't. As governors around the country began to respond and implement a version of our recommendations, I waited for President Trump to see that what we'd recommended was shutting down the country. This explosive response didn't happen. Instead, the virus pummeled New York City and the bedroom communities in Northern New Jersey and Connecticut and on Long Island just outside it.

Since February, I'd been envisioning a catastrophic scenario—lurid images born of what we'd all seen first in China and then in Italy—here on our shores. At first, it was a point in the distance. By March 16, that point had grown much closer and was taking shape in New York City.

On February 29, a man from Queens became the first confirmed case of community transmission in New York City. On March 1, the first confirmed case was identified in Manhattan. Two days later, there was the first recorded person-to-person transmission. Less than a week later, the number of cases in New York City rose to sixteen. Governor Cuomo had declared a state of emergency on March 7 for the entire state.

If any place in America needed fifteen days to slow the spread, it was New York City. The virus had staked its claim on an urban center that was a prime candidate for extensive community spread. Highly populated, densely packed, New York City was a place where people jammed into buses and subway cars, where even at the best of times your personal space was reduced to a tissue-thin layer. What we were beginning to see as a highly transmissible virus was in its optimal environment. Looking at the cities of Europe and doing the math, I feared it was already far too late to prevent what was going to happen in New York City. All we could do was work to prevent the hospitals from being overwhelmed. Indeed, later research into the virus antibodies of infected people would show just how vast the spread was in New York City: the virus had been spreading silently in the city at least since early February. For the coronavirus pandemic in the United States, New York City became the tip of the spear in a life-or-death battle that hadn't been seen in America in over a hundred years.

The weekend of March 14–15, while I had been preparing my pitch for what became 15 Days to Slow the Spread, New York City had closed its public schools and was already encouraging businesses to allow employees to work remotely. Mayor Bill de Blasio and New York City's health commissioner—who, just two weeks earlier, had encouraged people to go about their daily lives—were justifiably sounding the alarm.

On Monday, March 16, the day after our guidelines went into effect, I stood outside the Vice President's Office watching TV footage of Mayor de Blasio stating that the city might need to shut down and residents might have to "shelter in place." De Blasio expected to make the difficult decision over the next forty-eight hours. It seemed the message of the circuit-breaker, flatten-the-curve strategy had connected with his administration. Perhaps that's why I was surprised when, not long after, New York State governor Andrew Cuomo implied that no such thing was in the works—de Blasio lacked the authority to implement a shutdown, Cuomo said, and such a move was unnecessary. As Governor

Cuomo said himself, "I would have to authorize those actions legally. It's not going to happen."

At that time, I got the impression that the White House was surprised that Governor Cuomo wasn't taking more drastic action, especially given that, on the day we announced 15 Days to Slow the Spread, Cuomo had called for more federal leadership. But rather than taking our recommendations and guidance right away, he waited several more days, until March 20, to order that, statewide, all nonessential businesses had to close, effectively beginning New York State's shutdown.

Seeing this back-and-forth unfold, I was astounded. Here was a mayor who maybe could have moved a bit sooner, but who was now trying to do the right thing for his city, and there was the state's governor interfering. This squabble resulted in days of delay and put a huge number of people at greater risk. Losing those days of lockdown proved costly to the city, but what was even more detrimental would be the long-simmering rift in the leadership between the governor and mayor. A feud that predated Covid-19 by years suddenly had much higher stakes, now that human lives were on the line. Yet, not even a crisis of this magnitude could overcome their apparent political pettiness. This was the first real indication I had of the power governors possessed in such situations. Later, I used this awareness to appeal to governors directly, but at the time I was confounded (and still am) by why Governor Cuomo, at this early and precarious point in the pandemic, would use his power to undermine Mayor De Blasio and the people of New York City.

Almost overnight, New York City came to dominate the news cycle, and as the third week of March elapsed, I came to see the big role New York had played in why I hadn't received any immediate pushback from the shutdown. My presentation to the vice president, which was relayed to the president, had predicted that things would become dire. And now, I was proven correct—with wall-to-wall media coverage, no less. The tragic events in New York had, it seemed, shown the president and vice president the wisdom of the path they'd already chosen. When the picture I'd painted started coming true in New York City, it immediately had an impact on them—especially given that this president, born in Jamaica, Queens, was a native of New York City.

I recognized that when, in two weeks, I asked for an extension of the flattening-the-curve shutdown, tragic events beyond my control might again help my case. On March 15, when I'd pitched the shutdown to Vice President Pence, the United States had just shy of five thousand cases and seventy-four deaths; for most people in America, especially those in the White House, the virus was still fairly abstract. Something would be needed to stop people in their tracks—the CDC, the president, the media, the American public—and convince them once and for all that this virus was for real. With the rate of spread we were now seeing out of New York, my fear was that whatever was beginning there would be that thing.

FOLLOWING THE PRESIDENT'S MARCH 16 announcement of the guidelines, the frenzy that had characterized my first two weeks took on an even greater pace. Whereas my initial two weeks had been shaped by trying to put our slow-the-spread approach in motion, now I had to use my time as effectively as possible to gain any new ground we could, while also figuring out how to extend the slowdown.

A massive array of issues, constituencies, and areas of need had to be addressed, and as coordinator of the White House Coronavirus Task Force response, that was my job. Members of the task force spoke with grocery store executives, encouraging them to offer special seniors-only hours in places where home delivery wasn't possible. I reviewed the medical literature about various therapeutic treatments that might be effective against Covid-19 and studied the mitigation protocols that health officials in China, South Korea, and elsewhere had followed, to determine their efficacy and applicability in the United States. I prepared for and then participated in phone calls with physicians and nurses groups and public health officials responsible for tribal lands and prepped for media and press conference appearances. We reviewed the guidelines that the CDC was putting out regarding foreign travel, cruise ship excursions, child care facilities, long-term health care facilities, and hospitals and clinics. We spoke with five thousand state and local officials on a conference call—including governors, attorneys general, secretaries of state, mayors, city council members, country commissioners, and local public health officials. We were going as fast as we could to make the biggest impact we could.

On March 16, in tandem with the White House announcement of our guidelines, the task force began holding daily press briefings. These quickly took on increased importance and visibility, as events were moving at such a high speed, and we were conveying important information in the briefings. For each press event, I'd review the remarks that those in the Executive Office of the President or the Office of the Vice President had prepared. Following a brief preparation with the vice president, we would go down the hallway, past the Oval Office, to the Press Briefing Room. Once inside, Seema Verma, Bob, Tony, Steve Hahn, Surgeon General Jerome Adams, and I, and sometimes HHS secretary Alex Azar, would stand together for several hours not moving. We were the supporting players, with first the vice president and then the president as the leads.

No matter my role, I was in an unfamiliar setting. Cameras, still and video, were everywhere. The members of the press, who were carefully seated at a distance from one another, conducted themselves professionally. At these briefings, I always felt not quite put together as I gazed out at these perfectly dressed journalists—smartly put together even in the midst of a pandemic. When statements were made by, or questions asked of, the president, I had to figure out how to look serious and not react to what was said, no matter my inner feelings. I asked my colleagues how they were able to stand and keep their facial expressions so unmoving. They told me to focus on a distant point at the back of the room and think of other things while still paying attention.

Most of these press conferences were incredibly serious in tone, as we went through what we knew from Europe and our own country and what we could do together to slow the spread. When I had to speak, I carefully tried to use my time to convey what I was seeing in the numbers, something I thought was especially important to contribute. One evening, I spoke directly to Millennials, who, I believed, were critical to our slowing the spread. Millennials, usually the children of Baby Boomers, were uniquely positioned to communicate both to Gen Z about the risks of the virus and to their own parents and grandparents (Boomers and the Silent Generation) about protecting themselves. Frankly, I was worried about the Baby Boomers and how they might discount the risk of the coronavirus and continue to gather, but I knew they would listen to their concerned children. Millennials were the backbone of communicating many of the mitigation elements. Members of Generation X, those in their '50s and '60s, were a bridge demographic, sharing characteristics and points of view with Millennials and Baby Boomers depending on a variety of factors.

I wasn't always perfect at communicating data or what the data meant. One evening, while speaking about where the virus was and where it had been successfully prevented from spreading in the community using the recent measures we'd recommended, I turned to the U.S. map we'd set up nearby to point out Montana. My mind went blank. Which of those two large, rectilinear shapes was Montana and which was Wyoming? While I could name every country, capital, and most of the cities, counties, or provinces across the globe, especially in sub-Saharan Africa, I couldn't identify the state by shape. I turned to the vice president for help.

"Montana or Wyoming?"

"Montana, Dr. Birx."

I could tell he was surprised. I should have known this, and before the next presser, I made a point of learning not only all the states, their capitals, and their large metropolitan areas, but also most of the more than three thousand counties in the country. There were times I might not have gotten the pronunciation quite right—such as when I pronounced the x in "Bexar County" when it's supposed to be silent, like "bear"—but I learned, and I tried not to make the same mistake twice.

Limited testing for the virus remained one of my foremost sources of anxiety, as it would for some time. We still didn't have enough tests, we didn't know how to test thousands of Americans quickly, and we were already seeing problems with the supply chain for all testing supplies.

After our initial meeting during my first week on the job, I communicated regularly with the test developers and commercial laboratories. They were very responsive. Day after day and week after week, Abbott and Roche continued to provide me with daily information on where test kits were being shipped, the volume of tests being performed, the test positivity rate, and how many tests were going to large reference labs (labs that receive specimens or samples from other labs) versus hospitals. They weren't required to send me this data, but they did, filling in some of the gaps the CDC had created. The private sector was able to move at speed far better than the governmental agencies. Time and time again, it was the private sector that responded rapidly to save the country. They were the backbone of our ability to respond to the crisis in real time, whether with tests, PPE, therapeutics, or vaccines. Yet, nothing they did would overcome the months-long head start the virus had over our testing capabilities. We were still months away from having the volume of tests and the processing speed we needed.

In the meantime, we were coordinating with state and local governments to expand testing sites to sports stadiums, community centers, churches, and retail parking lots and getting mobile testing facilities up and running. We were preparing to utilize personnel to collect thousands of samples efficiently and were figuring out how to get the results back in forty-eight hours or less. From mid-March and into April, we had thousands of labs across the United States fully qualified to use FDA-approved testing platforms. To ensure that we had readily available data, we built in a requirement that test results would go not just to the client and their physician, but also to the state and federal health agencies. We'd also made headway on test production, with the FDA agreeing to allow commercial manufacturers to distribute their newly developed tests without an emergency use authorization. The FDA also issued new guidance that would allow states to develop tests independently, but it would take time to get those tests and sites fully functional.

This was the good news. The bad news was that stocking those facilities and making testing more accessible were still problematic. As a consequence of the testing shortfall, the case data we needed in order to see the true extent of the spread would rely on people voluntarily going to testing facilities, being evaluated at emergency rooms, or being admitted to hospitals. The shortages were so acute that HHS discouraged people without symptoms from getting tested unless they had been exposed to someone positive or had traveled to a global hot spot like China. As necessary as rationing tests was at this point, this emphasis on symptomatic testing had the unintended negative effect of appearing to confirm the CDC's bias toward symptomatic spread while simultaneously underrepresenting the true number of people infected.

Even so, with a testing data system up and running, and with required reporting, the daily testing rate was rising rapidly. Still, using the CDC's testing recommendations, I calculated that probably only 30 percent of the cases were being accounted for. As time went on, following CDC recommendations, hospital administrations and public health agencies began asking people to stay away from emergency rooms unless they were critically ill. Hospitals were being overrun, becoming hot zones. Again, as a result, fewer and fewer people were being tested at the community level, a fact that further obscured the reality. We'd eventually figure out how to test outside traditional medical facilities, but this early in the pandemic, at a time when I needed the most accurate numbers reflecting the breadth and depth of the outbreak, I was not going to get them.

PERIODICALLY, IN BETWEEN ALL the meetings, conference calls, and other responsibilities, I would stop by my data team's little conference room each day and reengage in the ongoing, rambunctious debate about just how much silent spread there was. I was still trying to chip away at the CDC's position, but more than that, I was trying to overcome my own team's doubts. Getting an extension of the shutdown depended on my winning over at least one of them.

In my efforts to get the CDC on board, my meeting with Tom Frieden on March 14 was a helpful start. I was mildly heartened four days later, on March 18, when I read a "CDC Daily Key Points" document that finally acknowledged that community spread of coronavirus was present in the United States. In reading the CDC's assessment of the initial major U.S. outbreak, at a nursing home in King County, Washington, I came across, and highlighted, this line in its analysis of why the response there had been less than optimal:

> 5. Delayed recognition of cases because of low index of suspicion, limited testing availability, and difficulty identifying persons with COVID-19 based on signs and symptoms alone [my emphasis].

I didn't know who had written this analysis, but it was clear that someone, either at the CDC or on the ground in King County, was acknowledging that relying solely on signs and symptoms alone wasn't working. Unfortunately, this acknowledgment still wasn't in general circulation at the CDC. All the right language was there—"community spread" and "difficulty identifying" and "based on signs and symptoms"—but they weren't recommending the actions needed to address the silent spread. What we needed now was massive, proactive testing, especially in hot zones among people under age thirty-five—those most likely to be asymptomatically infected. And we needed to limit all gatherings.

When I put my data team together at the end of my first week on the job, I hadn't been looking for Irum or anyone else on it to be a yes-person (not that Irum, Chuck, or Steve ever would be). In fact, I was hoping for the opposite from all of them. The team served as an important critical and scientific sounding board. We trusted one another; even then, we had each other's back. We didn't always agree, but we valued the critical thinking and insights each of us brought to the table. None of us was always right, and we didn't dwell on the mistakes of the past, but instead, ran forward on as little sleep as we could manage. They would listen to my ideas, evaluate the numbers I proposed would reflect the number of possible fatalities, and rebut what didn't make sense to them.

And this was where the lack of testing and the lack of data converged. Not only were our domestic numbers lower than what was actually happening on the ground, they were also less precise, lagging far behind in the complete demographic details that were needed. We were working on it, but commissioning new software would take time—time that, it was increasingly clear, we didn't have. Yes, the people at Johns Hopkins were doing their best to provide the most comprehensive statistics, but like everyone else, they were at the mercy of the virus: Hospital staff and agency staff at some locations were becoming overwhelmed with patients. When hospitals become inundated, data gathering, and ensuring that the data is complete and is shared promptly, becomes a lesser priority. As a result, the evidence that would make the best case for extending the shutdown was precisely what we lacked. The clock was ticking toward the end of the fifteen days, and I felt every second of it. We needed more.

On March 18, Irum convened a virtual summit with ten statistical experts from around the globe, including the United Kingdom, the Netherlands, and the United States, representing institutions such as Imperial College London, Columbia University, Harvard, and the CDC. Each expert produced two models, the first of which assumed no further intervention after the two-week shutdown ended. This so-called unmitigated response—which would constitute no further mitigation and would allow the infectious virus to move unconstrained across the country—produced some dire figures, with between 1.5 million and 2.5 million estimated deaths in the United States over the next several months as the virus ran rampant, unabated. Without an extension of the shutdown, this would likely be the scenario we'd face.

With their second models, the experts produced predictions of deaths based on a variety of mitigation assumptions and their level of effectiveness: school closures, social distancing, and a strict lockdown. Though the models produced different percentages, full lockdown was estimated to reduce the number of cases by 60 to 90 percent. This was with 100 percent compliance, a theoretical possibility but a real-world impracticality. A million and a half to 2.5 million deaths was horrifying to contemplate, but equally disturbing was that reducing that number between 60 percent and 90 percent would *still* result in a huge death toll, approaching 150,000 to 500,000, and this was just modeling this surge—not future surges.

We needed to refine these projections, as the death of 150,000 to 500,000 people was too much of a numerical spread to take to the vice president. The bottom line was that mitigation could be an overwhelming "success," and yet it would *still* result in more than 100,000 people dying over the next few months. By hearing what other experts had determined, by seeing these extreme figures, the task force and White House, I hoped, would be driven to further definitive action. These numbers, which represented deaths on a then-unimaginable scale, had laid bare just how far behind we really were.

As the case numbers from around the world and in the United States rose, I inched my team closer and closer to consensus on my 50 percent–asymptomatic figure. (While lots of research was being done at the time, a more exhaustive and conclusive study in May 2021, by the University of Chicago, would put that figure over 50 percent.) By the end of the third week in March, Italy had enacted its strictest measures, with 90 percent of the country shut down. On March 11, three days into the nationwide Italian lockdown, and on the day all bars and restaurants were ordered closed, Italy had 9,172 cases. On March 22, when the country closed all factories and ended nonessential production, they had 59,138 cases, a sixfold increase. They weren't close to flattening the curve—the worst was still ahead for them. Interventions take time to work, but we wouldn't have that data from Europe before we needed to ask for an extension of our shutdown.

In comparison, New York City, still the hardest-hit area in the United States, had 1,263 total cases on March 11. By the twenty-second, that figure was up to 33,073—a twenty-five-fold increase in eleven days. It wasn't just the absolute number of total cases that was troubling; it was the rate of the rise. New York City was going from 300 to 1,000 to now 3,000 cases per day. This hundredfold increase in cases was the kind of exponential growth, a day-over-day doubling, that lets you know community spread is rampant. This rate of rise couldn't be explained by symptomatic spread alone; most cases were being missed. New York City was in full exponential growth, precisely paralleling what we'd seen in Italy.

If there'd been any doubt before about the extent to which asymptomatic spread was a significant contributor to the rise in case numbers, the New York City numbers convinced the team that my estimate of at least 30–50 percent asymptomatic spread was likely accurate and possibly even conservative. Even though the testing issues prevented us from having the quantity and type of data we needed, a preponderance of numerical evidence, the picture was crystal clear. For those of us in that tiny room, doubt had been replaced by a communal sense of dread.

We asked the modelers to change the assumptions in the models to reflect a New York City–like outbreak, with an agreed-upon 30 percent of silent spread cases reflecting the rate of case increases in the New York metropolitan area. Irum and I wanted to model what the fatality forecast would be if ten to twenty-five of the largest metropolitan areas in the country (those with more than a million residents in the city and immediate outlying suburbs) were affected. It was important that our scenario closely reflect reality—these places would be affected *serially over time, not simultaneously*. Modeling for them would allow us to see how the pattern would impact the allocation of precious supplies.

I did not mention this "metropolitan area" approach to predicting the numbers in the task force meetings. Introducing this kind of complex modeling to the whole team would have been too fraught with potential bones of contention and arguments for one methodology over another. No real good could have come from our getting wrapped up in the minutiae of how New Orleans and its demography (population density, age makeup, racial breakdown, percentage of comorbidities) was different from, say, Detroit or Milan. Keep the focus on the most salient point—the serial nature of the viral invasions across the country and how to prevent it.

My numbers team and the other modelers were at work adapting the Italian data to fit with U.S. demographics to produce a projection. All the task force needed to understand was that the virus (and its disease) was progressing rapidly in the New York metropolitan area. Also we had cases in all fifty states, and that without intervention, all states and all major metropolitan areas could end up staring down the barrel of what New York City residents currently were. To paraphrase the old expression, as New York City went, so might go every major metropolitan area.

Strong as they were, these models and projections couldn't account with certainty for how willingly Americans would comply with the recommendations compared to Europeans. So many factors went into calculating compliance that we could use only an estimate. It was here that I best understood Tony's reluctance to rely on models, but I also believed that when it came down to it, what we were determining was the potential fate of hundreds of thousands or even millions of people. Better to go with the best calculable estimate than nothing.

I didn't want the press finding out about all of our different modeling scenarios. I couldn't risk having every pundit on TV preemptively tearing into the cornerstone of the case I was building for the president. It wasn't perfect—and the press would point that out—but I couldn't let the perfect become the enemy of the good. Whatever our differences regarding the extent of asymptomatic spread, I was fairly confident that Bob, Tony, and I remained solidly in one another's corner. I also believed that Steve Hahn, another ally and someone who truly understood data, would see the wisdom inherent in an extension of the fifteen days.

Yet, in the final week of March, Bob and Tony disagreed with some aspects of my strategic approach. We were all in alignment about the need for continuing the current mitigation. I was seeing Tony, Bob, and Steve in our daily task force meetings, and I emailed, texted, or was in phone contact with Tony nearly every day, frequently more than once. There was no aspect of the response on which Bob, Tony, and I didn't consult one another. We each served as the others' sounding board and support system when inevitable frustrations set in. We provided one another with both wise and effective counseling.

We went back and forth on the length of the extension; we were all in agreement that one was needed. I felt it should be thirty days, but when I broached the subject of asking the president for this, Bob and Tony both felt that it would be more prudent to ask for another fifteen, wait, and then ask for another fifteen after that. I disagreed. I didn't believe the president would have the patience or the political will to go to the

American people and say, *Here we are again, asking you to do this one more time*. But those thirty days would give us a chance to limit the kind of exponential growth we'd seen in New York from expanding across the entire United States. Anything shorter would be seriously inadequate, and we couldn't take the chance of the president agreeing to only fifteen more days and then stopping short of what was actually required.

In the end, I held firm to my gambit. Go for the thirty. Tony, Bob, and Steve accepted. Tony worked with me on the graphics and the wording. Time compressed, and anxiety levels rose—I didn't need any data to confirm that. I felt it in the thumping of my heart and in my worried, exhausted sighs.

WHILE I WAS GETTING modelers to help project the impact of mitigation, members of the Council of Economic Advisers were modeling the pandemic's effects on the economy for the second quarter. Jared Kushner shared these forecasts with me.

In one, Tyler Goodspeed, a member of the CEA, had written, "I am not a public health expert. But in discussion with the brightest epidemiological modelers in recent weeks [read: not us], it has become clear that while interventions may alter the path of critical cases, optimistically they can attenuate the cumulative case load by maybe 20%." In essence, he was saying that the most by which mitigation could lower cases was 20 percent. Once there was wide community spread, this could be true, but we were trying to prevent that initial community spread from occurring in the first place, to protect the main metros across the United States that weren't yet into community spread. Goodspeed went on to add that there was a need to be very clear-eyed about the staggering economic costs of mitigation and about what any mitigation might do from a public health perspective.

To the best of my knowledge, the CEA and the White House senior advisors who supported their view had been silent about our initial fifteen-day ask; I doubted they would be again. They'd planted their flag on the hill they'd chosen to die on, declaring (without supporting data) that any solutions the medical people produced would be only 20 percent effective.

In another email Jared shared with me, a member of the CEA noted that one forecast didn't account for the economic effects of shutting schools down, pointing out that education expenditures at the state and local level accounted for 3.35 percent of GDP. By March 18, 79 percent of schools in the United States had been shut down without federal recommendations. If that educational spending decreased by 79 percent over a three-month period, the annual rate of real GDP growth would be reduced by 10.2 percent. Later, more references were made to economic devastation, "an unprecedented economic contraction—more than double the magnitude of the worst quarter of the post-war period to date." You didn't have to read between the lines: The economic people didn't think a 20 percent reduction in fatalities was worth the cost to the economy.

These reports were having an impact on the president, I could tell. On March 24, at the halfway point of our 15 Days to Slow the Spread campaign, President Trump stated that he hoped to lift all restrictions by Easter Sunday, April 12. He wanted the country "opened up and just raring to go." Otherwise, the economic toll would be too great. Lives versus livelihoods were on the line, and he was siding with the latter.

I was shocked when he said this. With those words, it became clear just how far I'd have to move him in a matter of days.

In many ways, the economic figures felt just as scary as what we were modeling, but we could never forget that the figures we were tracking weren't just numbers—they represented *human lives*. I didn't have the time to think much about the assumptions being made in their models about the value of human life, but each time I walked into our little conference room, I had to remind everyone on my team that we all needed to be at our best, given the forces we were up against.

I was troubled by the sense that everyone in the White House seemed to have their own data stream and interpretation of that data. Some didn't understand the reporting cycles—over-the-weekend data reporting was always incomplete, and the Monday figures I provided tended to be lower than the actual case and fatality counts. Fatality reports were notoriously delayed by weeks in the United States. You didn't just need to use the data; you needed to understand the nuances of that data and how easy it is to under- or over-interpret the data at one point in time. Yet, every Monday, the economy people would report up the chain to West Wing personnel that things were improving. They'd cherry-pick a single positive data point and use that to produce a general, overly sunny forecast. I tried to align the data sources and their interpretation, but day after day, new emails would arrive with a summary of the state of the pandemic that I would then have to refute. Everyone was looking for "better" data—data that would make the pandemic look more "flu-like," less deadly, less of an issue.

During the first few days into the 15 Days to Slow the Spread campaign, I had begun to hear whispers from those within the CEA and elsewhere about my data.

Where is she getting her information?

Are these sources reliable?

I have different data than she does.

Of course, data was unstable, and we were still trying to develop comprehensive reporting for cases, hospitalizations, and deaths. The kind of data I was getting easily from European colleagues seemed to be elusive in the United States.

I didn't know the source of the whispers, but I knew that if I could turn those whispers into a normal conversation, so much the better. I decided to sit down with the other data-interested personnel inside and outside the task force to find out the sources for their statistics. I figured we could go through them together, to improve both mine and theirs. Perhaps they had a source I wasn't aware of. Did they have a better source for daily Covid-19 hospital admissions? Where were they getting their laboratory data?

We didn't want to dismiss their work out of hand, so Irum and Daniel Gastfriend met with Tyler Goodspeed from the CEA and Derek Kan from OMB to discuss it. If, when I later made my assertions, the economic-leaning members of the task force knew that some of what I'd calculated had come from their camp, they and others in the administration might be less likely to raise objections.

Meanwhile, Irum was continuing her work with the outside modeling experts. She and they were refining the graphs they'd created on viral spread without mitigation measures, so that they best represented the data and were accessible to the nonmedical eye. While I was satisfied with the work the modelers had done on this first task, the more important task was presenting the impact of a mitigated response. None of the models adequately represented what I believed the country's infection and mortality status would be with various levels of mitigation.

Consequently, I continued to work on these myself and engage in active debates with the data team. I'd assess and assess and assess what I was seeing happening in the United States and in Europe. I took the current mitigated Italian case, hospitalization, and death curves and applied these one by one to the ten to twenty-five largest major U.S. metro areas we thought could see community spread. I then adjusted each calculation to reflect the future weeks when each area would see a peak. Crucially, I was projecting a figure based on each metro area having its own curve serially, not simultaneously, displaced two to three to four weeks into the future. I assumed a ratio of hospitalizations to deaths similar to the Italian experience. I did that because the quality of the Italian health care system was similar to ours, but whereas Italians were an older population, we had more comorbidities; I believed these two factors would balance each other. I then combined this data and projected over the next two months to predict the possible impact on the United States. I produced a range, with a low end and a high end based on whether ten or twenty-five of our largest cities evolved as Italy had. Our domestic reporting was so inadequate and late that I didn't even have enough data from New York City to make clear projections; so I had to use Italy's data.

My first projections put us at eighty thousand to two hundred thousand people dead just during April and May, the first surge.

I had my sights on fully briefing the vice president on the weekend of March 28, just three days before the expiration date of the original fifteen-day Slow the Spread campaign. In anticipation of this, on the day before the meeting, I would reveal the projections I'd come up with to members of the task force for the first time—including to Bob and Tony. I needed the vice president to be ready to brief the president over the weekend, to ensure that we could announce the extension before the end of the first fifteen days.

Then, much to my sadness, other supporting evidence for an additional thirty days of restrictions began presenting itself in the real world. My initial wake-up call back in January had been the graphic footage from that Wuhan hospital; for many Americans, including those in the White House, similar images coming out of Elmhurst Hospital in Queens produced the same startling effect. Pictures and stories of what was happening in New York City began to appear in various media outlets in the days before my Friday meeting with Vice President Pence. One twenty-seven-year-old doctor there described the situation to the *New York Times*, characterizing it as "apocalyptic." Over the previous twenty-four hours, they'd had thirteen patients die.

Across the city, nearly 4,000 Covid-19 patients were hospitalized, and FEMA believed that, over the next days, all 1,800 intensive care unit beds in the city would be occupied and would likely remain so for weeks. Two city hospitals were already reporting that their morgues were full. Mortuary space was expected to reach capacity, and the state had asked for eighty-five refrigerated trailers from FEMA to receive the dead. Soon, those trailers would be parked outside various hospitals in the city. Online, photos showed nurses using plastic trash bags as personal protective equipment. Reports circulated of two patients being hooked up to the same ventilator due to the shortage of equipment. These kinds of visceral reminders of what Covid-19 could do visually reinforced the data, filling in the gaps and telling a story that numbers alone couldn't tell.

The president received a very personal reminder that week when he learned that Stanley Chera, a New York City entrepreneur, had been diagnosed with the disease, was hospitalized, and was faring poorly. Very soon after Chera was first hospitalized, he was placed in a coma on a ventilator. President Trump described Chera as a friend and began publicly remarking on how vicious Covid-19 could be. I saw this as the president's recognizing that people like him not just in age—Chera was only a few years older—but also in similar economic circumstances couldn't count on wealth as a form of immunity. Initially, I did not press this point in any of my presentations. I heard the president mention Chera several more times that third week. He also talked about Elmhurst Hospital; he knew that hospital. Suddenly, this pandemic was not abstract to him, but very real and personal. Hearing and seeing the president grow more somber, I sensed he might be more receptive than I'd initially thought to dropping his position on relaxing the guidelines by Easter.

Add to this the scenes being broadcast out of New York: the ambulances hurtling down abandoned streets, the cacophony of multiple sirens blaring, the lines of sick people outside hospitals, and the dire, excruciating tales of suffering, survival, and loss—the story and the numbers together presented a vivid picture not of what was to come, but of what was already here. Worse, the scenes in New York City pointed to what was likely to take place in other metro areas around the country. If we didn't extend the shutdown, a New York–type outbreak in every major city across the country would be our future—the kind of future that would see one or two million people dead by the fall.

I walked into my meeting with Vice President Pence on March 28 confident that I could make my case. As usual, Pence responded soberly to my presentation. His chief of staff, Marc Short, was unusually quiet, not pressing me on any points.

At various intervals, the vice president did ask, his urgent tone revealing far more than his poker face did, "Deb, do you really believe it's going to get this bad? Do you really believe that this many people will die? Do you really believe the hospitals are going to get in this much trouble?"

At each point, I told him that I was certain about what I had concluded.

It was hard for anyone to swallow the notion of the United States going from fewer than 5,000 deaths to between 80,000 and 200,000 over the next eight weeks as the best-case scenario.

My meeting with President Trump was scheduled for Sunday morning. Prior to meeting with him, I consulted again with Tony.

"I think you need to up your numbers," he told me.

I blanched at this. "Really? I've already told the vice president eighty thousand deaths at a minimum."

"Tell the president one hundred thousand. That's a big, round number and will resonate with him in a way that eighty thousand won't. I also think that's more accurate." He then went on to say that the upper limit should be 240,000, not 200,000.

Still, these were projections, and the difference he was recommending wasn't that substantial. After I met with Tony, who made great edits to the text to make the bullet points clearer, it was back to work, revising the slides and documents I'd present to the president. Tony might not have liked models, and he might have wanted a higher degree of rigor and certainty, but he recognized, just as I did, that we had to move with what we had, not what we wanted. Waiting for perfection would have doomed us all.

As I walked through the White House on Sunday, March 29, I immediately sensed a different vibe. Instead of the manic flurry of aides coming and going, the clamor of televisions from outer offices, the general hum and buzz of a busy office space, the scene was subdued. Its having been a Sunday contributed to this, but the place was by no means empty. The atmosphere wasn't quite somber, but watchful.

Before, I'd been able to walk the hallways and feel nearly invisible, just one of so many others whose degree of importance, influence, or insight wasn't particularly noteworthy. That day, though, eyes diverted from desktop screens to follow me. I wasn't sensing fear so much as anticipation.

Unlike in the past, I didn't have to wait long before being led into a room in the Residence known as the Yellow Oval. I was nervous. I had never been in the Residence before. I stood and waited for the vice president to signal where I should put myself. I sat on the yellow couch next to the chair the president usually occupied. The others, including Vice President Pence and Marc Short, either stood making small talk or sat. We waited.

I saw Tony Fauci looking pensive. This was a small group, not the task force. This was a different room, not the Situation Room. This meeting had been specially arranged, most likely by the vice president. I wasn't sure how much detail he had gone into, if any at all, in briefing the president. But our presence here in the Residence signaled that the vice president believed me and the numbers I had presented to him. I honestly didn't know what to expect from the president. A couple of days earlier, I'd landed myself in the media's crosshairs for praising the president's ability with numbers and data in an interview with the Christian Broadcast Network, but up to that point, he had taken my guidance and respected the data I'd given him. Now I needed him to do that again, and I had no idea if he would.

The president walked in. He was dressed in more casual clothes than I was used to seeing him wearing in the Oval, a pair of slacks and a polo shirt.

The vice president looked at me and signaled me to speak.

Despite my nerves, I plunged right into the deep end. Opening with my PowerPoint graphics, I said, "Mr. President, we need to take additional action immediately. I'm recommending that we extend the Slow the Spread measures by thirty days."

"What will happen if we don't do the thirty days?" he asked. He had cut to the chase.

I paused for a second, then decided to hit close to home: "If we don't, I'm certain that we're going to have fifty, a hundred, and potentially a *thousand* Elmhurst Hospitals. That means more trucks outside those facilities. That means more bodies inside those trucks. We're going to see city after city looking like what New York does right now. It will only get worse."

"I know that hospital," Trump said. "What's going on there, it's horrible." His eyes narrowed for a moment and his brow furrowed. He relaxed and tensed these facial muscles as he continued to take my words in. The sense I had had about the watchful atmosphere outside the Yellow Oval was now inside it. I'd been in the president's presence at only a few meetings by now, though many more press conferences. I'd noticed how he frequently held on to the lectern stiff-armed, his shoulders spread, making himself even larger a presence.

He didn't do this as he looked at me. He seemed to contract rather than expand.

"My friend is there. I'm younger. Don't weigh as much as him." He recovered quickly: "What are you basing this on?"

I explained to him how the United States was tracking as Italy had: we were two weeks behind where they were. I hit hard on Italy's case fatality rate, the toll it was taking especially on people over seventy, people with other health issues—the overweight, the hypertensive, those with known cardiac or other respiratory or systemic issues. I showed him one of the charts we'd created. It literally and figuratively demonstrated the graphic nature of the reality. It showed a steep and inexorable rate of rise of infections, hospitalizations, and deaths.

"That fast?" he said.

"Yes, Mr. President."

"How many?"

"One to two hundred thousand dead by the end of May. Best-case scenario."

Again, he seemed to deflate. "Do you mean that there will be body bags there? Refrigerated trucks? Just like at Elmhurst?"

"Yes, Mr. President. Hundreds of hospitals."

"Worst case?"

"If we do nothing?"

He nodded.

"Millions. Somewhere between one-point-five and two-point-five million."

He flinched as if I'd struck him. He looked up from the graphs I'd handed him and then back at me. Back to the charts and then back to me. "Are you sure?"

"I am, Mr. President," I replied without hesitation. I held his gaze.

"One hundred thousand to two hundred forty thousand dead even with another thirty days?" For the first time, his voice had lost its matter-of-fact tone.

"Yes."

Hearing those words come out of his mouth made what we were talking about even more real for me. I sensed then and in the preceding week that the president had a grasp of the enormity of what our country faced. I tried to imagine what it must be like for him and some of the others in the room to hear these numbers. I'd been operating on the front lines of the decades-long HIV/AIDS pandemic, which was still taking the lives of nearly a million people per year. He, and most Americans, didn't have my perspective.

We sometimes use the term *sobering* to describe the kind of Covid-19 mortality numbers I was talking about. For anyone without my background, projections like mine might have been seen as the kind of hyperbole spewed by a barfly in a rambling, incoherent denouncement of governmental malfeasance. Still, I could understand others being up in arms at the apparent impossibility of what I was suggesting.

Now, as I write this nearly two years later, the numbers are no longer impossible; they are our reality. Ultimately, with the additional thirty days in place, my forecast for the three-month period proved to be accurate:

March 28: 139,732 cases
   May 31: 1,889,000 cases
March 28: 2,844 deaths
   May 31: 109,058 deaths

No reasonable person, when talking hopefully in March 2020 about adjusting to a new normal, could possibly have imagined this included accepting that, by February 2022, as I write these words, we would already have seen nearly 6 million deaths around the world, with another nearly 10,000 people dying globally by today's end. In the United States today, there were "only"—and it breaks my heart to use that word—over 100,000 daily new cases and over 2,000 daily deaths. We have surpassed 930,000 Americans lost, with the potential to reach over 950,000 by March 1, 2022. Those 2,000 or so—who among us wants their lost loved one to be part of that "or so"?—brought the death count to an unfathomable and avoidable 950,000 Americans lost.

At the time of our meeting in the Yellow Oval, all these numbers were in the future, hypotheticals that none of us knew would come to pass, hypotheticals I wanted desperately to avoid. I had no idea then how inured the president would become to the growing numbers, nor that so many of us would come to see them as an acceptable consequence of a collective reluctance to do the right thing—whether that was to wear a mask, avoid large social gatherings, refrain from dining indoors, or get vaccinated.

As I sat there, waiting for the president to speak, the words running through my mind were *This has to happen. We have to get the additional thirty days.* As I sit here now, the words running through my mind are *This didn't have to happen. This shouldn't have happened. This can't go on happening.*

The president continued to sit, one hand covering his mouth. He dropped it to join the other one folded in his lap. I didn't dare scan the rest of the room. I kept my eyes on him, gauging. He wore the same expression he had worn while discussing his very sick real estate developer friend.

"Tony, what about you?" He turned to Dr. Fauci. "Do you see it the same?"

"Yes, Mr. President, I do."

The president nodded. "Okay, okay."

We thanked the president and rose to leave the Yellow Oval Room. As I gathered my things, the sense of relief was palpable. The last thirteen days had been the most stressful, anxiety-inducing, and busiest of my life. Making the case for 15 Days to Slow the Spread on March 15 had been difficult; creating this pitch for an additional thirty days had been harder by many orders of magnitude—and of course, the president hadn't even committed to anything at that point. But I read his "Okay, okay" as acceptance of the additional thirty days.

But the truth was, as much as I had wanted the extension, I also recognized that regardless of this win, the situation remained dire. We weren't going to somehow snatch victory from the jaws of defeat with a last-minute miracle. This was about limiting the damage, doing just what "flattening the curve" implies. No one likes to think in such pessimistic terms, but after effectively losing January, February, and part of March, this was our reality.

What mattered most today was that the president—perhaps surprisingly to many—had done the right thing, though I suspected his decision wasn't one that pleased most of his trusted advisors in the economy wing of the task force and elsewhere. Whatever my view of him as a politician or a person was immaterial. In this one instance, he had listened to the data, looked at the graphs and the evidence, and had made the only choice he could—and in doing so, he was helping us deliver a crucial message to the American people. These mitigation policies were needed. It wasn't a question of economic vitality, but of individual lives. I just hoped he had the political will to keep them in place.

CHAPTER 8

Who Are You Going to Believe?

"We will never shut down the country again. Never." President Trump's tone was emphatic, edged with agitation. Furrowing his brow, he concentrated his full attention on me. His pupils narrowed into hardened points of anger.

We were standing in the narrow five-by-eight-foot space just outside the formal White House Briefing Room, which was crammed with hard-working communications staff. It was the first week of April 3, mere days after the president had announced the thirty-day extension of the Slow the Spread campaign to the American public, and the ground had shifted suddenly and without warning.

I felt the blood drain from my face, and I shivered slightly.

A moment later, I stepped into another press briefing, swept along by the frigid wake coming from the president's broad back. I experienced that unnerving sensation of having crested a hill too quickly while driving.

As tempting as it was to believe that this was a belated April Fool's joke, I knew it wasn't. I didn't know what precisely had brought about his change of heart, or who had convinced him I was wrong, but his belief in me—in the science, the analyses, the graphs that had gotten the thirty-day extension approved—seemed to have disappeared nearly overnight. His stern look suggested I'd betrayed him, misled him somehow. I didn't take his anger personally—three years of watching him on television had trained me not to be dumbfounded by his words or actions. Still, the whiplash was intense.

After a month of positive and relentless forward progress, President Trump's harsh words and ominous tone immediately cast a dark shadow over the country's future. What interests had driven this new push? What behind-the-scenes actors had influenced him? More concerning: What were we going to do to prevent the calamity I'd predicted from coming to pass?

I assessed the situation: *Okay, we've gotten forty-five days. We're not going to get any more. How else can we protect people moving forward? What other ways can we push this rock back up the hill?*

We were less than a week into our second circuit breaker, protecting the metros where the virus wasn't in full community spread. While I'd always assumed getting another shutdown would be a near impossibility, having the door slammed shut so hard, so fast would leave us with few options if the situation on the ground continued to deteriorate. If we ended the shutdown too early and the American public's behaviors reverted to pre-pandemic devil-may-care, thousands more could die. These additional thirty days would be critical, and slow, careful re-opening down the line would be essential.

What I couldn't have known then was that this day would mark a permanent change in my relationship with President Trump. His about-face created a seismic shift in my ability to speak directly to, present data directly to, and influence him in person. Though, in theory, data-based decision making was still possible in this White House, from here on out, everything I worked toward would be harder—in some cases, impossible.

With the president's harsh words still ringing in my ears, I walked into the April 7 press conference and took my place on the dais, standing off to one side. I wondered if the makeup I'd been wearing to hide my exhaustion was adequately masking the "What-the-fuck-just-happened?" expression on my face. I had come into the White House knowing that having the president's ear would be crucial to my success. I'd never suspected how suddenly I could lose it, how impossible it would be to get it back, nor how sweeping the impact of his ghosting me would be for Americans.

With the cameras trained on all of us, I maintained my composure.

And with that, April began.

• • •

IN APRIL 2020, NEARLY everything came undone. So much of what went on that month codified the president's instincts about the virus and about the "cure" being worse than the disease. The ongoing shutdown was good news: we had preserved whatever positive strides the states had made in March. But behind the scenes, the administration was laying groundwork for a radical change. It didn't matter what I or any of the scientists thought: whether cases were growing by 10 percent or 300 percent, whether testing had improved, whether more people were dying. From the start of April, so long as Donald Trump remained the chief executive, the federal response would be different, and I would have to adapt to effectively protect the country from the virus that had already silently invaded it.

Since my arrival, the demarcation line between the economic and public health interests had been clearly defined. I always knew that President Trump was influenced by whomever he'd last spoken with. What I couldn't yet decipher was whether his new attitude was coming from the political side (Mark Meadows, Marc Short, Peter Navarro, Derek Kan, Derek Lyons, and Stephen Miller) or the economic side (Tyler Goodspeed and Kevin Hassett). To some extent, the question was one of semantics: with this White House, all politics was about the economy. Still, the distinction mattered. If I was going to watch my back, I would have to identify who the biggest threats countering my analyses were.

Tuesday, March 31, was Mark Meadows's first day in the office as the president's chief of staff. I hadn't actually met Meadows. He'd attended the 2020 Conservative Political Action Conference (CPAC) and had potentially been exposed to someone with Covid-19. Just before I went to a Sunday task force meeting on March 8, he called me from home quarantine. He was clearly unhappy about the CDC's recommendations. He saw little scientific evidence for the length of quarantine or for the incubation period of the virus the CDC had cited. The data set was too small at that point, he said, for such strict measures. To be clear: the data set was extremely limited. It was based on a choir in Washington State and didn't take into account that the members infected on day ten and fourteen may have been so from another source. Sequencing of the virus from each choir member hadn't been done, so it was possible those late cases were unrelated to the choir's exposure.

First impressions are revealing, and right away, I put Meadows in the "the cure is worse than the disease" camp—but with a twist. Unlike Marc Short, who too easily dismissed any claim the scientists made, Meadows wanted to see the full data set and hear the rationale behind our interpretations. He was willing to have an open debate. In later conversations with him, he often asked me for the raw data used for our recommendations. I was always suspicious about whom he shared these figures with, what additional analyses were done, and how they were used. Still, he was the president's chief of staff; if he requested something of me, I assembled the data. At the time, sharing numbers was part of my job. I couldn't yet see how it would also be a part of my undoing.

On April 3, just minutes removed from President Trump's emphatic declaration, it was impossible to dissect his message with great assurance, but the trail of breadcrumbs seemed to start with the projections I had produced to get the additional thirty days: the 100,000–240,000 deaths. I suspected that as far back as the European travel ban debate, when I asked the economy people where *their* numbers were, someone within the administration had been reinterpreting the data sets I had used for my projections. Their goal was simple but dangerous: to let the president believe I was wrong in my calculations and therefore couldn't be trusted.

On April 8, I received a memo on which I was cc'd. It came from the CEA's Tyler Goodspeed—whose earlier memo, you'll remember, had stated that any interventions we made against the viral spread would be only 20 percent effective. Kevin Hassett had asked Tyler to turn his attention again to the response to the pandemic. Hassett had once served as the president's CEA chairman. He'd left that position but was then brought back to President Trump's White House when the pandemic response began. Just like Secretary Mnuchin, both Goodspeed and Hassett had demonstrated a brilliant understanding of economics. Collectively, they and others at the CEA had built trust and respect within the West Wing. They had credibility—something I'd apparently earned and then quickly lost. Yet, neither had dealt with a vast public health crisis of this nature before. For all Kevin Hassett's excellent ability to integrate economic data, he and Tyler were in new and complicated territory.

With Tyler's April 8 memo, I had my answer for who within the senior administration was at work countering my data. Kevin Hassett,

Tyler Goodspeed, and other members of the CEA team had put together an independent analysis of the data I'd used for my computational assumptions, using similar case curves they'd developed. The problem was that they used vastly different assumptions than the ones I had used. Italy and many European countries had reached a peak for new infections, Tyler wrote. "We then adjust these projections to U.S. population level and apply country-specific Case Fatality Rates (CFR) to cumulative population adjusted projected cases. Specifically, extending each country's curve using a cubic model and scaling it to the U.S. population." He continued:

> Approximately 104,000 U.S. deaths, assuming the highest observed CFR (Italy, 12.9 percent), with a high estimate of 147,000 deaths in the event the U.S. were to track Spain in cases and Italy in CFR. Assuming the actual CFR observed in the U.S. to-date suggest substantially fewer cumulative deaths (approximately 26,000).

Their conclusion: in this first surge, 26,000 people would die by Memorial Day. We had predicted between 100,000 and 240,000. Clearly, this was an enormous difference. My projections were nearly four to ten times greater. The CEA and I saw two very different futures. I saw a pandemic of historic proportions; they saw a fairly average year of seasonal flu.

On the day I received the memo, April 8, the United States had already had 18,000 fatalities. Did the CEA really believe at this stage, given how the numbers were trending, that only 8,000 more people would die in eight weeks? By April 12, literally four days after the memo was written, the country surpassed the 26,000 deaths Tyler and his team had predicted. On April 8 alone, 2,234 people perished from Covid-19; the spring surge topped out on April 21 at 2,725 fatalities per day and stayed above 2,000 for seventeen more days. Obviously, this was still a long way from Memorial Day.

The underestimation and underselling of the seriousness of the outbreak's progression was obvious to me. They were trying to meet data with data. But they weren't using the appropriate data assumptions to arrive at their figure. Not obvious was who was responsible. I did what I usually do and checked Tyler Goodspeed's credentials. He had published widely on banking and financial regulation. While the memo had come from him, I knew from meetings and conversations that the driving force behind it was Kevin Hassett, the expert economic forecaster. To be clear this work was not from Larry Kudlow or Secretary Mnuchin. They never contradicted my numbers or projectinos and instead used them.

They had used the same data I'd assembled, and they'd done the math right, but they'd used very different assumptions to come to conclusions that were very different from mine—and much more palatable to the Trump White House.

Their model wasn't accurate for these reasons. First, the CEA presumed incorrectly that viral outbreaks in the major metropolitan areas would occur simultaneously and not serially. Second, they treated the demography of each major metropolitan area identically. And finally, and most crucially, they had grossly underestimated the United States case fatality rate, failing to take into account both the delay between infections and fatalities and between fatalities and their reporting.

Their projection was misinformed. I had used the Italian data because I was constantly in contact with that country's public health people. From those discussions, it became apparent that Italy had more comprehensive and up-to-date data. Their figures included the most essential demographic information (age and comorbidities), and I was able to use their fuller data to project that our outcomes would closely match theirs. Our data-reporting system was always going to skew projections primarily due to lateness of reporting, and I was able to account for that. Because I knew our death data was delayed by weeks, the fatalities from today's cases would happen and then be reported weeks in the future. Today's reported deaths were from weeks prior when cases would have been exponentially lower.

The economic team's faulty assumptions would always produce an underestimate. By using the wrong CFR, they would continually compound this fundamental miscalculation, day after day, case after case. They sent their report out with those wrong assumptions. They never bothered to consult with me. That was frustrating enough, but believing that a total projection of 26,000 dead when 18,000 had already perished was mind-boggling.

They had failed to understand a fundamental point I had tried to get across to them repeatedly: I had used the data from Italy to make clear

that we would experience something like what Italy had experienced—
but crucially, the United States wasn't mitigating the spread as aggressively as Italy was, so our fatalities would assuredly be higher. Also,
Americans suffer from comorbidities at a higher percentage than Italians, again contributing to a higher CFR. When you factored in these
two variables (lower mitigation measures and poorer population health),
the U.S. case fatality rate would absolutely be worse than Italy's.

At first, I wanted to immediately dismiss the CEA's forecast due to
those fundamental errors. The flaws were obvious to me, but it would
be harder for people without an epidemiological background to see
them; they didn't know that Italy reported deaths regularly, and we
didn't. You have to know your data sources. You have to know their
limitations.

I suspected that the CEA's faulty analysis and numbers had gone to
the president and other senior White House advisors, like Jared Kushner, Marc Short, and Mark Meadows. The CEA had done its analysis
and filed its report just days after I presented mine to the president and
convinced him to authorize the thirty-day extension. Just as I'd been
preparing daily reports, models, and projections, so had the CEA.

As gatekeepers for their respective bosses, Short and Meadows had
control over who and what went to the vice president and president.
The Hassett/Goodspeed analysis must have made its way immediately to
the Oval Office. This president wasn't going to care about the subtleties
of the analysis. The more cynical of the White House advisors would
have bottom-lined it for him. They'd likely reduced it to this:

*The doctors say a hundred thousand to two hundred forty thousand
dead.*

Your economists say twenty-six thousand.

*Debbi Birx used the wrong case fatality rates—she used Italy's. We
used the right one, the United States'.*

You know we are better than Italy. Our hospitals are better.

*You know she isn't very good. She's just another civil servant in over
her head.*

Reading between the lines, then:

Debbi Birx is wrong, really wrong.

She has overestimated, by four to ten times, the number who will die.

*Debbi Birx intentionally misled you to get you to do something that was
never needed, shutting down the country for another thirty days.*

The number will be under thirty thousand—almost eerily close to the seasonal flu.

This will be an acceptable level of loss to the American people.

No need to worry. No need to treat the pandemic aggressively.

Trust us, not her.

We know better. We are your team. Together, we built your economy.

We know numbers. Our numbers are right.

While I had been open about my forecasts being based on the best available data, they had presented theirs under the guise of its being much more consistent with the *real* numbers, the *American* numbers. In fact, their flawed methodology did not, and was never going to, reflect reality.

It helped the CEA enormously that their data happened to jibe with a typical seasonal flu death count, magically matching what the president and many in the administration had been saying all along. As time went on, and especially in retrospect, I saw that as soon as the Hassett/Goodspeed report was circulated among the Oval Office leaders, the president had all he needed to confirm his initial bias that the novel coronavirus was just like the flu. With this confirmation in hand, President Trump, I believe, simply stopped focusing time and attention on the public health side of the pandemic response. Mistakenly, he no longer believed he needed to support any of the mitigation efforts that were key to slowing the spread. He'd moved on. A set of flawed numbers from his "best" people supported the notion that he'd been right all along.

In the moment, the importance of the 26,000 versus 100,000–240,000 discrepancy didn't fully register with me. Over time, as the deaths continued to rise, the CEA would up its estimate, eventually getting to 66,000. That was still well below my projections but near enough to mine that we could have engaged in dialogue. I could have shown which of their errors explained the difference, how they had employed a similar model but with wrong assumptions. But no one initiated that conversation.

In April, my attention was being drawn to so many other areas that, regrettably, I didn't try to reclaim the president's attention. And I suppose it was too much to expect, given the atmosphere in this White House, for his most senior advisors, to say, "Our original estimate was wrong. Dr. Birx did a better job, and her figures are more representative of reality." Admitting to an error is the right thing to do, but not in this administration.

I don't think the CEA or Mark Meadows ever returned to this issue, and the flawed numbers were left uncorrected. I still wonder: What if I had demanded an audience with the president and Hassett? What if we had sat down and gone through the differing projections, side by side? Would doing so have changed the president's mind? Would he have understood that I really did know what I was doing and could be trusted for objective facts and figures? I might have been able to alleviate the administration's worst fears by suggesting that the optimal way forward wouldn't necessarily "shut down" the country and that we could, in fact, maintain significant economic activity while still protecting Americans.

I don't know. Today, I regret not having tried.

This admission doesn't mean I did nothing to address the grievous differences in our respective projections. During the travel ban deliberations and the discussions over shutting down for fifteen or thirty days, the two wings of the task force (economists and medical professionals) had been urged to present our cases and arrive at consensus. Once I received the CEA report, I wrote to Tyler Goodspeed and Derek Kan. I pointed out that their math was good, but that we needed to discuss their underlying assumptions. I fully expected that we'd get together and that each side would present its case. This never happened.

When advisors cherry-pick data, without fully understanding the data or its sources, to paint a contrary picture of a pandemic to the president of the United States, it is not only intellectually dishonest, but also morally negligent. When the CEA didn't respond to a request to work out our differences, they sent a powerful signal: *We don't care about your numbers, we care about supporting the president's wishful thinking*. With that single memo, they created the first of many inflection points in the president's engagement with the pandemic response.

I saw this ongoing data discrepancy as a harbinger of future instances in which others would either step forward openly or work behind the scenes to undercut the scientists' influence on the president and on policy making. At this stage, none of these efforts were overt. Mostly, they manifested themselves in the president's no longer engaging with me or the other scientists on the task force. This was not the case for the vice president and the heads of the other agencies: They, and other members of the task force, stood together. They remained grounded in using data for decision making and, critically, in supplying public health information, equipment, and therapeutics to save as many lives as possible.

I didn't think then, and I don't believe now, that Steve Mnuchin or Larry Kudlow instigated the Hassett/Goodspeed report and the CEA's inadequate projections and incorrect assumptions. During the debates on the European travel ban, and for a short while after the debate over the duration of the shutdown, they grilled me hard about our projections. They seemed generally convinced, if not by the numbers themselves then by the level of rigor we had applied to produce them. (They hoped the economic damage could be limited—as we all did.) After that, my interactions with Steve Mnuchin were limited to his participation in task force meetings, because he was off working on the legislative agenda to support the American people economically through the pandemic. I believed then and I believe now that Steve took the pandemic very seriously and understood the risk of Covid-19 disease to Americans. He also understood the economic impact it would have on people, and he worked 24/7 on a series of bills, and on the penultimate CARES Act, to get funding efforts through Congress.

When he was about to make press appearances, primarily on the Sunday shows, Larry Kudlow always came to me and said, "I'm going out there. What can you tell me about what's going on?" I believe he presented a balanced approach. He would talk about what the administration was doing to respond to the economic situation through its policies and legislative agenda. He would also talk about the importance of the personal behaviors and measures needed to stem the community spread.

To further counterattack the effects of the economic advisor's inaccurate projection, I sent Irum to work with the CEA folks on their daily report. No matter what she did, no matter how she advised them to address the three main problems with their analysis, she met resistance. Just as they believed that the president wouldn't listen to us, they weren't going to listen to her. They'd spoken. They'd supported the president's view, and that was now that. He'd heard "twenty-six thousand," and there was no way to make him un-hear it.

It wasn't surprising that the CEA and those who believed their projections avoided having an actual discussion on the numbers—especially once my daily reports and summations of the data clearly showed that we surpassed their 26,000 deaths a mere four days after I received the Goodspeed memo. Tragically, by May 25, Memorial Day, nearly 100,000 Americans had lost their lives to Covid-19. Our projections were accurate; the CEA's were not. But who among them was going to step up, in that environment, and admit they were wrong?

In late July, when Kevin Hassett returned from working on another assignment, he asked Tyler Goodspeed to again engage in this area in which he didn't have any great expertise: modeling public health projections. Tyler did the right thing: he declined. I believe he'd seen how far short their figures had fallen and what had resulted from the CEA's faulty report. After recognizing the consequences of his first attempts at pandemic modeling rather than economic modeling and predictions, Tyler didn't want to repeat the same mistake—unusually humble behavior in this White House.

I sensed at the time of President Trump's "never again" remark, and even more strongly with the CEA memo, that the brief window of opportunity I'd used to make my case for shutdown—the president's fear for his own health when friends and contemporaries were on ventilators or dying—had now closed. A unique set of circumstances, a moment of vulnerability, had nudged me closer to the front and gotten the United States thirty more days of mitigation.

Somehow, from the time he agreed to the thirty extra days, the president had convinced himself both that he was physically invulnerable and that if he didn't do everything he could to get the country (that is, the economy) back up and running at full speed, he would be politically vulnerable come November. The general election was just around the corner, and a robust economy was his ticket to four more years. Once the president was again numb to the devastating effects of the disease he'd seen ravage his friends, he was off and running in the other direction, leaving me in his dust. Even after falling ill himself, he would never again return to that Yellow Oval "Okay, okay" moment I'd witnessed.

I suspected then, and am now absolutely convinced, that by the time I got the CEA memo with its flawed assumptions, I had already been many days behind in the race for the president's attention and trust. I am sure that in every internal senior advisor discussion, the president was reminded of just how wrong I had been in my projections. It never mattered that my projections were right then and continued to be right throughout 2020 and 2021; being right apparently didn't matter. I know I find little consolation in it.

After my initial success with slowing the spread, I often felt I was just one chart, one statistic, one direct meeting with the president away from getting him back on my side. I believed that I was always *this close* to getting him to use clear data to drive mitigation efforts. If I had, perhaps he would have been willing to more forcefully advocate for basic mitigation efforts like masking and reducing the numbers of people dining indoors. He might have told the American people that gatherings of friends and family were one of the main causes of infection and that frequent and strategic testing could prevent a worsening of community spread. Reality, of course, was much crueler. We would have slowed the spread of the virus and significantly reduced the number of people who perished before Labor Day. Not being able to demonstrate to him how close the two sides, economic and medical, actually were in our numbers was, and remains, heartbreaking.

AS HARMFUL AS IT was to the public health response, the CEA team was just one front in the White House war on the medical professionals on the task force. As April would reveal, the marshaling anti-shutdown forces looked for any opportunity to undermine the data and metrics we were using to justify the seriousness of the situation—often pitting science against science.

Around that time, Jared Kushner shared an email with me. Up to that point, Jared had been mainly a peripheral figure, moving at the edges of my vision as he managed some of the essential logistics and supply issues related to PPE. He had understood the need to go directly to the suppliers and began what became the air bridge that brought essential gloves, gowns, and surgical masks directly from China, Malaysia, Thailand, and Vietnam throughout the late spring. To my mind, he was an effective means of getting to the president. I had no direct interaction with him on which to base any other assessment.

The memo Jared shared with me cited a National Institutes of Health analysis of the Italian data I'd used as a baseline for my projections for the case fatality rate. The memo stated that the NIH had determined that only 12 percent of the Covid-19-related deaths were directly attributable to the virus. Twelve percent? This was a ridiculous number. Now it seemed someone was pitting a trusted scientific organization, the NIH, against the task force coordinator. More significantly, they were hoping to erode the bedrock on which our case for shutdown had been built: the very reliable, precise, and timely information the Italians had provided. I immediately called Tony. He hadn't seen the analysis and certainly didn't support its conclusion.

I had had no idea the NIH was doing this study. If not for Jared's heads-up, I could have been blindsided. Like Joe Grogan, he was warning me. In this instance and others, Jared Kushner let me know what was going on. Indirectly, through him, I was able to gain insight into what was taking place outside the task force, in the hallways and behind the doors of the president's private dining room.

As he did throughout my White House tenure, Jared listened to the data I presented in support of my recommendations. I believed then, and I still believe, that he and the vice president advocated for my analyses and recommendations with the president—whether it was the need to expand testing, the importance of masking, or the critical message about viral spread among friends and family.

In a White House infamous for loud voices and backdoor meetings and where a general lack of discipline ruled the day, it was difficult to remain objective. When faced with so much randomness and uncertainty, could anything anyone said be trusted? And whom was the president trusting? The importance of what you had to say didn't seem to matter. It was about access—access to the president. It was about who could get to the president and who had the last word.

It was also about where people were getting their information. When I reported fatality figures, I integrated the data from multiple reliable sources in the field—from health ministers across Europe; from the myriad Covid-19 data-tracking sites, like Our World in Data and Worldometer; from Johns Hopkins; and from hospitals and nursing homes. Stephen Miller, the White House director of speechwriting and one of the president's most influential policy aides; Stephen's wife, Katie Miller, Vice President Pence's communications director; and Devin O'Malley, the vice president's press secretary, all repeatedly claimed that the figures I'd provided were wrong: they were miscounts, they said; unverified accounting, they insisted. The Executive Office didn't have its own source of data, so whom were they relying on to make this claim? What were they basing these claims on? They never presented me with evidence to support their positions.

Well, for one thing, there was no shortage of baseless claims pinging around the internet, and likely these fantastical stories were at least partially responsible for the "data" points the Millers and O'Malley had used to support their position that there were fewer Covid-19 cases and fatalities. Conservative radio host Wendy Bell, streaming live on Facebook, told her listeners that due to a change in death certificate procedures, "there's a huge chance that Covid death numbers are exaggerated, to the tune of 94 percent." This false claim mutated from there.

An immigration hard-liner, Stephen Miller used my insights and data when they overlapped with his focus on stricter border control. When I pointed out the increase in cases in Imperial, California, and El Paso, Texas, and other cities on both sides of the border, for example, Miller wanted to use this data to restrict border access. He believed that so-called illegals were responsible for the increase in cases. The CDC did a deep dive with city and county officials and the hospitals along the border where increased cases and hospitalizations occurred—and determined that American citizens and those with dual nationality residing in Mexico seeking care in the United States were driving up the numbers. But this bit of truth didn't seem to matter. Miller and others used it as a wedge to further the divisiveness in the country over immigration and distract the public from the real problem at hand.

Mark Meadows also frequently challenged the hospitalization and fatalities figures I provided and questioned my sources. He listened to me, and we engaged in dialogue. At one point, he stated that many of the hospitalizations that had been coded as "Covid-19" admissions were actually the result of a SARS-CoV-2 test being administered after the patient had already been admitted because of an auto accident or for some other health reason. A Covid-19 diagnosis was an incidental finding. As such, those "after the fact" admissions shouldn't have been counted toward the total number of Covid-19 cases.

Others suggested that hospitals added a Covid-19 code to their billing only to get the increased funding allocated to Covid-19 inpatient treatment. Later in my visits with hospital administrators, I learned that no hospital wanted more Covid-19 patients to care for and overwhelm their beds and ICUs. When this happened, the hospitals were forced to shut down elective procedures—the true moneymakers for them. Caring for Covid-19 patients is a complex, highly nurse- and physician-intensive struggle that uses up far more human resources than could ever be fully reimbursed. Even after I brought this information back to the task force, correcting their errors, social media postings continued to be seen as equally legitimate, factual sources as my on-the-ground on-site, truth-based findings.

On their own, these cases of incorrect and faulty data could easily have been dismissed as being endemic to our age and the rise of the internet as a source for all kinds of information, good, bad, and indifferent. But these claims were coming from people in the president's inner circle, people influencing decision making at the highest level, decision making that would determine whether Americans lived or died. It was deeply troubling.

At about the same time Jared's and others' emails were crossing my desk, another influential group of researchers chimed in with results that further eroded confidence in my projections. Researchers affiliated with Stanford University and the University of Southern California had conducted a study using antibody tests to assess how many people diagnosed with Covid-19 disease based on symptoms had actually been infected with the SARS-CoV-2 virus.

When you are exposed to a virus, your body produces antibodies to fight it. An antibody test (as opposed to a diagnostic, swab-up-the-nose test, like the ones we've become accustomed to, which detect active Covid-19 infection) determines whether your immune system has produced antibodies specific to that virus. If the test is positive, you can be said to have been infected by the virus, whether you are asymptomatic or very mildly symptomatic.

In the study, researchers sought out volunteers in one county near San Francisco and others living in Los Angeles County. How those volunteers were selected and their reasons for participating could have created sampling bias, particularly if those included already suspected they had been infected and were looking for confirmation. The results in both counties "showed," through the presence of SARS-CoV-2 antibodies in the blood, that far more people were infected with the virus than the diagnostic testing and data gathering had shown.

This sounds, on the surface, like a good thing for my case that silent spread was highly prevalent—and it was. News outlets picked up that portion of the story. But what the researchers had determined also undermined the conclusions we had drawn about the number of deaths that were likely to occur. One author of the study, in an interview with the *New York Times*, discussed their methodology, the results, and the conclusions drawn from them, and said, "This is very consistent with the fact that the virus is very common but not killing at the rate we thought." Gina Kolata, a well-respected science reporter for the *Times*, wrote, "But the new data suggests most adults will experience milder to asymptomatic infections." She went on to draw a reasonable conclusion that, if the study was accurate, then the fatality rate was more in line with the seasonal flu "than [with] a pandemic of profound lethality."

Despite the study's confirming that more people were infected with SARS-CoV-2 than previously believed, many in the White House chose to focus only on the much lower death rates and the flu fatality comparison and not on the silent spread that drove community spread and thus led to those deaths. They looked at this data and concluded: *Covid-19 is not a big deal and won't be a big deal. It is no worse than the seasonal flu, and we don't lock down the country for the flu.* They added this highly problematic study to the pile of evidence that *Debbi Birx and, by association, Tony Fauci are wrong in their projections and views.*

When the Stanford study was published in *The Lancet*, a highly regarded medical journal, this was another nail in the coffin of our projections and status. Stanford is a highly respected institution; so is Harvard Medical School, which publishes *The Lancet*. The message appeared to be: *Trust them. Don't trust the White House medical scientists.*

I absolutely believe that all ideas and theories need to be questioned and debated, and the court of public opinion is one venue for airing these discussions. But the fact that all these pieces of research contained such fundamental flaws in logic and methodology damaged the cause of public health at this crucial moment in the pandemic. Together, these studies created an opening for an intellectually dishonest assertion, one that suggested that no one really knew anything about the virus, so anyone could be right and anyone could be wrong. What this perspective crucially left out is the fact that science operates on the principle of "Let's test these ideas to see how right or wrong they are and under what circumstances these judgments are valid."

Despite the failings in these studies, their sources possessed impressive enough pedigrees and carried enough reputational weight to cast the studies as definitive, giving them the power to muddy the waters just enough, to sow just the right amount of doubt. Perhaps because I had been outside the United States for so many years, working for PEPFAR, I was initially a half step behind in recognizing the harm that could be done by weak science such as this. In a post-truth America, it seemed, some in the White House, and pockets of Americans, would use any shred of evidence, even at this early stage, to undermine the public health case of the seriousness of Covid-19 to specific vulnerable groups, to make the objective somehow subjective.

I saw then (and I see now) why the public was confused by the conflicting messages coming out of the White House and the public health agencies. Some of the senior advisors in the West Wing had deliberately sowed and were quickly harvesting a crop of disinformation that outgrew and overshadowed the deliberative, careful data collection and analyses we had done. Just as these theories and justifications took root in the White House, so they did across the country, creating alternative interpretations of the science that would have profound implications for future mitigation efforts—from masking to testing to reducing indoor gatherings.

Once doubts over the data-driven, science-led response had crept in, the floodgates opened for exploiting ambiguities within the data and creating a parallel "data-driven" alternative reality. A pervasive attitude of "No one knows for sure, so do what you want" soon spread around the country.

Meanwhile, as these flawed but well-pedigreed studies arrived, other, subtler misrepresentations of the science and data had been pouring in from all sides. White House senior advisors repeatedly inundated me with published reports of various tenuously related pandemic topics from many sources. I believe many, like General Kellog, provided these to be helpful, to ensure I was seeing everything, but others used them to specifically undermine what I was saying and what I was asking them to do.

As the CEA report had shown, I was no longer the only one in the room armed with data to support her arguments. But not all data is created equal. This kind of blurring of the lines between complete data, warts and all, with its explanations for gaps and biases, and cherry-picked, incomplete data specially assembled to support a preconceived idea or theory, was the most dangerous area for a scientist to sink into.

An effective pandemic response has to build on a foundational bedrock of truth or, at a minimum, a shared understanding from which each side builds its argument. In the case of this pandemic, everyone had to agree that Covid-19 was a major risk to the health and well-being of the American public. Simply put, sizable influential elements in the Trump administration did not believe this about the virus in January, and they did not believe it about Covid-19 disease in April. This was in spite of what had happened in Italy and in spite of the devastation currently reaching a fever pitch in New York City and New Jersey, which we were all witnessing in real time. When we should have been debating the finer points of strategy, most of the Trump administration—and crucially, President Trump himself—were again arguing that the problem wasn't greater than flu. The faulty studies and the memos from the CEA successfully gave my opposition more to work with.

The Trump administration hadn't believed in the risk before I arrived. Over the course of March, I'd done everything I could to build consensus about the substantial risks to specific groups of Americans. I viewed getting two successive periods of shutdown as evidence that, on some level, I'd been able to break through with this message. But these new developments laid bare that any consensus that had been achieved was fleeting. Despite mounting evidence to the contrary, many in the Trump White House hadn't changed their mind about the risks; it had just taken time for the opposing side to figure out how to counter our public health arguments and approach.

While there were undoubtedly people in the White House who viewed Covid-19 as a major risk, they were in the minority. I saw the entire NSC take the virus seriously (as did the vice president), and I believe to this day that Jared Kushner and his team saw the reality of the pandemic. The NSC had seen the early reports out of China and Asia before my arrival. Indeed, through Matt Pottinger, it was they who had recruited me to the White House to reinforce their warnings. While they didn't play as active a role as Vice President Pence—he always listened to me and made every call to governors on my "need-to-call" list—to ensure they took the threat of the virus seriously. They also saw that solutions existed but that combating the misinformation and divergent points of view was as much a battle as trying to contain the pandemic. How can you combat something effectively if you can't agree that it is actually a threat?

Despite these many frustrations, at least we were getting validation from others in the scientific community that our projections were accurate. At some point in the midst of all this back-and-forth, Tony passed along to me an email he'd randomly received from a highly credentialed statistician who'd run the numbers himself and come to the same conclusion we had. It was helpful intellectual support, reinforcing that we were correct, but it didn't lessen the feeling of menace I sensed around me.

Tony and I connected nearly daily, week after week. I made sure he was seeing in the data what I was seeing. It was critical to me that we saw the same evidence the same way. This was true for Seema Verma, Bob Redfield, and Steve Hahn, too. The doctors' group met three to four times a week, and I spoke with Bob and Steve day after day. I don't think they ever wavered in grasping the seriousness of the pandemic or the need to do everything we could to battle the misinformation both within the White House and without.

In some of our discussions in the spring and carrying over into the early summer, Bob and I addressed a topic that was much discussed and debated from the first revelation of the outbreak in Wuhan. Because several research facilities located there were actively studying coronaviruses, much speculation went on about whether or not the wet market was definitively the point of origin of the outbreak. Bob and I would note that this virus, unlike other SARS strains, was unusually fit to adapting to a human host. Often with zoonotic viruses in that first jump from animals to humans it takes it awhile to adapt to its new host, but SARS-CoV-2 was unusually adapted to humans with high infectivity in the first surge unlike the prior SARS-CoV-1. Its unique characteristics presented us with real challenges but I wasn't engaged in answering one of them—determining its origins. Once it was out and active, my priority was to save lives and though we had the genetic sequence (not the samples themselves, mind you) from the Chinese later than we would have liked, once we had it using that information took priority over any other consideration.

Both Bob and I were aware of other unintentional leaks or contamination issues arising out of labs around the world. While not common, they did occur despite safety measures being in place to prevent them. Whether that was the case here, we couldn't say for sure, but it was possible. As I later told a house subcommittee investigating the full scope of the pandemic response, I didn't have a definitive answer to their question regarding the origin of the virus. I did tell them that it was possible to get that answer, but it depended upon the Chinese releasing the very first samples of the virus taken from those who were first infected. It would be necessary to study the evolution of the virus in those first moments and study all of the original strains. Without having those samples of the original strains of the virus, it would be very difficult to ascertain with any degree of certainty whether it came from the wet market or a lab.

In either case, I don't think that people were exposed to it intentionally. Entering that debate would be another distraction.

Steve Hahn and I bore the brunt of the intellectual assault that was hydroxychloroquine. Studies regarding its potential use as a Covid-19 therapeutic were passed along to both of us. These came out of France, China, and the United States, both from doctors and from people who had access to the internet but not a firm grasp of scientifically determining the drug's efficacy as a treatment. Peter Navarro and television's Dr. Mehmet Oz were among those pushing for us to back their belief that a drug used primarily for the treatment of malaria, and a sister drug, used to treat systemic lupus, rheumatoid arthritis, and other autoimmune conditions, should be used to treat Covid-19 disease.

In a crisis, taking proactive measures is important, and at the beginning I saw the FDA's "emergency use authorization" (EUA) of hydroxychloroquine as falling into that category of proactive measures. So was continuing to definitively test to see if the drug would prove to be, in time, a therapeutic we should adopt and use more widely.

While the debate about the use of the drug sucked up too much of my time and attention, it did far more damage than that. Instead of the president's delivering a consistent message about what we knew were effective mitigation measures and effective treatments, he was more preoccupied with touting the benefits of unproven, untested, potentially counterproductive drugs that had been brought to his attention daily by his inner circle. On March 21, the president tweeted about using hydroxychloroquine in combination with the antibiotic azithromycin, calling it a potential "game changer" and urging that the malaria drug be put to immediate use. A day later, he contradicted Tony, who had replied "The answer is no" to a journalist's question about hydroxychloroquine's effectiveness. Tony had gone on to state that the evidence for the drug's being effective was "anecdotal." The president said he felt good about it—that's all it is, "just a feeling, you know" and "we'll see what happens."

Not everyone was able to sidestep hydroxychloroquine. Dr. Rick Bright, who, before being ousted, was in charge of vaccine development at the Biomedical Advanced Research and Development Authority (BARDA), a division of HHS, claimed that he had been dismissed for pushing back against what he called "misguided directives" to advocate for the use of hydroxychloroquine and chloroquine. He cited this among other examples of how the federal pandemic response was being interfered with.

Trump's public cheerleading for hydroxychloroquine blurred important distinctions in some cases and caused direct harm in others. An Arizona man, having heard the president's message on television, took a product that contained a form of hydroxychloroquine and died. The American Heart Association, the American College of Cardiology, and the Heart Rhythm Society warned that the combination of hydroxychloroquine and chloroquine might not be appropriate for patients with existing heart problems.

Later on, *The Lancet* issued a report on the efficacy of a variety of drugs that could be used to treat Covid-19 disease, including hydroxychloroquine. I was worried about this study's validity due to the difficulties in global data collection I had seen. When I was asked about the *Lancet* piece, I said that it had some value. The study clearly clarified who was suffering and dying the most from SARS-CoV-2 infections: people with comorbidities. I thought my statement was clear, but the press later falsely reported that I supported the use of hydroxychloroquine. Why? Because I had said the *article* was of value—but in this, I had been referring to the demographic analysis showing that it was older patients and those with comorbidities who died from Covid-19 disease. I never said I thought the paper offered definitive support for or argued against the use of the drug or any other treatment. Such studies were under way in the United States and would produce definitive results down the line.

Within weeks of publication, the *Lancet* report was retracted. Additional reviews of the data showed that it was unreliable with regard to the interpretation of treatment. What wasn't retracted was the claim that I supported the use of hydroxychloroquine. Understandably, this created more confusion. Did the White House and the Coronavirus Task Force believe in the effectiveness of the drug or not? The president and some of his advisors may have, but the rest of us were waiting for a definitive study we could trust.

The FDA approved the provisional use of hydroxychloroquine in hospitals and clinical trials only, stating again that the drug hadn't been shown to be "safe and effective for treating or preventing COVID-19." Despite this qualifier, twenty-two U.S. states stockpiled nearly thirty million doses. A May 11 study of more than fourteen hundred Covid-19 patients hospitalized in and around New York City found that those who took hydroxychloroquine with an antibiotic were twice as likely to experience cardiac arrest.

By mid-May, the president announced that, despite the concerns and warnings that hydroxychloroquine should be used only in hospital settings, he was taking the drug preventively and had been for ten days. Upon completing this two-week regimen following a viral exposure, he proclaimed not just that the drug hadn't killed him, but that it had received "tremendous, rave reviews." Though I was used to hearing hyperbole from the president, this felt particularly egregious. The reviews that mattered—what doctors and researchers were still working on—hadn't come out yet.

By June, Steve Hahn and the FDA had reversed their position on "hydroxy" and revoked their EUA. It's important to note the language used in their announcement: "Additionally, in light of ongoing serious cardiac adverse events and other potential serious side effects, the known and potential benefits of chloroquine and hydroxychloroquine no longer outweigh the known and potential risks for the authorized use." By anyone's standards, this was a reasonable and measured response.

Peter Navarro, one of President Trump's trade advisors, who assisted in distributing the drugs, made even more plain what the administration thought of scientific rigor in commenting on this FDA rethink: "This is a Deep State blindside by bureaucrats who hate the administration they work for more than they want to save lives."

What the FDA and Steve Hahn had done was exactly what one does in the scientific community: work to confirm the truth or falsity of a position. You examine the results. You make a decision based on the data.

What Navarro was actually saying was *This administration hates all data that doesn't support the unproven contentions we believe in that come from non-peer-reviewed anecdotal reports.* Peter was knowledgeable and passionate about manufacturing and the need for the United States to reestablish the manufacturing of critical PPE and essential medication—an important point—but he also brought that same passion to the hydroxychloroquine debate. Peter believed so strongly in the data supporting hydroxychloroquine use, that he came to task force meetings armed with sheet after sheet of studies supporting his position. The first week in April, he got into a loud argument with Tony. Standing behind me and leaning over my shoulder, he angrily waved a stack of papers over my head toward Tony, shouting, "Here is the evidence this works! You're ignoring all the data, and it's killing people!"

There were other moments like this, when people passionately engaged with one another. Sometimes the disputes were based in evidence; other times, on peripheral unsupported positions. Some people in the White House seemed to believe that when reason failed, passion might carry the day. Peter Navarro crossed a line.

Steve, Tony, Bob, and I (and others) never took any of this personally; nor did any of us dislike any members of the administration. We hated the lies and distortions that were supported. We were deeply concerned when they used small, poorly designed investigations to arrive at definitive answers. We were enormously frustrated when "debates" took time away from the hard work of getting the data to avoid continually confusing the public.

Even with all the evidence in hand, the FDA continued its studies, concluding, on July 1, that "serious heart rhythm problems and other safety issues, including blood and lymph system disorders, kidney injuries, and liver problems and failure" were a consequence of using this drug. Tony and the NIH had done their own study and come to the same conclusion. Assistant Secretary for Health Brett Giroir also stated that he couldn't recommend the drug's use.

Meanwhile, President Trump's conclusion was this: "Hydroxy has tremendous support, but politically it is toxic. If I would have said, 'Do not use hydroxychloroquine under any circumstances,' they would have come out and said it's a great thing."

Because Steve Hahn, as FDA commissioner, was the one directly involved, he did the talking for the rest of us in rebutting the president's disdain. Speaking on ABC News, he said, "This is about science and data. There are randomized trials that show it doesn't work."

Real science and data are great things when they're used transparently in open debate.

CHAPTER 9

The Enemy of the Good

Losing President Trump's support for the countrywide measures so early in April had other detrimental consequences—mainly, that he and others in the administration quickly turned to reopening the country as quickly as possible following the end of the thirty-day Slow the Spread extension period.

As early as April 7, at the direction of the White House senior advisors, some on the task force and many others in the administration were at work on a parallel mission: to develop a plan to reopen the country as soon as possible. As task force coordinator, I was to be part of this "tip-of-the-spear" staff work group. We were given until April 15 to draft and submit to the Executive Office the plan that became known as "Opening Up America Again."

Not only was the timing tight, but how we were moving forward exemplified the problems inherent in any group enterprise, particularly at the level of the federal government. Simply put, too many chefs in the kitchen. Initially, at least three separate groups were duplicating efforts to develop this guidance: HHS, OMB, and the task force. In addition, OMB was being aided by an outside agency, the Boston Consulting Group. Wisely, Joe Grogan saw that this hydra-headed approach, the kind that had failed before in other endeavors, needed to be streamlined.

To avoid interagency squabbling and turf wars, Joe and I worked to consolidate the three working groups into one, placing specific people I trusted in leadership positions on core topics and assigning tasks to the various groups. In our first meeting, I shared with the larger group which tasks each subgroup would be responsible for. No one pushed back on their assignments; it was obvious I was trying to strike a balance in each group between the public health–minded and the economic/politically–minded—the latter with high political interest in reopening businesses, schools, churches, and other institutions. I also placed Bob in this group, along with White House senior advisors, to ensure that the guidance would be embedded in public health messages.

This was one of the important things I could accomplish in my role as coordinator. I wasn't the secretary of HHS, I wasn't the director of OMB, and I wasn't a cabinet member. The heads of the agencies were still in charge of their people in the larger working group. Though I couldn't directly persuade everyone around the table to agree with me, I could at least ensure that knowledgeable personnel acting upon strong, evidence-based data were present in each work group. We had less than a week, so we all needed to work independently and full-out. I also internally organized my data team to ensure that we moved forward deliberatively based on the best information, data, and science at the time. Remember, we lacked a historic road map to show us what had or hadn't worked in the past.

If I stacked the deck in my favor in any way, it was by putting myself in charge of drafting the "gating" criteria for a deliberative, safe, data-driven reopening in the states. I did my research, relying heavily on the work of Drs. Tom Frieden, Scott Gottlieb, and Zeke Emanuel. At this point in the evolution of the response, we'd be moving past the emergency "flattening-the-curve strategy" that had blunted the initial surge and toward the crafting of guidelines that reflected where we were currently in our understanding of the virus and the specifics of proactive mitigation to prevent immediate viral surges across the country. My goal was to make the reopening of the states a data-driven exercise: as each state moved toward reopening, it would have to pass through one phase, or gate, before moving on to the next, thus ensuring the greatest degree of public health safety. This was my way forward within the context of the president's "never again."

To move from one phase of reopening to the next, states would have to demonstrate a fourteen-day continuous decline in new cases and test positivity, clearly demonstrating that its community spread was under control. Within my gating criteria was a clear call for weekly "sentinel testing" (that is, testing across a community, even of people who appear well) to serve as an early alert of asymptomatic community spread. This was very important to me, as I fundamentally believed that community spread began silently among the eighteen-to-thirty-five age group, who were more social and more likely to gather together indoors. Seeing the first signs of silent spread and responding with mitigation would protect the vulnerable and ensure decreased hospitalizations and fatalities. Early mitigation with masking, reduced indoor dining, and limiting friend and family gatherings would ensure that businesses, schools, and critical community engagements could continue operating and that the virus would not reach the vulnerable. I was scrupulous in making sure the benchmarks for proceeding from one gate to the next were as stringent as possible. Any early setback would drive a state back to an earlier gate and prevent further deterioration.

Along with the hard-and-fast fourteen-day rule and the hard stops, hospitals had to be absolutely capable of handling patients. They had to have enough PPE and staff and a robust testing program in place for their health care workers and the community, to detect early community spread among their staff. With these stopgaps, my goal was to ensure that most states wouldn't get to phase three (fully open) until late summer. That way, we could use the summer to prepare for the devastating fall and winter I foresaw.

I needed direct control over this dimension of the guidelines. With these gating criteria, governors had a simple, straightforward, usable, and modifiable template to guide them and to adjust to their unique situations. States without significant community spread in April would be able to move forward cautiously in May and would serve as an early road map for what was and was not working. We could continually monitor the on-the-ground situation for any evidence of community spread and mitigate it immediately where needed.

Each reopening task group worked on drafting a portion of the two documents. The first was a brief list of bullet points that would live on the White House website and that the president would use during the press conference announcing the reopening guidelines. We'd refer to this as the framework document. The second, a CDC document, would be a far more detailed guidance document paralleling each aspect of the bullet points and would be posted to that agency's site. Also, I wouldn't have the final say-so over what was included in either document. The framework document would have to go through the Office of the Staff Secretary (Staff Sec), which managed the stream of documents that flowed into and out of the Oval Office and was considered by many the nerve center of the White House.

Staff Sec was run by Derek Lyons, an attorney admitted to the bar in Texas and DC. Derek and I engaged in a lot of back-and-forth over the framework document. While he was very helpful in streamlining and clarifying the language, I met with strenuous resistance from him in mandating that masks be worn everywhere indoors and in all outdoor public places where social distancing wasn't possible. I'd submit the framework document with the mask mandate included, and he'd send it back with it removed. I'd reinsert it, and again he'd reject it. Back and forth we went, until eventually, I got an approximation of what I wanted and the American public needed. It was frustrating. "Strongly consider using face coverings while in public, and particularly when using mass transit" clearly isn't a mandate, but a suggestion. It shouldn't have been that difficult to get the administration to recommend simple masks, but it was.

While I had been dealing with the White House's push to reopen, the CDC, after weeks and weeks of discussion, finally provided the guidance on masks we had needed back in February. On April 3, it recommended that the public wear cloth masks. Perhaps naïvely, I thought that the CDC's having come out in support of masks would make mandating them much easier. Masks were critical to any chance of our reopening the country, making indoor spaces safer and ensuring that schools and businesses could fully open. Yet, that same day, in a briefing announcing the recommendation, the president said, "It's going to be, really, a voluntary thing. You can do it. You don't have to do it. I'm choosing not to do it, but some people may want to do it, and that's okay. It may be good. Probably will. They're making a recommendation. It's only a recommendation." In equivocating in this way—instead of making a strong statement about the CDC and its experts being world leaders in public health care—the president essentially gave the American people carte blanche to ignore the guidance, as he would be doing.

At the same time, President Trump, without using the terms *asymptomatic* or *presymptomatic*, urged all people to behave as if they were infected, even if they exhibited no signs of being so, and to exercise a level of care accordingly, a practical, straightforward way to open safely and remain open. This was the closest we got to a formal acknowledgment of silent spread. If only this type of statement had continued to be made day after day, offering a practical, straightforward way to open safely and remain open.

Instead, the overall message was muddied. For one thing, the president's stating that he wasn't going to wear a mask became a huge problem then and throughout the response. Also, the CDC guidelines made no mention of masks preventing the wearer from infection. Summarizing the rationale for wearing simple cloth masks, the president said, "In other words, if a person has the disease without knowing it, wearing a very basic mask can help prevent that person from infecting others." This was true, but it was also confusing, as many Americans I met over the coming months would tell me.

Americans have common sense, and the CDC recommendation on masks did not pass the commonsense test. How could the material absorb what they exhaled but not what they inhaled? If there were two layers, and the layers were the same on each side, how could it be that only one side was effective? I believed from the beginning that masks could partially prevent the wearer from transmitting and receiving the virus. How effective were they at doing both? Simple tests could be done to determine this. For weeks in task force meetings, we'd discussed the need to get these inexpensive fabric tests performed (especially on the Hanes masks being made in coordination with HHS and ASPR), to ensure that we were distributing effective protection. If the country didn't have a large enough supply of surgical masks, we needed to know how well cloth masks worked—at least in the laboratory. It would have taken a matter of mere hours to get such tests done and analyzed.

Regulations prevented the White House (to which the task force was attached) from entering directly into contracts with vendors. As coronavirus response coordinator, I therefore wasn't empowered to execute or fund such tests. Worse, I couldn't direct anyone else to get them done by private laboratories, either. I could ask, and I did. I asked the CDC, the NIH, ASPR, and the FDA to move on testing the efficacy of various cloth, surgical, and N95 masks so the American people would have the clear, definitive information they needed to inform their actions.

Unfortunately, territorial stickiness again gummed up the works.

The CDC was the logical choice for carrying out such tests, but the agency seemed paralyzed on this point. Why it didn't devise and then execute tests on masks remained a constant frustration, and a mystery to me. I prodded and prodded, but never received a satisfactory answer.

As for the NIH, ASPR, and the FDA, they didn't believe that pursuing this kind of testing was part of their job. So, definitive tests on masks didn't get done by the federal agencies. Instead, university labs, private-sector labs, and even CNN had the tests performed—but these tests weren't systematic enough for the CDC or the task force to make data-driven recommendations. Without a definitive statement one way or the other about the effectiveness of masks as two-way protectors, the public became confused, frustrated, and, ultimately, angry and distrustful.

I don't know even now why these simple tests were not done and the public informed of the results so they understood that masks worked in both directions. This clear information gap and lack of definitive data allowed others to fill the void with their own interpretation of mask effectiveness, which continued to create confusion.

In October 2020, a Japanese team of researchers did what we had been asking for since March. They studied the effectiveness of various kinds of masks and published the results, finding cloth masks 20–40 percent effective and N95 masks 80–90 percent protective for the wearer from infectious viral particles. The percentage increases to 50 percent and 99 percent in preventing escape of viral particles from the infected wearer. So if both the uninfected and the infected wore masks, the protection from infection was magnified. The CDC briefed the HHS secretary on this at the beginning of November, hours before the general election and months past the spring and summer surge of the virus, and weeks into the third and deadliest surge. He immediately brought it to the task force's attention, but this whole episode was months too late. If the CDC had followed the model of the Japanese study, the data that might have emerged could have changed America's perception of masks.

Americans wanted and needed clear, evidence-based guidance on masks, and they weren't getting it. The cumulative effect of this? While plenty of people took the CDC's guidance as a call to start wearing masks, that guidance also cast doubt on masks' efficacy among much of the American public. Inadvertently, it fed the fire of those in the camp who believed masks were unnecessary and uncomfortable and required too much thought and preparation and who were otherwise predisposed against wearing them.

In contrast, the FDA, in the face of an enormous public health emergency, was allowing compassionate use of safe, promising therapeutics that allowed doctors to try new and old medication to combat the deadly virus. The FDA also issued emergency use authorizations without complete and final data. It lowered the level of definitive evidence required to make decisions, while still carefully balancing safety needs. The CDC needed to be just as aggressive in its prevention science and data collection leading to evidence-based advice. The agency needed to lead America in the study of prevention interventions, from masks to indoor physical distancing to vaccines. Americans were willing to learn with us, but our lack of evidence and our inability to tell the whole story led to confusion and distrust then and today.

The CDC's mask guidance also undercut the critical message on silent spread that had already been so hard to convey. When people read or heard about the mask guidelines, they likely thought, *I'm not infected. I don't have any symptoms. I'm not going to infect anyone else, so why do I need to wear a mask if it isn't going to keep me from getting infected?*

Beyond being counterintuitive, the mask guidance also appealed to people's self-interest—but in the wrong way. People are more willing to take precautions they believe will protect them. The idea of wearing a mask to protect *others* while doing nothing for them made the guidance much easier to ignore. Of course, ideally, we want people to act foremost in the interest of others, but this, sadly, does not always align with real-world behavior.

For all the legitimate issues I had with the CDC's new position on masks, pushing back against it in public or in the media would have been equally fraught. The highly respected CDC had spoken; attacking its guidelines as insufficient or incomplete seemed too risky. Its reputation largely insulated it from criticism, especially any emanating from this White House—in the face of any criticism with a whiff of politics to it, the media's fallback position was to side unquestionably with the CDC. Besides, it had taken so much to get the CDC to put out even this muddled guidance on masks that the drawbacks to further undermining the guidance were far too great. The CDC wasn't clear enough, it didn't pass the commonsense test, but at least it had made a small step forward. Perhaps the clearest indication of this was how many people started wearing masks in specific areas of the country. Overnight, groups and individuals sewed masks for themselves, their families, and their communities. Even if it wasn't as dramatic or uniform across the country as it could have or should have been, progress was progress.

Still, an unfortunate downside to the new guidance was that when I pushed the White House to include a more aggressive masking mandate in the reopening guidelines, I lacked the CDC's full backing; without definitive data, it could only be "recommended," which made the mandate even more of a long shot. The pushback from Derek Lyons on this point was especially forceful.

I thought I would get significant pushback on the phased reopening described in the framework document, with my strict gating criteria for states. I doubted the White House would approve the criteria, but I was helped by an unlikely source. Another modeler not associated with the task force shared her projections with Derek Kan and the senior White House advisors. Even before we'd "reopened," as many as forty states, she said, were already in phase three, the green, most open phase. According to her calculations, those states could already reopen fully. This was music to the ears of most in the White House, and the message became: *Go ahead and set up your system. It will look good but will be essentially meaningless and unnecessary, since only ten states will be in the most restricted phase.* It was great that they thought this was the case, but it wasn't. Because no one in the administration had bothered to look closely at the gating criteria, and with the modeler's "forty states" misinformation fed to them from outside our group, they had failed to understand fully how the gating system worked. According to the now-approved gating criteria, all states had to go through a mandatory two-week period just to prove they could start phase one.

But because of this failed understanding, I was able to get our strict gating criteria and framework plan approved and posted to the White House website as official policy. Just as important, we'd effectively extended the Slow the Spread campaign for two additional weeks, to mid-May, as each state demonstrated its control of the virus in their state.

The CDC's own "Guidance for Implementing the Opening Up America Again Framework" (not to be confused with the framework document posted to the White House website) was bound up in bureaucracy and oversight issues. While the briefer, White House document essentially provided a list of *what* needed to be done for the states to reopen safely, the CDC's document would provide the even more critical component: the specifics of *how* those things could be done. The states were relying on this more detailed document to establish the best practices and protocols for reopening safely. The CDC document was to be posted to that agency's website as a single complete document—but this never happened, and it went up piecemeal over weeks, providing the states with fragmented, disconnected guidance.

This delay was hugely problematic. Because the CDC didn't get this document done quickly enough, governors and state health officials had to scramble to find answers, frequently relying on the task force for any kind of guidance that should have come from the CDC. The CDC blamed OMB for the delay, but I'm dubious of this; drafts of the CDC document were still being circulated well after the April 15 deadline. I suspect that some of the CDC's lateness was a result of its typically cautious approach and of the glacial pace at which internal processes and reviews moved, exacerbating these institutional tendencies. Some of the delay was due to editing related to asymptomatic spread. Perhaps there were other reasons, but bottom line: posting implementation guidelines for reopening weeks after the reopening officially "began" limited the guidelines' effectiveness. It was one more misstep that eroded state-level confidence in the federal support they needed.

But I had other concerns with the CDC's guidance. In the early drafts, I edited one section of their document to push to make clear, again, that silent spread was at work in producing early silent community spread, highlighting the importance of testing. I placed bullet points to that effect front and center. When the CDC finally posted the guidance, these explicit statements had disappeared from the opening paragraph. Though a close reading of the document's entire one-hundred-plus pages and various addenda would have explained the role of silent spread, it should have been explicitly stated at the beginning of the document, highlighting the importance of proactive testing of sentinel cohorts, as it appeared in the White House Framework. Instead, it was implied, and, consequently subject to possible misinterpretation.

This guidance wasn't just for state health officials; it was for the governors and their staff, nonmedical people who would have benefited from a clear explanation of my strong, foundational beliefs about testing that would allow us to act proactively and not reactively to prevent a hot spot from expanding into a larger hot zone. When we saw test positivity rates rising in a specific area we needed to expand the "sentinel cohort" testing program. That meant intentionally routinely testing those under thirty-five—community college students, nursing students, EMS students, all health care workers, and potentially high school students. People who were in the community but also at sites where they could be routinely reached and tested. If any of those asymptomatic people tested positive, they would isolate for ten days, thus cutting off the transmission route that might eventually lead to the most vulnerable group—the elderly and those with comorbidities or compromised immune systems. Testing those with symptoms was helpful, but it would never effectively work as well as sentinel testing in speedily reducing the chances of spread throughout a larger community.

The discussions in task force meetings and within the reopening work groups remained contentious on this point. The FDA, HHS, and the CDC were aligned in disagreeing with aspects of my position. They all believed, to one degree or another, in testing more, but they remained highly focused on, and prioritized, the symptomatic individuals who present later on in community spread.

But with tests approved by the FDA for symptomatic individuals only, the Centers for Medicare and Medicaid Services was able to pay for testing only of people with symptoms, people who had been in close contact with someone with symptoms, or people known to be infected. The CDC then recommended testing priority to only these cases, making it hard for the states to proactively test for the early asymptomatic cases often found in those under thirty-five. It came down to money. With no payer, either the person being tested had to pay out of pocket or the states had to fund the testing directly. The states needed to use CARES Act monies or other funding to perform this critical step. Eventually, some clever young Americans learned to pretend to have symptoms or exposure, just so they could be tested to protect their family members. This should never have happened, and yet it continues today.

Short of mass available testing at the first evidence of community viral spread, we needed to test weekly all long-term-care facility (aka "nursing home") workers, those employed in communal transport, and those who lived in multigenerational housing. Seema Verma would do precisely this with a strategic testing "experiment" to prove the efficacy of weekly testing in identifying the role of asymptomatic spread. Because of her position as the head of the Centers for Medicare and Medicaid Services, she was able to mandate weekly testing of staff and residents in long-term-care facilities based on the level of community spread, ensuring we isolated the first cases and prevented onward spread among the most vulnerable. Because these facilities received public Medicare funding and were used to scrutiny by the agencies overseeing them, Seema was able to apply both the carrot and the stick. And with control over policy, she could be proactive. As a consequence, the viral fires that ravaged many long-term-care facilities in the early months of March to May 2020 were eventually brought under some control in many places due to this proactive testing approach. This kind of data-driven policy making would become the model we all needed to follow, and exactly what I was pushing the CDC and others, in mid-April, to enact—with little success.

ON APRIL 16, PRESIDENT Trump announced the "Guidelines for Opening Up America Again." The messaging on masks wasn't perfect, but at least the gating criteria would be highly effective. We had provided a framework that could be applied as is or adapted within each state and county.

As with the Slow the Spread campaign's measures, the Trump administration was fine with not having full authority over the reopening criteria and gating—although, when he announced the guidelines, he did say he would point out any states that opened too quickly. Three days before announcing the reopening guidelines, he'd said unequivocally in a press conference that he had the power to lift stay-at-home orders in the states. As he put it, "The president of the United States calls the shots." I was in that press conference, again biting my tongue. It wasn't my role to step in and correct the president about federalism or how, under the direction of the vice president, the working groups on reopening had been conferring with the governors all along. Three days after that "the president calls the shots" statement, the president pivoted on the authority issue, acknowledging that it would be up to the governors, with federal support.

Ultimately, it is true that we couldn't enforce a strict federal requirement for reopening. The United States of America is a collection of states with independent states' rights. As much as the president some-

times promoted the far reach of his authority, the states would determine their own reopening. We could only offer guidance and support.
But that guidance needed to be clear and consistent, and the support
needed to be continuous and adaptive to meet the needs of Americans
and the states.

By the time the reopening guidelines were announced, popular resistance to the Slow the Spread campaign was already evident. Initially, I
saw the protests to mask wearing and social distancing as an inevitable
by-product of the spring surge having affected some metropolitan areas
to a greater degree than others—the virus had not yet reached some
of the protestors' more rural counties; many didn't personally know
anyone who'd been affected yet. This was actually evidence that the
Slow the Spread approach was working. I sensed that some saw the virus's absence in their area as evidence that they had never been at risk
and would never be at risk. Many viewed the virus as a plague on the
large cities—places where they had chosen not to live. I was sure those
complaining and actively protesting the mitigation measures felt that
they had sacrificed their Main Street to fight a virus that survived and
thrived only in places where they didn't live and would probably never
visit: Chicago, New York City, Detroit . . .

What I didn't expect was that, the day after he announced his own
administration's reopening guidelines, President Trump himself offered
support to anti-shutdown demonstrators in Michigan—and later in
Minnesota and Virginia, all states that had Democratic governors. As
April dragged on, his grievances and frustrations with a science-based
approach to combating the pandemic only grew more publicly hostile.
On April 17, echoing the protest cries, he tweeted, "LIBERATE MICH-
IGAN!"

With these words the leader of the United States began encouraging protestors to take back their states and ignore local public health
guidance. And the resistance to our message increased throughout the
pandemic. His words also signaled an end to the sense of our "all being
in this thing together," to seeing our sacrifices as a shared burden we
shouldered to produce a shared benefit. Rather than uniting us around
a collective cause, the president was exploiting the differences and divisions between us. It was stunning to see that, only a day after announcing his own administration's guidelines, Trump was supporting

constituents whose chief goal was to undermine them. The juxtaposition infuriated me. It was as if he were determined to erode any progress the reopening guidelines produced.

More practically, the protestors and the president were wrong, of course. "Liberating" Michigan would only have liberated the virus just at the time when Detroit was exploding and the virus was being effectively mitigated, preventing spread throughout the state. At the end of March, Michigan ranked third, behind only New York and New Jersey, in Covid-19 deaths. By mid-April, it was the third state to reach the then-shocking figure of twenty thousand cases. At the time of President Trump's tweet, cases and deaths were still rising in urban areas. Instead of making an inciting remark, he could have used the opportunity to inform the public. He could have explained why Michigan was in the condition it was in, why the reopening phases were so essential. Instead, the president sought to undo that state's hard-fought progress.

It was at this point that the "never again" message he'd delivered to me in private finally entered the public square. Having received two divergent narratives about the pandemic, one from me and one from the CEA, President Trump had been ambivalent about what he thought of the policies he had originally endorsed. Now, in siding with the protestors, and in other actions and statements, he had shown where his ultimate loyalty lay. He believed the CEA memo. Never again would he side with us and the data I was providing. His ambivalence had hardened into categorical resolve.

AS APRIL STRETCHED ON, I began to express more of the profound sense of confusion and frustration I was feeling, the alienation—the outsider on the task force; the politically inexperienced one; the less polished, less practiced media type who was used to speaking her mind plainly as often as was possible and practicable.

During one of the informal "doctors' meetings" we felt were necessary to convene as April progressed, Tony, Bob, Steve Hahn, and I took a moment to reflect on the progress we had made. Yes, we had moved the president to the 15 Days to Slow the Spread, the 30 Days to Slow the Spread, and a careful reopening of America. We were moving supplies and medical personnel to where they were needed, and the New York

City surge was beginning to abate. The wonderfully upbeat Steve said, "So, how do you think it's going?"

I looked him in the eye and said, "We'll be lucky if we survive this." I read confusion in his face. I went on: "We're going to be hated. We will be the ones who'll be blamed by both sides. We will be hated by the right for not agreeing with the president and hated by the left for staying and trying to support as comprehensive a response as possible under an administration they loathe. Guilt by association." I let the thought hang there.

Steve nodded.

In encountering all these advances and retreats, I had a gut feeling that so much else was going on beneath the surface. Bob, Tony, and Steve tried to dissuade me. But in that moment, I felt I existed in a nearly overwhelming moment of contradiction. I worried that the four of us wouldn't survive in this environment, while at the same time I was convinced that we would absolutely need to see this through. Despite the dominant negative impression of us in the press and within the administration, I knew one thing: You don't abandon the battlefield in the midst of a fight because it's hard or because it could damage you personally. You stay. You redouble your efforts. Our hope now was that all our efforts to produce the reopening guidance would pay off in fewer cases and fewer deaths. We also needed to ensure we were ready for the fall. We needed better treatments, a full federal stockpile of PPE, therapeutics, tests, and vaccines.

I knew I'd lost the president's attention. What I couldn't know was that, after April 21, I wouldn't have another presidential briefing on the state of the pandemic in the Oval Office until the summer surge was fully raging. I would occasionally, after the CEA report, be called into the Oval from my perch outside in the hallway, to make a point with the president prior to press conferences, but it always felt like he didn't see me. He didn't acknowledge me verbally, except for his "never again" reminders, and later he stopped doing even that. I became a nonentity. He appeared to find nothing useful in what I had to offer—which was a realistic assessment of the situation and a reliable, accurate forecast of the disease's trajectory and toll. If I wasn't of use, I ceased to exist— except for a few moments as a prop in meetings where the so-called optics required I be present. I didn't know it at the time, but the infamous "disinfectant" press briefing fit into this category.

• • •

"AND THEN I SEE the disinfectant, where it knocks it out in a minute, one minute. And is there some way we can do that by injection inside or almost a cleaning?"

It was April 23. The president was at the lectern in the White House Briefing Room. I was sitting to one side, against the wall. Trump was not speaking to me, but to the Department of Homeland Security's undersecretary for science and technology, William Bryan, sitting two chairs down from me, closer to the door the president and White House staff used to access the room.

The president moved on to discuss the benefits of sunlight and of disinfectants that were effective against the virus outside and on surfaces—a justification for encouraging people to get outdoors and children onto playgrounds. Then he mentioned the use of light and disinfectants as treatments for Covid-19.

What was happening?

I sat there, unmoving, my hands clasped tightly in my lap. I watched the president's exchange with Bryan and then looked at the floor, knowing my face would betray me. I kept my breaths slow and shallow. I tried to control my body's reactions to what the president was saying. My brain whirred as I attempted to track his reasoning. How had he gone from sunlight sterilizing children's playground equipment and the use of disinfectants on surfaces to using light and disinfectants—*injecting disinfectants*—to treat the human body?

I wondered where this was going, why he was on this tangent. I would soon get my answer.

This isn't often reported, but prior to this press conference, William Bryan attended our usual daily task force meeting. We had earlier talked about wanting children to be active, to be able to go outside again, use playgrounds. Many cities and towns had closed their playgrounds and parks, limiting families' ability to socialize and engage outdoors, especially in cities. We wanted families going to playgrounds and parks. I believed the outdoors was safe, and we had asked the DHS to study the impact of sunlight and disinfectants on surfaces to determine if playground equipment could be made safe and to encourage mayors and governors to reopen these outdoor spaces, parks, and playgrounds.

They could. Sound research had said so.

I had thought the matter was done and dusted, the evidence clear, the conclusions sound. But somehow, between that task force meeting and this press conference, everything that had been made clear had grown foggy again.

After the task force meeting, as I later learned, Bryan was pulled aside and taken to the Oval Office. This kind of post–task force gathering with the president happened infrequently now. In this case, I was not present; nor had I been invited. Instead, as had been so often the case by then, I was already outside the Press Briefing Room, about thirty-five feet from the Oval Office, waiting for the president to arrive. I waited and waited. Minutes passed, then a half hour, and then nearly an hour. What was going on?

During their tête-à-tête in the Oval, as I would later learn, Bryan briefed the president about the impact of sunlight and disinfectant on killing the virus. The president then asked him about the potential use of sunlight, heat, and disinfectant, not to clean surfaces in support of outdoor activities, but as a *treatment*. Reportedly, Mr. Bryan politely listened to the President's ideas and said he would discuss them further with his team.

Fast-forward to that awkward press conference moment. Bryan and the president, it appeared, were continuing their off-the-cuff Oval Office discussion, but now it was happening in front of the media and before the eyes of millions of Americans, millions of people around the world.

As the president continued his musings on the use of heat and light, I looked down at my feet and wished for two things: something to kick and for the floor to open up and swallow me whole. With the fifteen- and thirty-day campaigns, we had finally shown that this administration was taking the viral threat seriously and that the American people should, too. The president's disinfectant remark could unravel all that, and at the worst possible time, when the emphasis should have been on the slow, deliberate, careful reopening of the states. I knew the media would pick up this ball and run with it, discrediting weeks of careful work.

At the end of his long dialogue with Bryan, the president finally turned to look at me, in my chair against the wall, to ask about the benefits of heat and light as a treatment. I replied, "Not as a treatment." Then, pivoting from the bad science to the good, I pressed on, talking about how the body naturally defended against invaders. I was employing, in my response, a commonplace truth of immunology: the role of fever in the body's defense against viral infection. I hoped to get the conversation back on track, saying that heat and light couldn't be a treatment per se and denouncing the tenuous connection between the disinfection of an inanimate object and a true antiviral treatment. But doing so only made things worse for me.

Many people thought I should have run up to the microphone and shouted, *Not a treatment!* That I should have leapt from my chair, hurled myself between Bryan and the president, and shouted, *This is nonsense!* I know myself, and I wouldn't do such a thing even today.

To say that I was caught completely off-guard by the absurdity of what was being said in that room that day is an explanation, but not an excuse. Yes, I should have done better. I should have said "Not a treatment" more forcefully, several times over. I should have ignored my deeply ingrained, military-honed instinct not to publicly correct a superior and followed my instincts.

Instead, in that moment, I thought, *Correct what needs to be corrected and move this along.*

It wasn't my reputation or standing I was worried about at this point. The more time and attention this disinfectant nonsense received, the bigger distraction it would be from what we needed to focus on right now: safely reopening the country. The important thing was to get the right message out there: sunlight and disinfectants were not treatment options, but a means to get children outdoors and playing with friends.

In the end, my facial expression, one of worry and disbelief, did the work my words could not. What the president was suggesting was nonsense. It wasn't worthy of the nation's attention. The vast majority of the American people, I believed, would see it for what it was: an impulsive statement, as improbable as it was improvised. It said more about the state of the speaker's mind than about any fundamental scientific truth.

I left that briefing furious. In a matter of minutes, with the uttering of a few words, our credibility had been blown up. None of what the president had said about disinfectant made sense. He needed to recant and recant immediately, before this spun out of control.

Because I didn't have direct access to the president, I asked Jared Kushner, the next best thing, to make sure the president knew that what he'd implied in raising these bogus ideas with Bryan undermined the serious science we were using to communicate to the American people. I told Jared that the president needed to immediately correct the statements he'd made in the press conference. I know Jared went to the president.

The next day, the president said he had been "joking."

The media and the public saw through this. Many of us have used the same excuse after saying something hurtful or unwise. But there can be consequences to some "jokes," especially when they come from the mouth of the leader of a country. They can't ever truly be unsaid; the damage remains done.

Having the president take an idea from one context and apply it to another fit into a larger pattern I had witnessed. I had the sense coming into the White House that the president's team had struggled even before the pandemic to filter the information the president so voraciously consumed. Whether or not the information was credible; whether the person who espoused a particular belief, point of view, or "fact" had based it in sound reasoning and evidence, hardly seemed to matter. With the pandemic dominating the news cycle, social media, and other areas of the internet, the White House was awash in conspiracy theories, speculation, and a panoply of perversions of science posited by nonexperts.

The president's team couldn't deal personally with all the phone calls, emails, texts, and random mentions the president came across while viewing his four screens or the information his friends fed him. Because of this, his senior advisors and the vice president's communications team, Devin O'Malley and Katie Miller, often passed these people on to me to speak with. These outside influences ranged from TV doctor Mehmet Oz to pillow salesman Mike Lindell—and they gave oxygen to all manner of ideas that had no business making their way into the White House. The inclusion in the president's body of information of even the most baseless, suspect ideas created the illusion that all information, no matter the source, was equal and equally trustworthy.

Understanding science can be hard. Nuance is important. Ideas and statements get distorted. And you can't just "consider the source"; you have to think more deeply and not accept anything at face value, no matter the source.

The disinfectant debacle brought about a fundamental change for the task force. For one thing, the daily press conferences ended. This meant it was harder to communicate, to get our data, concerns, and solutions out there directly to the American people. More than that, though, it codified the shift that had taken place throughout April. The doctors on the task force, especially me, no longer had an audience with the president. He was no longer hearing our analysis of the data, nor the solutions for combating the virus. We were effectively cut out of the Oval Office discussions and the decisions that would come from those discussions. Trump's senior advisors and outside sources could present their views on the pandemic unchallenged. The days of presenting graphics and analyses to the president were over. The people who wanted to neutralize my influence had won that battle, but the larger war was still raging. While we all had roles to play, and while there was still a lot of good we could do while there, our days of being able to shape White House communications on a large scale were over. We needed to find another way to influence the critical decisions and policy making—in the states.

From that point forward, whenever I felt we needed a press conference to communicate with the American people directly, it would take weeks and weeks of begging the vice president and Marc Short to schedule one. What press briefings were held—which I often wasn't a part of—had the feel more of campaign events. In them, the message was: *Here's how great we're doing on testing or therapeutics or vaccines.* The White House would never return to the serious message needed on this pandemic: where the virus was and who had to take precautions.

IF APRIL WAS WHEN the economic forces in the White House regrouped against the science of the virus, May saw us increasingly marginalized, solidifying President Trump's, and his administration's, resistance to our efforts. As the tide turned, the mood grew darker, more sinister. Death threats and warnings of attacks (sometimes sexual in nature) against me and my daughters had begun in April and were occurring with a frequency that was hard to ignore. I took the first set of them to the State Department, since they were my current employer. They were used to these types of threats to ambassadors around the globe. They asked me to continue to track them. Tony got protection, and I got to do more collecting.

To be honest, I couldn't read the hate mail. The sheer volume of it was overwhelming. I never reported it again to the State Department. I reminded myself that I had been in much more threatening and unsafe areas in my life and had survived. I would survive this. I did get help from the Secret Service, though. They were amazing—at their posts day after day, protecting all of us. I greatly appreciated their work. Once, they did something very kind. Like many households in America, we had a family telephone plan that dated back decades ago, when my daughters were in high school. But now we were getting threatening, sexually violent texts. So, the Secret Service called our provider and got my phone locked, and overnight, the texts disappeared, never appearing again until I left federal service. I will always be grateful for this kindness shown me. Those messages were deeply unnerving to my daughters, and I was so grateful when they stopped. The mail threats didn't, but at least I didn't have to open those and read them.

Tony and I had become the avatars for science, both for the White House and for those in the public who disagreed with the shutdown. Science became the enemy, data-driven debates dissolved into threats and expressions of hatred.

On May 6, without fanfare or bluster, the president casually said the White House Coronavirus Task Force would soon be disbanded. In its place, a new task force would be created, one focused on getting the country's economy fully recovered.

At first, I had no idea what was happening. No one had said anything to me about these changes. I had to find out through the media, when the rest of America did. I couldn't believe it. Here we were, in May, two months into my coming on board, and the president was essentially declaring victory over Covid-19. Though he'd be keeping Tony and me on board as the lone survivors of the public health wing of the disbanded task force, the economic wing would be getting not just the green light, but the keys to all the cars and permission to detour around the unpleasant realities of the pandemic—all to get to the Promised Land of a revitalized economy.

I stayed—knowing that if I didn't, no one else would be brought in to replace me, leaving Tony alone to flounder in the wake of the president's now disregard for public health. I spoke with Tony; he was equally upset. We had looked forward to the day when there would be no need for a Coronavirus Task Force, not to its being prematurely ended as another false signal to the American public that all was well.

Looking back, I see that my public responses to events like this were often muted. But things happened so fast in that environment, and with such great frequency, that instead of flinching or raising your hands to protect yourself, you simply had to let whatever was incoming bounce right off you, so you could focus on the task at hand. Given the summer surge in 2020, and the surges that followed, I don't like to think about how much greater the loss of American lives would have been had we not done what we had up to that point. Without the Coronavirus Task Force:

There wouldn't have been the hospital database that tracked vital supplies, health care personnel needs, and critical therapeutics, helping get them where they were needed.

There wouldn't have been the PPE early alert systems that identified and helped resupply every hospital across the United States short on masks, gloves, and gowns.

There wouldn't have been the amazing behind-the-scenes work of so many to get the message out through local media.

And without the task force, the governors almost certainly would have had to fend for themselves. Instead, they were able to speak directly to us with great frequency, and the governor's reports we sent them each week gave these leaders—especially of red states, whose populations were more resistant to mitigation measures—the cover to say they were only asking for what the White House had advised. Without those reports and the perception of White House backing for the mitigation measures they provided, there would have been dozens more Michigan-like protests, more open defiance of mask wearing and social distancing regulations. The governors would not have been able to hold the line.

Fortunately, the day after saying the task force would be disbanded, the president tweeted that it would remain in place indefinitely. He also mentioned how many people had contacted the White House to express their positive opinion of the work the task force was doing. He said he had had no idea how popular we were.

I wondered: *Popular? I have never been popular. I'm the data nerd. I'm used to doing the hard work behind the scenes. Popular? Does that really matter? Isn't it about how effectively we're doing our jobs?*

I can't be sure precisely what went into his changing his mind, especially given how marginalized we'd been for the last month. Perhaps he had realized that in spite of his distaste for the work we were doing, the political optics of disbanding the task force at this stage would not have been good.

He kept the task force going, but he would continue to undermine us publicly just as he had been. Perhaps it shouldn't have been surprising, then, when he came out against widespread testing. On May 15, he said, "Don't forget, we have more cases than anybody in the world. But why? We do more testing. When you test, you have a case. When you test, you find something wrong with people. If we didn't do any testing, we would have very few cases."

Rightfully, many people in the media seized on this statement, excoriating the president for his incorrect explanation for why the United States' Covid-19 case numbers were so high. If test positivity is rising despite increased testing, it means that there is ongoing, uncontrolled community spread. Together, both test positivity and tests per capita are the most meaningful metrics for determining the effectiveness of a testing program. (Remember: some people were having themselves tested again and again; while others weren't being tested at all.) When it came to this important measure, the United States actually trailed Canada, Russia, and other countries in Europe.

I heard what the president said on testing and dismissed it as just another example of his desire to reframe the narrative so that the negative picture (the high *case* rate) could be replaced with a rosier, more voter-friendly one. I also saw his remark for what it was: President Trump tossing red meat to his base.

As misleading as this rhetoric was, I would have been more concerned about his distortions on testing if his words didn't stand in direct contrast to what his administration was actually doing behind the scenes. For all his anti-testing statements at this point, the reality was that the task force, under the leadership of Vice President Pence, was pulling out all the stops to rapidly *expand* testing. Through the agencies, and with clear White House approval, we were spending billions of dollars on tests and testing supplies. We were aggressively testing in nursing homes and asking universities to use testing to bring their students back safely in the fall. We were flying in swabs from across the globe so that testing could be constantly expanded. We were using the Defense Production Act to streamline the manufacture of supplies and raw materials for the private sector, to build more manufacturing facilities, and to expand testing. We had ensured a constant funding stream for the purchase of tests through the ups and downs of the surges to make sure the private sector would continue to expand testing. We were working to increase testing of the most vulnerable, including at the Federally Qualified Health Centers and tribal nations. And we continued to work with the private sector and states to increase testing. And the White House had posted the guidance I had written to ensure broader, more strategic testing—finding the silent spread. Week after week, testing increased, and continued to increase throughout the months that followed. The majority of reported testing was primarily with the PCR nucleic acid test. By January 2021, we had reached nearly two million of these tests per day, and were also sending out fifty million free antigen tests monthly to the states and institutions that needed them most.

At no time did anyone tell us to stop this acceleration in testing. The president was a master at saying one thing to appease his base while his administration did another in support of combating the virus. This was the dichotomy then and throughout the months that followed. One thing was said, while a very different action was taken.

The direct effect of this particular instance of bluster by the president was negligible on me and the task force. However, there was plenty of collateral damage, specifically in its effect on public perception. How people viewed the pandemic response and what personal actions they took—these were the true casualties of Trump's language. As we moved through the summer and into the fall, we began to see more profoundly the results of the American public's not having the kind of leadership from the man in the White House that could have led them to do the right thing.

In a town hall on CNN, I was asked if I thought more testing was a bad thing—a question designed to drive an even bigger wedge between me and the president. I wouldn't rise to that bait; I kept on message. Testing to detect asymptomatic spread was good and necessary. Testing didn't increase positivity rates. Aggressive, strategic testing allowed you to spot the silent spread early. This kind of testing, in fact, would decrease the number of confirmed cases over the following weeks, because it would stop community spread.

I didn't want to paint the president's comments in a simplistic way. In fact, questions on testing were complicated by far more than merely what the president had said. The long-standing "test more or test wisely?" debate was ongoing and expanding. For months now, "testing czar" Brett Giroir, the CDC, the FDA, and I had been unable to agree on the best approach: Whom to test? How many to test? But there was also another wrinkle: Which tests should we be using?

From the outset, there were two different testing options, each with its merits: the nucleic acid test (also known as the PCR test) and the antigen (rapid) test. While the nucleic acid test was the more accurate, it provided a positive result long after the viral infection was cleared— that is, long after the infectious period. The antigen test could be processed and read much more quickly. This difference in turnaround time was no small thing; antigen tests could produce results in a matter of minutes, and could therefore be used to prevent ongoing transmission to others, as opposed to the days and sometimes weeks it was then taking nucleic acid tests to be turned around. The only problem was that the antigen test results were not as reliable. Still, while antigen tests did have acknowledged deficiencies in detecting seasonal flu, I believed the new SARS-CoV-2 antigen test was more reliable for detecting the virus than the previous flu antigen test.

In addition to believing we should test only those people with symptoms, the CDC and the FDA also felt we should use only the more reliable nucleic acid test for diagnosis, even with its longer turnaround time. The CDC and the FDA were okay with antigen tests, but not for official diagnosis and not for diagnosing asymptomatic cases. Meanwhile, asymptomatic testing was exactly what we needed the antigen test for. If you got a rapid answer that you were infected, you could immediately take measures to avoid infecting others. Getting an immediate answer, one that developed right before one's eyes, would appeal to younger people, who were more likely to be asymptomatic, and would encourage them to increase their personal testing. The CDC thought the antigen test result could be considered only presumptive. This caused incredible chaos in reporting and required reporting. In other words, its results would have to be confirmed by a nucleic test—thus requiring people to double dip, spending more of their time and energy. Also, there was no reason to report the antigen test results, as they weren't considered definitive. Once again, the perfect became the enemy of the good.

There were also issues of cost. Because of the CDC's and the FDA's positions on the antigen test, the Centers for Medicare and Medicaid

Services couldn't pay for them, and thus, private insurance refused to pay, too.

Similarly, states didn't want to pay for antigen tests because their costs couldn't be reimbursed; they would use the antigen test only if they were given them for free. Because of all this, antigen tests came with an out-of-pocket cost, whereas nucleic acid tests would be covered. This automatically put antigen tests out of reach of those who needed them most. Brett stepped up and worked to align the fifty million antigen tests sent out per month with the sites that needed them most through the fall.

At the same time, despite Bob and me having pointed to their advantages in phone calls, public health officials across the country united against the antigen tests. It's difficult to fault the state agencies and public health officials. They didn't want to break with CDC guidance. But this refusal to strategically embrace antigen tests led to even more community spread.

To advance my case, I devised a strategy for the use of our current testing supplies, one that would increase the rate at which the tests could be processed and that wouldn't strain our inventory. Early on in the pandemic, the availability of tests had been a major stumbling block. We had wanted to preserve as much of the testing materials we could for use in diagnosing the more vulnerable age groups. From the outset, the message had been that testing younger people widely was not a priority. I didn't anticipate that "test the older, not the younger" would become set in stone even after the supply strain eased.

For weeks within the doctors' group of the task force, we worked on a blueprint for strategic testing involving the antigen test, especially since the federal government had purchased 150 million of them. Tony was fully supportive, agreeing that my plan to use antigen tests for screening for silent spread was the right approach: We had a ready supply of them, the rapid tests were likely to be more appealing to younger people, they were proving to be reliable in this application, and by using them instead of the nucleic acid tests, we could reserve these latter tests for the most vulnerable age groups and other higher-risk people. To bolster my argument, I used Seema Verma's success with long-term-care facilities as a real-time experiment to see if lots of testing would prove effective at identifying the infected (who could then be isolated) before they were symptomatic.

Of course, none of this was different from what I had been saying all along. It's just that, before, most of the evidence supporting the need for widespread, strategic testing to find asymptomatic spread in the community had come from other, mostly Asian countries. Now we had increasing amounts of evidence from within our country supporting this strategy—evidence that showed clearly how tests could be used to track asymptomatic spread in a community.

And still, it wasn't enough. While Brett Giroir, the CDC, and the FDA understood the rationale for my position, I struggled to get everyone to shift gears into a proactive mind-set. Their objections were nebulous, insubstantial, nothing I could use to refine my case for a different testing strategy—essentially boiling down to a lack of will and follow-through. The FDA would not approve antigen tests for asymptomatic diagnosis until July 2021.

We should have made these faster tests universally available. This failure put the United States farther behind Europe in testing and made us more vulnerable to surges. I admit that the antigen tests were not as reliable as the PCR tests, but not using them cost us lives. Unknowingly infected, untested sons and daughters spread the deadly virus to their vulnerable parents and grandparents. I suspected—and later on, I confirmed this in personal encounters—that many young people in the eighteen-to-thirty-five age group who were unknowingly in the silent superspreader group hadn't wanted to wait the twenty-four to forty-eight hours it took, best case, to learn the results of their nucleic acid, or PCR, test—especially when that "best case" often turned into more than a week. They wanted an immediate answer, and frankly, they were right. They wanted a test they could use right before they saw their elderly aunt or their grandmother or their parent receiving cancer treatment. They understood being negative in that moment of interaction was what was important to protect others.

We had in our possession millions of antigen tests and nucleic tests that could have been distributed to more sites so people wouldn't have to wait for results. We had a quicker test available, one that would have told them within minutes, not days, that they needed to isolate or mask indoors when with their vulnerable family members. Knowledge is power, and we denied Americans that critical power by deciding we needed the perfect rather than the good.

CHAPTER 10

Find a Way or Make One

We'd gotten through a horrible April, but even as the cases we could see through our limited testing continued to significantly decline, I was still uneasy. In smaller meetings, when it was just Bob, Tony, Steve, and I, I'd express my gut-level concern.

"What are we missing?" I'd ask.

As we looked at one another, one answer became obvious: sleep. We all looked drawn and pale. The three of them had been at this for weeks before I came on board, and with no weekends to recharge, the toll of twelve- to fourteen-hour (or more) days, seven days a week, had become evident. My husband, whom I'd married only a few months before, had taken on all the household responsibilities. I'd given up walking to work, and he was my chauffeur. I'd also given up on cooking, something I loved, but I couldn't stop gardening. Every day, in the few moments I had between rising at three thirty in the morning to review data and getting out the door by seven to start my workday, I planted and pruned, weeded and watered, all in the dark. With so many in the neighborhood having walking the streets as their only outings I wanted our gardens to be welcoming and something to make them smile. Some things were still possible. Rain. Sunshine. Flowers.

This was also my thinking time, and during it, the question *What are we missing?* became part of my breathing—regular, insistent. Those thoughts competed with another set of thoughts: We were making some headway in critical areas. During the first weeks in May, the level of the outbreaks in the major metropolitan areas, like New Orleans, Chicago, and New York, was subsiding. The darkest days of March seemed to be behind us. It certainly wasn't time to declare anything like an overall victory, but like the flowers in the garden, they served as a reminder that there was hope. But given my natural and professional inclination toward worrying, I kept hearing that other voice asking, *What are we missing?*

At that point, as I weeded, and planted, I knew how important it was to stay vigilant. Along with those big picture improvements in the metropolitan areas, I was pleased that some of the CDC issues we'd been dealing with had improved. Irum's-led data team was scouring every source possible to learn of any increase in the rise of cases. Tracking and tracing down to the county and zip code level had proved effective in picking up the first signals of a possible larger outbreak. For example, in early May, isolated hot spots were emerging in Iowa, Nebraska, Kansas, South Dakota, Colorado, and the Navajo Nation. They weren't statewide; they seemed to be isolated to certain counties. When we picked up that flash of rapidly rising case incidence, we worked closely with the CDC, sending them the counties we were concerned about. The CDC took that data, and its investigators called local public health agencies where those sparks and new small fires were seen. The CDC was very good at doing this. This matched with the procedures and skills they used to track foodborne illness. Applying those same procedures to SARS-CoV-2, the CDC investigators were able to track down the source of the outbreak to precise institutions within the county.

In early May, the vast majority of viral spread was most often isolated cases at meat-packing plants, LTCFs, and prisons. Penetrating deeper, in some cases the CDC's team was able to find a single individual who was the first person to be infected and who then passed it along. Almost without exception, that person had visited a major metropolitan area that was still confronting high cases, and came back to their local community, likely asymptomatic, and infected others. From this, the CDC personnel, using data they received from the local authorities, tracked the spread and entered into the database a code that indicated the exact source of the infection—so for instance, when cases began in a prison, they were recorded as a *P* for *Prison*, since they arose in that particular type of institution. The CDC would support the state in testing everyone at those sites, revealing that hundreds of cases had developed over a week. The CDC would then continue to monitor the situation and continue to test in partnership with the state and local authorities; our data team did as well; soon, data showed that the flare-up had diminished in intensity and was nearly extinguished.

That approach worked well, but it resulted in an unintended consequence. Even though these outbreaks were isolated to a specific site, on some media outlets the entire state would flash bright red as the cases rose dramatically in this one county and then the next week bright green signaling the rapid decline in any additional cases in this county. I was worried that this would give Americans the false impression that in the future we could expect immediate control of the virus across broader geographical areas that quickly. We were controlling facility-based outbreaks and watching for evidence of community spread, but larger-scale outbreaks would definitely take more time to evolve and for our response to evolve along with it. Jumping the gun and raising false hopes too quickly was another aspect of the messaging response we had to guard against. Making sure that what was being conveyed in the media matched the reality our evidence base was showing was essential to managing the public's expectations and getting them to buy into needed behavioral changes.

It was the "nearly extinguished" that fertilized my sense that we were missing something. So far, in those first few months, with the majority of the cases in densely populated, cramped cities and their surrounding communities, the case incidence curves produced there were large—picture a nearly standard bell-shaped curve, something that has an obvious shape to it. The curves weren't complete, especially with still inadequate testing, of course, but a clear trend was readily apparent. In early May, along with those obvious curves, we were also seeing something like blips on an oscilloscope, small isolated bursts and tiny waves, which we identified and worked with the states and the CDC to immediately control, but was there something else lurking beneath the surface.

Based on the available data, for those institutional outbreaks, additional testing was done in the household associated with the worker. This broader testing in many cases also showed little community spread. Consequently, it seemed as if that outbreak was contained. (In addition to coding exactly the type of institution, the CDC also used *C* to indicate where community spread was occurring.) With these spikes at the start of May, that community component accounted for less than 10 percent of the total codes in counties where these institutional flare-ups occurred. Slowly, incrementally, but still perceptibly, that number increased for the next two weeks, rising at first to as much 30 percent of the counties' codes, attributable to a mix of institutional and, in some counties in the South, community spread.

What was the source of that spread? Overall the situation seemed to be in hand, even at this local level, but still the thought nagged at me that we were missing something. As May went on, my growing sense was that some seed had been planted beneath the surface, and had germinated, sprouted, and grown from a single seedling into a field of invasive weeds.

While I was wondering about that missing component, other aspects of the ongoing struggle to wrestle with so many challenges became more obvious.

The first week of May, I learned that both our task force data team and FEMA/HHS were running parallel data teams tracking the pandemic, including hospitalizations. That figure was critical in determining how the limited U.S. supply of remdesivir, the antiviral medication that was proving highly effective at treating Covid-19 disease, would be distributed. Given that we had the promise of approximately only a million donated doses over the next months, with a six-month wait before new doses could be produced, allocating wisely was top priority. Having this kind of parallel systems and duplicated data could potentially cause problems and wouldn't do anyone any good.

The remdesivir would come to us in weekly batches of twenty thousand to twenty-five thousand doses. The plan was to monitor the situation and adjust the allocation numbers based on rising new hospital admissions of Covid-19 patients. The CDC had improved its reporting system, increasing from 40 percent to 75 percent of the hospitals in the country reporting. My internal data teams were calling hospitals and tracking *new* Covid-19 admissions; the FEMA/HHS data team was tracking *total* hospitalized Covid-19 patients. The FEMA/HHS figure wouldn't help us determine where the remdisivir shipments needed to go. To be effective, this intravenous drug must be administered immediately after a patient is admitted to the hospital—at worst, within the first twenty-four to forty-eight hours.

Since my arrival to the task force, I'd been asking the CDC (for many reasons) to get all six thousand hospitals to report daily new Covid-19 admissions, current Covid-19 patients, and newly admitted intensive care unit patients; this was the only way to identify current needs as well as project future areas of increased need so we could be proactive in our responses rather than reactive. With the game-changing remdesivir, whose efficacy depended so much on the timing of its administration, we needed new admissions data now more than ever.

The problem I'd feared resulting from having two different data streams materialized. Despite my instructions that shipments of remdesivir go to hospitals with the highest new admissions, the initial week's supply didn't go to these locations. Rear Adm. John Polowczyk at FEMA had used the data it had: total Covid-19 inpatient status. As a consequence, we wound up sending supplies of remdesivir to facilities where patients, weeks into their illness and with long-term complications, still lingered. They would never benefit from the drug. Newly diagnosed patients who urgently needed the drug didn't get it. Shipments were sent to hospitals that lacked the refrigeration units needed to preserve the drug, and therefore ran the risk of spoiling what we'd sent them. In one instance, a batch of the drug went to the wrong "Columbia Hospital" because no one had checked the address.

I was furious.

The next day, I walked into the White House at 7:15, sleepless and unsettled. At 8 a.m., I attended the daily morning meeting held by Chief of Staff Mark Meadows. This meeting was an opportunity for all the major players in the administration to convene to devise top-level strategy on various issues, including the pandemic. I had only been asked to attend over the past couple of weeks while we were drafting the reopening of America guidelines. Also present were Joe Grogan; the deputy chief of staff, Chris Liddell; Jared Kushner; White House senior counsel Pat Cipollone; Marc Short; Katie Miller; two members of the president's communications team, Kayleigh McEnany and Alyssa Farah; and the stuffed pheasants that appeared to be a prominent part of the décor in Meadows's office.

When it was my turn to speak, instead of doing my usual daily data summary, I recited, chapter and verse, the story of the remdesivir debacle. In closing, I said passionately, "This is the kind of unbelievable level of fuck-up that ends up killing people. We can't keep doing this!"

"Dr. Birx!" Mark Meadows's tone and volume nearly matched mine. With my attention grabbed, he spoke more calmly: "We understand. We need to move on now."

I remained livid. I knew that if I continued in this way, I'd become *the bitch who couldn't let things go* instead of someone with the balls to tell it like it is.

As the meeting resumed, my anger at the remdesivir mess-up and at being shut down by Meadows simmered. Afterward, I went to Vice President Pence. I told him about the remdesivir problem and recommended that he integrate the FEMA/HHS and the task force data team parallel data streams into one single system that we could all use for decision making. This wasn't just about remdisivir allocations and distributions. This was a significant miscommunication, I reminded him, that resulted in the wrong hospitals getting the potentially lifesaving drug, and it needed to be fixed. Throughout all levels of the response we had to struggle with inefficiencies, duplications, and parallel data systems like this. We couldn't continue to make mistakes by using the wrong data or wrong data source. We needed all agencies to use a single comprehensive integrated system, of which Irum would oversee the continued evolving development and implementation. The vice president agreed that was what was needed, and we added that to our list of responsibilities. More immediately, Vice President Pence contacted Secretary Azar to get the distribution of remdesivir on track.

I tried to focus on the positive. With Irum in charge of that data centralization issue, we'd avoid repeating the same mistakes. Bob Kadlec and John Redd and Rear Adm. Polowczyk, after that single misstep, worked beside me week after week to make sure the drug got to the sites that needed it, and together we saved lives.

That night at home I listened patiently as Meadows went off on me over the phone. Angrily, he was demanding to know why a shipment of remdesivir he had requested be sent to a Florida hospital had not arrived. Six days earlier, he'd promised Democratic congressman Ted Deutch the supply, and his request still hadn't been met.

Dismayed and disappointed by another remdesivir delivery problem, I let him vent. He had a legitimate gripe; new admissions were rising in Florida, and I'd already taken steps to resolve the larger issue. After Meadows ended the call, still fuming, I checked the distribution sheets and then contacted Steve Hahn and Bob Kadlec, who verified that the shipment had gone out that day and would arrive the next. I then relayed to Mark Meadows the news he was hoping to hear. Another fire put out, I retired for the night.

The next day, I was called into the chief of staff's office. Mark cut straight to the point: "We have someone in that group leaking stories to the press. I'm certain of it." He was referring to the attendees at yesterday's daily chief of staff meeting. "I'm not certain who." He paused, as if to let the implication sink in. "I think you understand how damaging it would be for everyone if what you said yesterday ever got out publicly. In case you don't understand, hear me now: You will never say anything like that again in any meeting."

I was dumbfounded. Instead of fixing the leak problem, I was going to have to censor myself? In light of what had just happened with remdesivir and how I'd managed to find an effective solution to the problem, Meadows was still focused on managing personalities and leakers. Why not allow people to speak freely in a crucial internal meeting? I knew the answer: because appearances mattered. I imagine Mark Meadows was also worried about appearances this close to the election. He didn't want the administration to look like it wasn't on top of things—but what about addressing the real problem? Remdesivir distribution was in disarray. That's what the focus should have been on—optimizing operations, not optics. By this point, I had long understood that, for a lot of people in the White House, political concerns outweighed specific public health concerns. This most recent incident was merely a variation on that theme. My priority was clearly not aligned with theirs.

In the end, how I spoke in future chief of staff meetings became a moot point. My admin, Tyler Ann McGuffee, told me that the daily meeting had been "canceled." Since the meeting would no longer be held, my presence was no longer required. Of course, a new invitation to the meeting went out to all the prior attendees but not to me. Business went on as usual in there, I supposed, but losing the one time I felt I had an opportunity to speak frankly about the complexities of the pandemic and the federal response with senior White House leadership who could make things happen and, critically, had access to the president was yet another obstacle. The doctors and the task force had already been marginalized enough, and most doctors weren't free to openly communicate to the American people. Losing one more chance for direct access to the president's chief influencers would further marginalize me from the president's inner circle. So be it. I'd figure some way around that and concentrate more of my energies on the main task at hand—the COS meeting covered so many other topics unrelated to the pandemic, and I could use that freed-up, non-Covid discussion time to strategize ways around the other roadblocks.

On the other hand, I couldn't let go of the thought that if one of the men in the room had similarly gone off as I had, he would have been looked up to for his passion and fiery commitment. I was a liability—one that they thought they could easily dismiss by exclusion from a meeting. They underestimated me. I wouldn't be on their list of attendees, but I was always going to be attendant to sharpening the response at the ground level.

I understood from the outset that with this president, with this combination of senior advisors, during this election year, my impact on the executive branch was never going to directly produce the results we all wished for. Believing that I could directly move the president, the CEA, Mark Meadows, Marc Short, Derek Kan, and others not just to take the virus seriously, but to immerse themselves fully in understanding how data-driven messaging and mitigation practices would be effective was an impossible errand. I had spent decades moving presidents and prime ministers across the globe to enact critical policies—policies that many of them not only didn't personally believe in but also perceived as detrimental to them politically—to ensure the most vulnerable and marginalized residents had access to lifesaving prevention and treatment services. But unlike those other leaders I was able to convince, some of the West Wing group were proving to be implacable, immovable. Now, after all this time, I saw better which people were inert, and to what degree. I couldn't focus most of my energy on trying to move the unmovable; the strategy needed to shift. Just as I knew that in a pandemic getting people to change their behavior was very hard, that was true of those shaping the pandemic response as well. Better to leverage those who could help me impact the response than those who resisted me. There was Jared and there was the vice president, and those two men would be my go-to people in the White House, then and for the next nine months, to move the pandemic response forward.

Outside the White House, I had other allies. Those county outbreaks the CDC handled gave me greater confidence in our ability to coordinate task force efforts and theirs. FEMA was critical to all of our efforts as well. Bringing that kind of coordinated effort to scale was possible, and we were working toward that. I also heard in our calls with governors that, by and large, regardless of party affiliation or partisan divide, they were very committed and, I believed, willing to take additional guidance to ensure that citizens of their states were safe and their economies could recover. (I didn't usually do this directly myself; I'd create call lists for the vice president and provide him with talking points to ensure that the states were getting consistent messages.) In the military, you look for places where a force multiplier can have the best effect—in other words, any factor that can produce an effect greater than its size or apparent strength might reveal, to fight on a par with a larger force. I was always on the lookout for those and when I could I'd deploy them. This was the kind of thing I'd been doing for years with PEPFAR and before. I felt comfortable having that level of engagement. I didn't need to be among the generals or the chiefs of staff; working with the lieutenants and sergeants might prove to be more effective. I was learning.

The story the virus was telling continued to evolve. As we moved deeper into May, some of those blips on the radar of minimal spread into the community increased in frequency to match the rise in cases. Now nearly 40 percent of the counties' rising new cases had a component of community spread. Though we weren't seeing explosive growth on the scale of what we'd seen in March and April in the metros, case incidence and hospitalizations in those pockets were growing. In messaging directly to specific governors, in the report to all governors, and in my daily report internal to the White House and agencies, I sent out the alert: this was new; this was community spread in rural counties, not just large metro areas. Unfortunately, for several reasons, this didn't carry as much weight as I would have liked. For one, previous numbers coming out of the metros were large. That set a kind of bar then. If we weren't seeing those kinds of figures, then that likely meant that things were still better. And they were better in the metros, but not as better as we all hoped for everywhere.

All states have areas of population concentration. The data coming out of them carries a disproportionate weight when looking at state averages. I always tried to make this point and was frequently frustrated that after reopening, on the national and state level reports a troubling game of red light/green light was being played. When using those statewide averages, and applying them to the status maps that became more prevalent in the media, it became common to see states flipping from one of our color-coded indicators to another with great frequency. Those maps were misleading because they didn't reflect what was happening in individual communities, including rural towns. Our internal maps were finally being produced with color-coded counties. But in May and into early June, too many were still at the state level rather than for individual counties. I spoke with some media off the record, this time asking for them to move toward a more countywide account of the state status in the graphics being used. I never wanted a distorted picture of the reality we faced to perpetuate a false sense of security.

To one degree or another that's what a WHO and a subsequent CDC announcement about asymptomatic spread did. Once again, we were back in the debate about the level of asymptomatic spread. Dr. Maria Van Kerkhove, the head of the WHO's emerging diseases and zoonosis unit, announced that it was "very rare." The WHO's statement created a mini-firestorm of controversy in the scientific community, which challenged this assertion. The WHO tried to walk it back, but by that point it was too late. It could not unsay what had been said.

Shortly after this, the CDC stated silent spread was occurring, but it accounted for only 5 percent of the cases. The CDC did say that asymptomatic spread was "plausible" and that it "meaningfully contributed to ongoing community transmission," but these vague, cautious terms indicated that it still didn't believe, as I did, that silent spread was responsible for as much as 50 percent of the cases and the majority of infections in those under thirty-five. Testing wasn't as robust as it needed to be and children, for the most part, weren't being tested at all because many didn't have symptoms. Consequently, many still thought that children weren't being infected, or if they were they would only develop mild disease, and that they played a minimal role in transmitting the virus. Without more comprehensive data, the CDC wasn't fully accounting for the breadth of asymptomatic spread, and its 5 percent figure was far out of line with what I was seeing and projecting. I could see it in the trends and the numbers. I could see it in the rising test positivity in young people without any symptoms or clinic visits.

This difference in opinion had existed from the start of my tenure as response coordinator, and now, here we were at the beginning of June, with nearly two million Americans infected, and the public health agencies had yet to break out of their bubble of testing only those with symptoms or exposure to a symptomatic individual.

I had to do something to counter the too prevalent perception the administration was creating that Americans could reduce their level of vigilance. In line with what I'd told myself earlier about utilizing my go-to guys, on June 10, I sent Jared Kushner an email with the subject line "The One Thing the President Can Do that Will Drop Cases by 4 July." In it, I told Jared that I could see the storm clouds gathering across the South: we needed the president to come out strongly for masking. We could have an impact if we aggressively mitigated now. The next day, I spoke up in the task force meeting, making the same point: we needed mask usage in public indoor spaces to be at 100 percent. This would require a cultural and personal shift, and White House leadership needed to set the tone. I can't say I heard crickets in response, but it certainly wasn't a chorus of voices chiming in with their support.

Ultimately, Jared heard me, and I believe it was he and Hope Hicks who schemed to get the president to wear a mask for the cameras when he visited Walter Reed. That was one visual. But we needed such visuals consistently, day after day. Masking in the White House would set the example that masks worked—but only if they were worn regularly. I would have to see if my attempts to reinforce that main pillar of the response would be effective.

I turned my attention to another of those pillars—testing. I knew that the WHO's and the CDC's statements would come back to haunt me; they did. President Trump continued to assert that testing led to more cases. Like a trial attorney who makes an assertion knowing the judge will ask the jury to disregard it but also that those in the box couldn't unhear, on June 15, the president repeated his May 15 assertion that it was more testing that was the cause of higher case rates. Some in the administration used tools like this, misdirection, and labeling things "fake news" while producing their own, to effectively change public perception. Call it bizarre, call it brilliant—whatever you called it, the intention was always the same: hope that the doubt or alternative reality would find fertile ground somehow, somewhere.

Here's the truth I experienced in this schizophrenic environment: that doubt may have flowered in the public sphere, but those anti-testing statements fit another pattern. The president would say one thing, but in the pandemic response in general, and in this case specifically, testing efforts weren't interfered with. Neither the president nor anyone in the administration, to my knowledge, ever exerted undue pressure to limit the number of tests being performed or critically interfered with the production and distribution of testing supplies. Instead, the White House, through the task force, aggressively expanded testing using the Defense Production Act, and Brett Giroir used all mechanisms to increase tests and testing supplies. No one ever advocated for closing or limiting the number of testing facilities; instead, the reality was that we funded and supported new testing concepts, helped implement new testing strategies, and then moved them to state-run sites. Throughout the months we continued to scale testing and we continued to buy and distribute supplies across the country. I may not have always agreed with all elements of the specific testing strategy and the oversized priority given to testing those with symptoms; I also wanted to do more specific cohort testing, as universities and sports teams did to find the earliest infections.

Even if Trump wasn't actively taking steps to reduce testing, the confusion his tweets and public comments sowed on the issue was damaging. As evidence of the power and influence Trump wielded, many people were left with the impression that what the president was tweeting about testing was true when, in reality, it was the opposite. The more tests performed would drive the test-positivity percentages *down* as infections were rapidly identified and community spread and household spread mitigated. In fact, more testing, and more awareness of one's status, leads to a more effective response to that status, driving down transmission and therefore cases.

As simple as that math is, it was surprisingly difficult to make it understood. Time after time, throughout May and June, both Democratic and Republican governors would report their state's increase in cases was due to increased testing. But because their positivity rate was rising, it meant not just increased tests but increasing viral spread. While Trump's words on testing may not have altered the federal execution, they were absolutely detrimental to public perception.

I believed then and still do now that the president's assertions on testing were a by-product of his belief that the CEA's initial estimate of twenty-six thousand deaths was still accurate. Obviously, we had exceeded that number of deaths, but those extra deaths could be explained away through a number of right-wing theories floating around:

The reporting structures weren't efficient. Doctors were mislabeling non-Covid-19-related deaths as Covid-19 deaths. Testing more made rates of infection go up. Just as testing drove up cases, testing in hospitals drove up cases of incidental Covid-19, not real Covid-19 disease. And so on . . .

The fact that President Trump himself apparently didn't want to know what was really happening spoke volumes about how far down into a facts-don't-really-matter hole we'd fallen since I'd convinced him to announce the fifteen and thirty days to slow the spread. He was getting his information on the pandemic from others, others who were cherry-picking data, others who were convinced this virus was not severe and we were overreacting. Telling him the Covid-19 hospitalizations were exaggerated. The deaths were exaggerated. He was hearing from doctors who told him we were lying and purposely misrepresenting the pandemic to him to do damage to him and his reelection. Words word words. But the actions actions actions to combat the pandemic continued.

MORE WORDS CAME AT me. In early June, I was provided with the "White House Coronavirus Task Force: Report to President Trump." A summary report of some considerable length that I was to review, it was essentially a victory lap for the White House Task Force on its handling of the pandemic crisis. Probably initiated at the time when the talk was to end the current task force, this was the kind of document you might expect to get when a project is winding down or has already ended, not in the middle of a crisis that's still actively being managed. It had been clear back then and was clear now that the end wasn't in sight.

Just two days before receiving that report card, I'd noted in my daily report to the White House that seventeen states were seeing rising rates of infection, and I highlighted that seven of those (North and South Carolina, Arizona, Arkansas, California, Texas, and Florida) were experiencing the highest case numbers since the beginning of the pandemic. Things were getting worse, not better. The spring had been awful, chaotic, and we were scrambling. Now, with remediation recommendations in place (some of them well considered, others less so) and a summer surge that would lead to an even worse fall and winter, some in the White House were actually looking at this ongoing calamity as if it were over, as if they believed that if they said it was over enough times, perhaps that would make it so.

The report was another reminder that the testing pillar had to be fortified. Even if the president wasn't directly attacking it, he was engaging in a kind of propaganda effort to win popular support to his side. He could achieve three things: he could make a claim based on a wrong interpretation of the data that things were better than the testing evidenced; through his public messaging campaign he could get fewer people to test to drive down the numbers; and he would still not directly interfere with testing, in order to demonstrate American superiority. It was a marvel of a communications strategy, to demonstrate that demonstrably contradictory things could all be true. While that is sometimes possible, in this case it wasn't.

To a great degree of certainty, tests don't lie. They do produce false negatives, but that's not intentional, as a lie is. The data we gathered from testing didn't lie. It was subject to different analyses, and it was my job to counteract the false narrative(s) being told. That was a battle I had to wage on many fronts, and I believed that being as data driven as I was, I was the best person with the best team to win that intense skirmish. As immovable as some people in the administration were on some aspects of the response, I was equally immovable on my belief that testing and the data it revealed was essential to showing us the way forward. The president, that report, could use whatever words they wanted to, but the evidence told a truer story.

The testing kerfuffle, and the administration's premature claim of mission accomplished, was a case of collective farsightedness. While images of the awfulness of late March and April consumed my thoughts, the administration had its eye on election victory parties and the start of a new term. To make those visions real, they needed to get the American people's minds off the pandemic and onto to the administration's economic achievements—anything else but the rising death toll. This was a vision that many, regardless of party affiliation, shared. The White House and many members of the public mirrored each other's desires. They wanted to redirect their attention from the pandemic elsewhere— to Memorial Day weekend; to the Fourth of July; to November and the election; to adjusting to the "new normal" so many were talking about.

The sad truth, one I was still struggling so hard to make clear, was that the new normal would look a lot like the very recent past. We were on the edge of a new surge, a product of our collective Memorial Day start of summer activities.

I was still working hard to communicate what the present rising numbers of test positivity actually showed: they always projected what we would see *three weeks later*: illness and hospitalizations; and six weeks later, deaths. Still, with the CDC's and the WHO's pronouncements regarding asymptomatic spread echoing down the halls and in the minds of so many, the White House wasn't listening to me and remained stuck playing defense, reacting rather than attacking. If this report were released it would be akin to claiming victory in the middle of the first quarter.

It also became even more difficult for me to publicly counter that message. Ever since the disinfectant debacle and the president abandoning the daily press conferences, the Trump administration had effectively banned me from speaking publicly for weeks. I kept pressing for the resumption of regular press briefings. I believed the comprehensive presence of the task force members presenting on the state of the pandemic and the actions needed by both the federal government and American citizens was important. Making it clear what we knew and what we didn't; what we were learning together about the virus. I still believed it was important to connect directly with the American people, where I could cut through some of the clutter and confusion the White House and CDC messaging had created. Then there were the individual pressers that Bob, Tony, Steve, and I were doing. Most of these were stopped. None of us thought that wise. In a perfect world, CDC guidance and task force recommendations would have been in perfect alignment. Because they weren't, having a means to clarify our position was crucial. The CDC had its public platform, and now we were being denied our own.

I never knew who was responsible for this change. In my case, I believed that Marc Short and the vice president's communications team were primarily responsible for my being silenced, but it was never clear from whom this was coming. Had I been allowed to do more national press when states reopened throughout May, my message would have been largely unequivocal: Follow the gating criteria in the guidelines. We are seeing outbreaks in specific counties that the CDC is tracking carefully. We are looking for evidence of community spread. At the end of May I would have said: Here is where we are starting to see increasing community spread in the rural counties in Mississippi and Alabama. We see significant community spread in the Navajo Nation and we need to increase support there across the three states. Be more cautious. Protect the vulnerable in your household. Expand mitigation in these areas. Test more strategically, looking for those who are unknowingly infected as well as those with symptoms. The outdoors is safe for you to gather and have your children play. Instead, I could say little to nothing publicly.

Of course, I could have taken my thoughts to the air the way so many people in the Trump White House did: by leaking them anonymously, telling the public through other people what was going on with me inside the White House. For ethical reasons, I would never do that.

Eventually, the president's communications staff stepped in and cleared me to do more *local* media—revealing the likely motivation behind the virtual gag order: They hadn't wanted me shaping or influencing the *national* conversation in any way. My words would almost certainly have made it harder for the president and them to keep turning the page, effectively closing the book on Covid-19. I had, it seemed, benefited from a combination of benign neglect and "out of sight, out of mind." I sensed the president and some of his inner circle had already moved on from me and the elements of the pandemic response I was coordinating, making it easier for his communications team to grant approvals for media requests—as long as I didn't appear on the screens he was watching.

ON JUNE 16, VICE President Pence had an op-ed piece run in the *Wall Street Journal*. It was published with the unfortunate title "There Isn't a Coronavirus 'Second Wave.'" The subtitle was, "With testing, treatments, and vaccine trials ramping up, we are far better off than the media report." It was true that we were making progress in all the areas the subtitle mentioned. Although I wasn't involved in the decision to do, or the writing of, the op-ed, I did see it only in passing in hard copy, literally hours before it was submitted. I can only surmise that when the op-ed was composed, we were primarily seeing the institutional outbreaks that were being rapidly detected and controlled with rapid engagement of the CDC throughout the majority of May. But after Memorial Day, in retrospect, the second wave was silently moving through the Sun Belt undetected then exploding in the second wave by the beginning of July. The vice president had to endure heavy criticism for presenting an inaccurate assessment of the current situation. This op-ed moment stood in contrast to the seriousness he brought to the summer surge response, including in that time pushing us to provide the Sun Belt states the support they needed. He would also support and accompany me on visits to the region that would begin just days after the op-ed.

Some in the administration's wishful thinking about the pandemic's end overlapped with the start of the summer surge I'd anticipated and warned about in the White House. My *What are we missing?* question regarding the isolated outbreaks throughout May that the CDC was diligently addressing in state after state transformed after Memorial Day, due to widespread travel from areas in the north of the United States where the virus lingered, into a full community spread across the Sun Belt. We had underestimated the depth and breadth of Americans' movement. The resulting viral spread that went with them, primarily with eighteen-to-twenty-five-year-olds and then twenty-five-to-thirty-five-year-olds, created a new phase in the invasion. In days we went from isolated specific institution outbreaks to broad community viral invasion. Even with all the guidance we'd created and the numbers we had provided to help the governors best manage their states, some were still far too slow to react.

This was especially true for the states that had reopened in late April and early May and did not see a substantial initial rise in cases. After several weeks open, they saw their situations deteriorate rapidly following Memorial Day. Studying the numbers carefully, I saw clearly that states like Arizona, Mississippi, Louisiana, Alabama, and Florida were on the precipice of their own New York–style catastrophe in July. But it would be worse in those states than it had been in the Northeast. Their populations had higher rates of comorbidities like obesity, diabetes, hypertension, and heart disease. Critically, the virus was moving into rural areas and rural hospital systems, which lacked equipment and had fewer health care providers and less capacity. The very places that needed stringent mitigation steps immediately (mask mandates, reduced indoor dining, among others) had shown themselves to be the least likely to impose them. They hadn't experienced the destructive forces of this virus in the March to May time frame and thought they wouldn't now.

Up to this point, the vice president had been an important conduit for communication with the governors. Though I didn't have as much access to him as I had when I first arrived, I adapted. I asked for ten minutes of the vice president's time to make certain he understood the gravity of the situation As always, he was receptive to the idea of making direct calls to those governors. I got around the Oval Office roadblocks, but the vice president ran into another.

With every call, independent of party affiliation, governors all reported that their states were fine—their hospitals were not filling. Most equated viral spread with rising hospitalizations as they had seen in the media from New York City. As a result, they would wait until hospitals began to fill to act with increased mitigation. They still weren't getting the message about the three-week lag time, that testing of younger cohorts and finding asymptomatic infections was important, that small outbreaks could be mitigated but that the silent spread that was a part of them was still at work. There were earlier signs and signals of viral spread in their states, and not just the later-rising hospitalizations.

But calls from Vice President Pence to the governors alone would not be nearly enough. Increasingly, as the doctors on the task force ran into the limits of what could be accomplished from Washington, DC, I saw a need for more ground-level intervention that could carry our message directly to the decision makers in the states and cities—not over the phone or on Zoom, but in person.

During my first week with the task force, I had noted down, "It was clear that states needed more on the ground support and wanted us to travel and leave a fulltime team behind to support the states in their response and address any issues immediately." Someone from the federal government needed to go out into the field. I believed that was the role of the CDC.

I sent an urgent message to Bob Redfield, asking him to send CDC personnel to North Carolina and Arizona, two states at most immediate risk of a surge, to speak with local health authorities and the community. It has long been my experience that a situation looks very different up close than it does from behind a desk or from a computer screen. I was continually frustrated by the CDC's apparent unwillingness to put more people out in the field for long periods, not just a week or two. These surges were lengthy, and we needed boots on the ground continuously for months, not days, to understand what was working and what wasn't from messaging to supplies.

They could be the force multiplier. Outside the United States we had found that CDC effectiveness was greatly increased when actively embedded with and in close coordination with the Ministries of Health. Working right alongside the public health officials in each country day after day. Out of the office in the embassy and into the offices of the host government. I believed this was critical for the United States. CDC health officials needed to be permanently embedded alongside their counterparts in the state and county health offices. Learning from each other, listening and adapting best practices learned from other states and rapidly translated to other states. Using the CDC network to actively move information in real time and actively learn and share.

Early on in the pandemic, those scientists were informed that, if they chose, they could volunteer their efforts for thirty days to Covid-19, and a portion of them did so in some capacity as a part of their usual responsibilities. Many, many more did not. (The same applied to uniformed public health personnel at the NIH.) Their employment agreements with the CDC, in my mind, should have stipulated that, in the event of a national public health emergency, they would become a first responder, deploying to various places across the country.

So, in response to my urging Bob to deploy the CDC personnel, he said his hands were tied. He couldn't force anyone to go to one of the Covid-19 hot spots, or anywhere else for that matter. He had to ask for volunteers. I wanted people from both agencies to join in an all-hands-on-deck effort to manage this crisis. Nothing but their goodwill could compel them to do so. I applaud those who did.

Brett Giroir and the Public Health Service did answer the call, as did the Department of Defense, sending health personnel out to the states not for a few days, but for weeks and months at a time to support testing, hospitals, and nursing homes. The real secret sauce in many states, though, was the National Guard. They answered every call, took on every mission, and saved many lives.

To this day, though, I can't imagine why there isn't a provision in the public health safety plans to deploy every available health professional, both those in uniform and not, from all the federal agencies in the event of a pandemic. When we face a military threat, those who work for the Department of Defense, including our frontline soldiers, don't have the privilege of opting in or out. They are told where to go and what to do, and they go there and they do what they were instructed to do.

These structural and personnel limitations to our federal health agencies inherently constrained our reach, negatively affecting what we could accomplish in Washington. As the Trump administration's focus drifted from the pandemic response to the economy, those limitations had been revealed. The states where Trump's attention would have helped the most, those that leaned conservative, were precisely the states less inclined to accept our mitigation message from afar. We needed to meet these governors, mayors, and lawmakers on a more level playing field. Simply making pronouncements, placing phone calls, and filing reports was not allowing our ideas to break through.

With the hindsight of the nearly two years since June 2020, it's easy to forget just how much confusion there was back then about how SARS-CoV-2 behaved, who was at risk, how the virus spread, and why it was so hard to detect. At the time, it was difficult for us to discern from Washington what was behind the objections from the governors. Sure, it could have been political, and in some cases it was. But it could also have been that they, understandably, were having a hard time reading the muddled and at times contradictory statements coming from the Trump administration or the confusing and complex guidance from the CDC. The task force would say one thing, President Trump would say another, the CDC would put out an ambiguous statement loosely supporting the task force but leaving itself some wiggle room. Meanwhile, Tony was on national media trying to push back against the contradictory messaging, to break through the noise; and President Trump was tweeting his own personal truth. Honestly, I can understand why some governors were legitimately pulled in different directions, facing opposing impulses about how to move their states.

We needed to act. To enhance our communication with the governors, we created a weekly executive summary report to provide *every* governor with county-level data, showing where their state was at present in the pandemic and what that meant for the near future. With these reports, I was trying to simplify things for them: using the type of language and graphics that would make the picture as clear as possible, so they could get ahead of what was coming not to neighboring states, not to the region, not to states a thousand miles away, but to *their* state and specific counties within their state—as well as a whole-of-country analysis so they could understand how this virus was moving around the country.

While we received some positive responses to these state-specific reports, the sheer distance between the governors and Washington, and the lack of a full understanding of the virus and how to read the data, made it difficult to chart a course for each state. The economy-versus-health conversation defined this uncertainty. Lost in the debate over economic recovery versus public health response was any sense of nuance regarding the kinds of mitigation strategies appropriate to a given state. The media, often led by President Trump himself, seemed to frame the issue as a zero-sum game—either you're entirely on the side of the economy, or you're entirely supportive of public health. Either everything is open, or nothing is. You impose mask mandates, or you ban mask mandates. When it came to actual implementation of these policies and ideas, where the rubber met the road for the governors and their states, the reality was far more complicated, and it deserved a more in-depth and nuanced conversation. Unfortunately, no one in the administration, the CDC, or elsewhere seemed willing to go directly to the states in person and have that conversation face-to-face.

Ever since I'd been okayed to do local media, I had been trying to spread the message in individual states that way. Over time, I'd learned a few things about the media and its role in the pandemic—most notably, that while many people turned to cable networks to see reality painted with a broad national brush, a lot of them relied on their local affiliates. Generally, when I appeared on local news, the journalists there were interested in getting accurate information specific to their community. They weren't trying to make headlines themselves. They weren't asking "gotcha" questions. It didn't matter if they were broadcasting in a red state or a blue state; they wanted to be of service to their viewers, who were interested more in protecting themselves, their communities, and their loved ones than anything else. As I began this smaller-scale outreach—appearing on dozens of news programs in early summer—I began to feel that bringing my case directly to the people, state by state, city by city, could actually work. It wasn't efficient, but I felt it was having an effect.

Still, the list of moves the White House was making to project victory and return to normal continued. I heard the ticking clock sounding the approach of a very, very difficult fall, while the CEA pushed to have the European travel ban lifted. I wasn't the only one who pushed back on the CEA's idea, and the vice president, Jared, and the task force did listen to reason and the data. Partially as a result, restrictions on leisure travel from Europe wouldn't be eased until late fall 2021.

Besides the looming summer and then the fall/winter surge, another specter was on the horizon: the general election. As much as the vice president remained fully engaged with me, took an active interest in what I was reporting, and did his best to support me, he would soon be on the campaign trail. The window on his constant daily availability to me was closing. As an indication of the White House's shift in focus, instead of meeting every day, our task force sessions had been cut to two to three times per week.

The election represented an entirely new variable, one that had me even more concerned. Indeed, if made public as seemed to be the intention then, the administration's "victory lap" report fit a narrative tailor-made for election season. With the election on the horizon, it would be that much harder to get the White House to raise the alarm about anything related to the pandemic. Any message that could interfere with President Trump's reelection message would be scrutinized, watered down, and weakened potentially to the point of irrelevance. As much as President Trump's and the political elements of the administration's focus had waned since April, the fall would represent something far worse. Not only would they be distracted by the campaign, but the virus would be poised for its most lethal stage yet.

AND SO, ON THOSE meditative mornings when I tended to my garden, I considered other options. I knew myself and my natural inclination to always hang in there and do more. Find a way or make one. I knew that, with one phone call, I could have the biggest megaphone in the world for a second through which to shout that the Trump administration had lost its sense of urgency; that it was moving on from the pandemic response; that a specific state had reopened too fast; that Americans choosing to travel across the country revealed that as far as they were concerned the pandemic was over; that we stood on the brink of a summer surge that would hit many states incredibly hard and kill a hundred thousand people, some unnecessarily.

I played out the going to the press scenario more fully. I would have done it eagerly, if I had honestly believed it would have produced some kind of course correction from the White House; I would have sacrificed my job and returned to my PEPFAR position in a heartbeat. It was true that Tony, Bob, Steve Hahn, and I had agreed that if one of us was fired the others would resign in protest. We were all still on board with that promise and were all still on the task force. I was neither hoping to be fired nor willing to resign independent of the others being let go.

The problem was, I didn't believe my departure would change anything for the better. Accusing the Trump administration of negligence wasn't going to suddenly produce a different response to the pandemic from them. And it missed the point that there were people every day within the very walls of the West Wing who were helping me get things done, moving policy, and ensuring action. So, it was never that simple. To the outsider it may have looked that black and white, but behind the scenes critical progress was being made using data to drive federal support. The machinery of the federal response was more focused and more data driven in decision making week after week. Progress could still be made. We needed to respond to the summer surge and be ready for the fall.

It was also obvious that the American public could already see that President Trump had staked out a position that opposed specific science, that opposed masks, that opposed testing. But behind the scenes we were still moving the ball down the field and mobilizing the vast federal resources to the right intervention. We still struggled with communicating effectively to motivate the American people to do what was needed when the virus came to their community. Communities would need to take steps to protect their larger circle. They had to act together and not divisively.

The evidence for the president's obstinance had always been in full view—he'd tweeted much of it himself. Leaving wouldn't bring much to light that wasn't already visible. My message of warning would last a news cycle, two at most, and then everyone would move on—except the people suffering and the states on the verge.

Moving policy is always more complicated than it appears. In the case of this pandemic, navigating the states' versus the federal government's role made things inherently more complicated. What constituted "support" and what was "interference" wasn't always as clear-cut as some liked to believe. The states would determine which mitigation policies to enact.

At first, in the beginning of the pandemic, the federal government's approach had been to focus on the whole of the country. Gradually, we began moving away from that to a more state-specific emphasis. This was a good thing, but I wanted to make it a much better thing. As much as we communicated with governors throughout the early spring, we couldn't address the feeling many Americans had: that the government was out of touch with how the pandemic was being experienced where they lived.

What overrode any other consideration was this: that other question in my mind about what was missing had now been answered. I knew that what we'd seen with those institutional outbreaks had now expanded beyond the confines of those buildings, beyond the people who worked in them, into their homes, and beyond into the community linked to our summer travel plans. I didn't want to spend too much time regretting not connecting all the dots, I wanted to do everything in my power to prevent that from happening again. We had not effectively foreseen the vast spread from the advent of summer. We needed to learn what we had missed so we didn't miss it in the fall. I had to get out of the White House and to the states. I had to cut through all the interference the president's words were creating. I had to look governors and state and local public health officials in the eye, read their expressions, know when I needed to provide greater clarification, and show them the data and the graphics to give them an evidence base to support the measures I had seen working elsewhere and that had been so effective in other places.

I might be starting from point zero with some of the governors, but our understanding of the pandemic and my own understanding of why so many people believed that this time would be different had evolved. The hope that we'd endured the worst of it was rooted in many causes. I'd have to do what I could to uproot them, but also learn from the experiences of states who'd been most effective in mitigating against the virus as well as the mistakes that had been made. We'd started with a whole-of-government approach; that would continue, but it had to be modified so that the whole could support the parts. That was what I was hoping to do and to discover, but this was moving into new territory, taking the approach I'd used elsewhere to see how effective it could be transplanted to American soil.

This on-the-ground effort should have been made by the CDC. For so many reasons, that agency was the right representative of the federal government to be leading such an effort. Yet, the CDC wasn't yet doing it, despite my urging. They were still only going to states for a week or two. They weren't meeting with most of the decision makers like the governors nor were they meeting with communities. So, rather than encourage them with words, I decided to show them with my actions. *Get out there! Do this kind of fieldwork. Learn from it.* Data at a distance was helpful. Data and experience and feedback from those on the ground were invaluable. A pandemic is a lived experience as well as a subject of scientific inquiry for future publication. Perhaps our visits to the states would help bridge the divide between the CDC's academic approach and our frontline public health one. We hoped they'd take notice of our example. The CDC and other agencies did have people out in the field, but those in the field often conducted their work remotely from hotel rooms, conducting Zoom meetings, creating another barrier.

I knew, from years of being on the road 40 percent of each and every year for decades, that I would be able to stay on top of my task force and coordination responsibilities and still visit the states. I broached the subject with the vice president, who was fully supportive. He turned the matter over to Tucker Obenshain, and the great team at the Office of Intergovernmental Affairs led by Doug Hoelscher and William Crozier.

So, the data team and I came up with an idea: Irum and I would travel together to states whose hot spots had put them in the red zone. We'd be risking our own health and safety, but face-to-face seemed the way to go. Just as it's easier to hang up on a telemarketer calling your phone than to close the door on the flesh-and-blood canvasser standing on your porch, we decided we needed to meet in person with the governors and others at the state and local level. We would take our message to them and learn what they needed, what was and wasn't working. We needed to see, listen, and learn.

I let Tucker know it was important that we get to Arizona, a real hot spot. It turned out that Vice President Pence and HUD secretary Ben Carson already had a trip to Texas scheduled for the last weekend in June. Irum and I could join them on the flight to Dallas. Tucker also arranged additional visits for us there, before Irum and I drove on to New Mexico and then Arizona. This was the blueprint for the months ahead. By December, with Irum, I'd travel to forty-four states, some several times, sometimes coordinating with the vice president's travels and other times hitting the road independently, covering thousands and thousands of miles visiting communities large and small. We would be on the road nearly 50 percent of the time.

Along with meetings with the governors, Tucker arranged sessions with state and local health officials, hospital associations, doctors, nursing associations, nursing homes, individual frontline workers, community members, and tribal nations. This first itinerary was a matter of convenience and happenstance, but ultimately, it would prove providential. We'd learn a lot in Texas, New Mexico, and Arizona, and I'd be able to use what we learned as Irum and I tracked hot spots and then went to where the virus had announced its presence. It wasn't enough to sit in state capitol conference rooms. We had to get out on the front lines and get a closer look at how this pandemic was being lived every day by people outside the government.

A big part of the reason I felt comfortable leaving Washington was I had recently supported the hire of Dr. Moncef Slaoui to head Operation Warp Speed (OWS), the public-private partnership to hasten the development of a vaccine against SARS-CoV-2. Dr. Slaoui would oversee Covid-19 vaccine development. His presence and guidance proved to be a strong hand the country could all grab on to, and he would carry us to the vaccine finish line.

Dr. Slaoui had retired in 2017, after working for thirty years at the pharmaceutical company GlaxoSmithKline. He spent several of those years as the head of its vaccine development division. The company had created vaccines against malaria and Ebola, something that had resonated with me during our interview. At the time, Dr. Slaoui was still on the board of directors at Moderna, but he resigned that post when he accepted the job to lead OWS, to avoid any conflict of interest. In so doing, he had signaled to all that private industry and the government could work cooperatively.

In the case of Operation Warp Speed, there was full White House engagement. Jared Kushner and HHS secretary Azar, invoking the Defense Production Act, made sure vaccine developers had all the critical raw materials and could conduct the trials and manufacture without delays. Over the course of the next months, we would see cooperation and mutual respect among all the groups involved. Had there been this kind of focused commitment in all areas of the pandemic response, including communications, the science of behavior change, ensuring that all Americans understood the science and data and were empowered to act, we might now be viewing our efforts as an across-the-board success, not just the patchwork one it was. With its unique partnership between the White House, HHS, and the private sector vaccine and therapeutic developers, Operation Warp Speed, I felt, was one instance where we didn't overpromise or underdeliver.

I was encouraged by one of the first moves Dr. Slaoui made, immediately bringing on board Carlo de Notaristefani, an expert in the vaccine manufacturing processes. It is one thing to develop a safe and efficacious vaccine; it is another to be able to make it in the quantity required. Proactive production and working out all the scaling from thousands to millions of doses would be key to the rapid rollout we all were hoping for and that was eventually delivered.

I was also pleased when another person with a long career in the military, particularly in supply and logistics, came on board. Gustave F. Perna, a four-star U.S. Army general, took command as chief operating officer of the federal Covid-19 response for vaccines and therapeutics. In July, he'd transition over to become COO of Operation Warp Speed, where he would coordinate logistics for the CDC's distribution plan developed in coordination with states for the yet-to-be-produced-or-approved vaccines. Getting out ahead of the approval process was a key move that was lacking in other areas. Like Dr. Slaoui, General Perna brought a wealth of experience and expertise to both those jobs. Similarly, Rear Adm. Polowczyk followed by General Stafford brought a similar data-driven, no-excuses, now-now military discipline to logistical supply chain matters and worked with Jared on the next-generation federal stockpile. Collectively, we were committed that no matter who won the election of 2020 they would inherit a fully stocked federal stockpile.

This whole-of-government approach that tapped into the expertise of the military and the deep partnership with the private sector was and is a new approach to pandemics and pandemic preparedness and should be integral to the full planning and execution. The depth and breadth of this type of partnership was unique to the Trump administration and not only saved lives but should transform future thinking and planning.

Technical advances certainly aided the rapid development of vaccines, but technology as a tool must be employed efficiently and thoughtfully. If there was one thing to be learned from vaccine development, it was that a shared vision, a balance between competition and cooperation, and an enormous investment of capital can do wonders in a pandemic. Operation War Speed is a road map of success for future critical vaccines and therapeutics.

As Irum and I prepared for our trip, I was well aware of what we had missed. That was the incessant pulse I felt beating in my chest. I was having my own "Never again" moment, vowing that we'd not miss any nuance in the data again, and that we'd translate this new learning into words and actions. Too much was at stake to do anything but. One more, two more, a dozen more times when we had to work through, around, or over whatever lay in our way, we'd do it.

I was pleased when, in late June, we made progress on the communication front. I was being allowed to meet with medical correspondents in the media, which gave me a chance to correct the president's message on decreasing testing. I could also utilize them to lay the groundwork for my upcoming work in the states. To counter the president's false claims about test positivity rates, I explained what the administration wasn't accounting for, "the two-week sequence of events." Two to three weeks after you see an increase in *positive tests*, you will see an increase in *hospitalizations*, followed two weeks later by an increase in *fatalities*. If you cherry-picked data from a particular point or from a specific state on this time line—you would be ignoring the longer-term trend. If fatalities from a past surge were decreasing in one area of the country, but test positivity was rising in another, what you would be seeing was evidence of *another surge in another region*. As the president preached his false gospel on the role testing was playing in the rising case numbers, he was ignoring what the data was showing us about the inevitability of more cases. More cases = more hospitalization = more deaths, as the virus always found its way to the vulnerable Americans living in communities and there was no vaccine yet to protect them from severe disease.

During these meetings with medical correspondents, I made the case that things were getting far worse for the summer and that this trend would continue into the fall and winter. One month after the April 15 release of the reopening guidelines, Covid-19 cases had increased by 125 percent and deaths by 174 percent. We couldn't yet see that, after the forty-five days we'd been granted had elapsed, by Labor Day, we would go from approximately 675,000 cases and 33,000 deaths to 6.2 million cases and nearly 200,000 deaths, combining the 100,000 fatalities from the spring surge and the 100,000 lives lost in the summer surge. Cases increased by 827 percent from April to September, while fatalities rose by 544 percent.

The summer surge was deadly, but if we had mitigated earlier with increased testing and masking and with decreasing social gatherings with the first increase in test positivity, lives would have been saved. But in talking with states then and now, every time the test positivity and cases began to rise they would say "this time the outcome will be different," "this time we have more tests," "this time it's a different variant," "this time we have vaccines"—yet the outcomes continued, and continue, to be the same. That will change over time but as long as there is widespread community spread and Americans who aren't protected, due to a poor response to the vaccine, due to age or immunosuppression, or due to the number of unvaccinated vulnerable Americans, lives will be lost.

I made the same case about the summer surge and the prospects for fall and winter in our doctors' group and with the larger task force: We needed to prepare for the coming heightened crisis. By the third week in June, I changed this message a bit: *The crisis isn't approaching. It's here.*

Tony wrote me privately to let me know he got it and that, in his view, "the wheels are coming off." The response from others was muted.

The president had publicly stated that he would intervene in pandemic matters as necessary. So, I tried to make the case through Jared that this was one of those necessary moments. During the last two weeks of June, I sent him several emails encouraging the adoption of universal mask mandates among members of the administration while in public, more sentinel testing (testing performed across a community, even of people with no symptoms), and the usual laundry list regarding supplies and therapeutics. He said he was working on it. I needed the president to talk to the American people and raise awareness and the need for them to act to protect their vulnerable family members. We had tools—we could blunt the impact of the surge.

As I counted down the days until my road trip with Irum, Vice President Pence updated me on his progress with the governors. I'd asked him to make clear that a significant surge was under way.

"Debbi I'm sorry to say this. I'm not having much success with these calls." He sounded discouraged. "They're hearing me, but I don't think they're really listening. They say they are fine, the hospitals are 'fine.'" But, I said to the vice president, the hospitals won't be fine—they will be overwhelmed again soon.

I read in his expression the same question I'd been asking myself for so long: *What are we missing in our communication with governors? We can see it—why can't they?*

It was the silent spread issue rearing its ugly head. Without demonstrable evidence that sentinel testing could provide, without hospital admissions rising, the waters appeared calm. But farther out at sea, rapidly approaching swells were rising. The tsunami was building.

We'd have our work cut out for us once we started to sit down with the governors.

Saturday, June 27, was spent packing. I'd gone to Jared Kushner, Tony was sounding the alarm, and the vice president was well aware of the situation. A surge was coming. Had he communicated it to the senior advisors and the president? I believed that he had, but I couldn't know for sure. Without direct access to the president, and with trust in me at a low ebb among his inner circle, I did the only other thing that I could.

In the top lines of the opening page of the daily summary document for that day, I heavily emphasized that if we didn't act now, the summer surge could see our reaching as many as one hundred thousand new cases a day—in the end, we reached "only" eighty thousand—and an additional one hundred thousand American lives would be lost. They were. The refrain I sang was the same: We needed to mandate the wearing of masks in public indoor places. We needed to close or vastly reduce occupancy indoors where people were unmasked, like in bars and restaurants.

As a kind of coda to that shared report, I wrote the doctors' group a postcard from home before hitting the road:

I have tried for 2 weeks to get acknowledgment of our situation and with current log phase. I am pushing as hard as I can. In local media I push for masks and testing but I am not scheduled for any national media and I know I will remain sidelined. Would be good to connect over the weekend if someone can get a line. Don't forward this to anyone.

In retrospect, I see how desperate that sounds. But when you can see the approaching tsunami and there are still people on the beach you need to run out to the beach and be clear about the warning. We needed to get to the beach. I continued to send out emails over that weekend. In one 5 a.m. missive to the doctors' group, I tried again to hammer home my point. At the beginning of the week, we were at twenty-six thousand new cases a day. Six days later, we were at more than forty-six thousand a day—the largest percentage increase in the United States up to that point.

We needed to act now.

We needed states to institute stricter mandates now—not wait.

I sent a separate but similar email to Jared Kushner. He was out of pocket, but he told me to contact his former college roommate Adam Boehler and Adam's associate Brad Smith. Both were young, successful contemporaries of Jared's serving in various advisory capacities within the White House and in other federal agencies. On a phone conference call for the four of us, I felt like a talking head, delivering again, at the top of the hour, the news of the day. I reiterated every point of the message I'd been trying to refine over the course of the previous weeks.

It's here.

It's going to be bad.

It's going to be *VERY BAD*

They offered assurances that they got it. They'd work on it. They'd get back to me.

I ended the call hopeful, but nagged by the Ghost of Overpromise and Underdeliver. A scant few hours later, I was back meeting with Vice President Pence, doing yet another version of my (legitimate) doom and (legitimate) gloom presentation.

The next morning, trusting that I'd done as much as I could to make clear what we were facing in the coming months, I set out. I'd entrusted the care and feeding of my plants and flowers to my husband. They'd be well taken care of, and I knew my family would take good care of one another.

Irum and I exchanged a look as we used disinfectant wipes on all the touchable surfaces of the car. I surmised that some spots were simply unreachable, but that didn't matter, you went after them anyway.

PART II

CHAPTER 11

Hitting the Road

Early on in what would be a run of visiting twenty-six states in just July and August, Irum and I frequently departed our hotels at 4 a.m. to the sound of downshifting semi-trailer trucks and the rattle of moths tap-dancing in the shroud of lights illuminating jam-packed parking lots. We hoped that our difficulties in finding available rooms and places where we could dine outdoors more safely and where more people than just the servers and other essential workers were masked would be the exception rather than the rule.

That first trip out of the Washington bubble and into the hot zones was both instructive and daunting. Irum and I armed ourselves as best we could with masks and disinfectant wipes. We had to be careful not just about our health, but also not to let our presumptions rule. Despite wearing masks, we had to be open. We had to be willing to learn. Some of what we learned was alarming. Some proved enormously useful in helping us shape the next approach to mitigation so a state could stay open safely.

We were on high alert. Mobility data showed us that though people in the DC area had decreased their movement by 60 percent, elsewhere—in Texas in particular, but also in Arizona and New Mexico, where we finished up our three-day swing—they were on the move. This meant they were carrying the virus from place to place, many of them unaware of what asymptomatic spread was. A few were heedless, unwilling to listen to anything they were told about the danger they were in or the danger they posed to others. We wanted to understand better: Had they not gotten the message? Had we not made it clear enough? More likely, we presumed, it was a combination of the two.

Too often, public health officials rely on reputation and scaling up the urgency with stronger adjectives or increased volume to change people's minds and behaviors. After years of working on the ground in communities and among the marginalized, we understood that meeting them literally where they lived and understanding the context of their lives would be essential to our moving forward together. Talking down to people through oversimplification, alienating them by finger-pointing, never works. It may feel good in the moment, when you are frustrated, but it makes the job much more difficult in the end. I had also found that, no matter where I was in the world, no matter the level of formal education, people were smart and had great common sense. They could understand the concepts and the nuances if you took the time to speak to them in clear language. Sometimes, as scientists and doctors, we like to make things complicated to sound really smart and communicate that we should be listened to because we are smart. This attitude and approach completely misses the mark. We should always strive to deliver a clear message, one that can be understood and acted upon by everyone.

We needed to keep everyone on our side because we had ideas we needed them to try. Throughout my months in the White House, I'd continued researching the virus, scouring the internet, reading scientific journals, and speaking with colleagues from around the globe. Irum had been doing the same, and she had discovered an unlikely but critical new model. We agreed that it could potentially serve as another way to actively mitigate and impact community spread without a shutdown.

Dr. David Rubin directs the PolicyLab at Children's Hospital of Philadelphia (CHOP). He is also a professor of pediatrics at the Perelman School of Medicine at the University of Pennsylvania. Rubin and other members of the PolicyLab staff conduct population and community-partnered research with a focus on the needs of high-risk children. As part of that work, they had produced a highly interactive Covid-19 forecasting model at the most granular local level. They would eventually use it to prepare school reopening guidance and in-person schooling recommendations.

Rubin used precision tracking of transmission and test positivity rates similar to what Irum and the rest of my data team were doing. From that, the UPenn/CHOP team independently produced a forecast that closely matched what we were seeing and projecting. Irum contacted him, and Dr. Rubin created models for us based on various assumptions we provided. We asked him to model the impact of a statewide mask mandate, significant expansion of outdoor dining while reducing indoor dining to 25 percent, and closing standing-room-only bars. We also asked that a model reflect keeping retail spaces open as long as masks were worn in all indoor public spaces. His models demonstrated the impact of these measures on the replication rate of the virus. Simply put, to be under R1 is a condition where every infected person infects less than one other person and community spread contracts. Actions taken to decrease viral replication to a point below R<1 results in controlling community spread.

This combination of mitigation measures, which spared many aspects of the economy, could very well be nearly as effective at preventing community spread as a full-scale stay-at-home-type shutdown. It had the impact of a "shutdown" but wasn't one. The president had given me that one directive: Do not shut down the country again. I believed we had charted a path forward that could work now (and, critically, in the fall) while meeting his directive.

The UPenn/CHOP model worked *in theory*. Now we needed to implement it at an actual population level and measure the impact on viral community spread. Many in public health preferred small-scale, controlled experiments, but a real-world implementation was more revealing because many more variables would be tested. We needed to get at least one governor to enact these mitigation measures statewide so we could study their impact in both urban and rural counties. In anticipation of our state visits, we had already laid the groundwork for this on a governors' call by having David Rubin speak to all the governors about his model and theories.

I believed the UPenn/CHOP model was one the states could use and everyone could follow. If it proved successful with one state, it could be effective as a template, an adaptable way to stay open safely, preserve both economic interests and public health, and ensure that America's children were in school in person full time in the fall. Every day, I carried with me the deep concerns about children's mental and physical health and their educational progress.

I was certainly aware that Irum and I, as representatives of the federal government, might not be welcome in the states, that our visits might be perceived as federal interference in their territory. Thankfully, my fears were unfounded. What I'd sensed while in Washington—that the states needed, *wanted*, more on-the-ground federal support—was reiterated a thousandfold by some governors and health officials. But the states didn't want just support. They wanted a clear and open dialogue in person to ensure the federal government was hearing them. They also wanted clear guidance like we'd created with the Opening Up America Again framework. They were dismayed that the Trump administration hadn't shown better leadership in communicating clearly, and they needed the president and others in the administration to model public health mitigation—to wear masks, to tell the truth about the importance of testing, to encourage Americans to stay the course for what was to be a long haul. In the weeks to come—whether it was a red state, a blue state, or a purple state—behind closed doors, when in face-to-face, private meetings with us, governors and other officials repeated some variation of this desire. Their pleas drowned out the sound of the semi-trucks and the fluttering moths. Those voices were what I heard as I fell asleep at night and what I woke up to in the morning.

GOVERNOR GREG ABBOTT OF Texas was our first stop. Flying out of DC on Air Force Two, with its standard airplane cabin preventing physical distancing, I reviewed the data on Texas and my previous interactions with the governor. Throughout March and April, he was among a group of governors—Phil Murphy of New Jersey, Gina Raimondo of Rhode Island, Michelle Lujan Grisham of New Mexico, and Doug Burgum of North Dakota—who frequently, during and outside the weekly gubernatorial conference calls, asked directly for additional guidance and for specific supplies to meet their needs.

Abbott had made many of the right moves in Texas—declaring a state of disaster, limiting dining at bars and restaurants, closing gyms, and restricting social gatherings to ten people. His state had adopted the White House reopening framework guidelines and had worked with me to tweak them as necessary. Governor Abbott adopted the three-phase gated criteria program we'd devised, and it was initiated in Texas on May 1. Like most states, Texas had ridden the waves of the virus, its crests and troughs. In mid-June, the state had eased indoor dining restrictions, and most businesses were open. Just prior to our arrival, the governor had to order bars to shut down and restaurants to once again restrict capacity to 50 percent, the lowest yet. He also banned any outdoor gatherings of more than one hundred people, unless the organizers got local government approval. Texas was experiencing the post–Memorial Day summer surge and was the first state to rescind reopening measures. Abbott was under a lot of heat from both sides—for not doing enough, for doing too much.

Texas was stuck in this feedback loop. We offered Governor Abbott and Texans a way out.

At the start, the governor seemed skeptical that doing three rather simple things—universally masking, restricting indoor dining occupancy, and allowing only smaller social gatherings—could make such a dramatic difference.

"We've done these things," he told me. "We can't seem to make any real headway. It's like we're on a roller coaster."

"You've done *two* of the three things, Governor," I told him. "Without mask mandates, you lessen the positive effects of the other two."

He frowned. "Masks? Nobody seems to agree on that one. It's a real hard sell. It's not going to make too many folks happy."

"It will keep more folks alive," I said. "That should make everyone happy."

I went through my explanation of the role silent spread played in the pandemic and said that, short of doing sentinel testing of specific groups, masks were the most efficient means to stabilize the situation. Abbott was tired of the back-and-forth; so were many Texans. Without a clear mandate from the White House and the CDC on the efficacy of masks, he had chosen not to mandate them. Considering what the UPenn/CHOP model demonstrated, and eager to do the right thing, something that would preserve economic interests and public health, he told me he would consider taking some masking measures.

Governor Abbott recognized that his state was in trouble, and on July 2, a few days after we left Texas, he issued a mask mandate. He devised a sensible, creative solution that took into account Texas's fierce history of independence: the mandate would be statewide but would exclude counties with twenty or fewer Covid-19 cases. This implementation of mask mandates based on the degree of community spread was a data-driven, tailored approach that we learned from him. The result of this approach was that it seemed less arbitrary to the state's residents. As was true in many places, many of the moves Governor Abbott made were viewed in various quarters as too little, too late, while others saw them as overreaching overreactions. As for me, I was glad to see that he'd added the third element of the UPenn/CHOP model: a modified mask mandate would help us see how effective the three measures could be.

Next, we had to wait for the results. Cases in Texas peaked around July 15, consistent with the masking mandate being added to Abbott's decreasing indoor occupancy and his clear public messaging. The UPenn/CHOP model was working! We would carry that message forward.

Later, as the governor was being attacked on both sides, either for being late to the game or for overstepping his authority, one state representative tweeted that Abbott thought he was king. But the governor was living out a dichotomy that existed in so many places. In the political and social arena, striking the right balance was possible, but it would take considerable time and a healthy dose of commitment from leaders and the general public to stay in that zone.

As would be true for many states, in the months that followed, state leaders would need to continue to engage in dialogue with those of us at the federal level, so that we might jointly analyze the data and mitigation efforts to review their impact and decide on next best steps. The task force continued this bidirectional learning until the last governor's report was issued, on January 18, 2021. We provided the governors with clear state-specific recommendations based on their local county data. More important, we listened to each and every governor, independent of party affiliation, and we learned from each and every governor, independent of party affiliation. This is what the vice president believed was critical, and we came to understand why.

Governor Abbott would remove much of the statewide mitigation in the spring of 2021. Yes, he should be held accountable for his decisions. But the federal government should as well. The agencies and the White House didn't stay directly engaged with every red state or blue state outside of the governers' calls. They didn't continue the dialogue we had initiated so that data and issues could be reviewed together. Without that dialogue, neither side was listening and learning. The federal government didn't revise approaches or provide direct state-by-state recommendations. The virus and the situation on the ground continued to evolve.

Governors require constant engagement and support. They need a clearinghouse of effective strategies that worked in other states. These are lessons we need to carry forward to be ready next time.

We learned two things from the Texas experience: First, Governor Abbott's decreasing maximal indoor dining occupancy to 25 to 50 percent of the state's existing fire code occupancy figures allowed for public-sector enforcement and consistent application. And second, closing the "bars" in Texas wouldn't have done much good. Nearly three thousand establishments that in other states might be classified as "bars" are, in Texas, classified as restaurants, better known as "roadhouses." As a result, if we used the term *bars* when referencing closures, the owners of these roadhouses could make the case to health officials that *We're not a bar, so that doesn't apply to us.* It may seem obvious that "bars" refers to places that serve alcohol, but this linguistic and legal distinction was one example of how our general guidelines were being interpreted and applied on the ground. During our weekly governors' calls, there wasn't time for any governor to raise unique concerns like this. For this reason, such critical nuances escaped us. In person, we got the message. The bar-versus-restaurant distinction alone made our Texas trip worth it. It proved that being on-site could yield positive results.

In Texas and elsewhere, we realized that we couldn't always trust what we were being told. Some state officials were reluctant to report bad news to the governor or to Washington. Also, some were using bad data as a source of such reports. The use of state averages rather than local county numbers was an ongoing problem. State averages provided a misleading picture of where the hot spots lay. While, for example, Indiana might have had a low case incidence rate at one point, test positivity in Brown County or in the zip code 46135, in Putnam, was exponentially rising. Working at this level of precision was a major key to an effective mitigation response.

The tyranny of averages played a role in a less-than-proactive response. While reports at the federal level indicated that our supply management was in great shape, we found out that some states were lacking much-needed tests and testing supplies (swabs, tubes, extraction media, pipettes), remdesivir, ventilators, and other crucial items. This information was helpful but disheartening.

Personally, we were finding solutions to some road trip issues. We learned where there were safe, secure places to pull off the road to change clothes discreetly. We even figured out how to do so in or near the car—wedged between the open front and passenger doors and, at least a few times, behind trees and Dumpsters. In minutes, we would go from car-comfortable stretchy pants to full business dress, stockings, and heels. And we developed a new appreciation for fast-food drive-thru windows. We also found that those working drive-thrus were a good source of information about how a particular community was doing vis-à-vis mask use and other aspects of mitigation.

We continued to read about the negativity and hostility flaring up around the country over masks and shutdowns, but we saw no real sign of it, and none was directed toward us. We were moving around incognito—after all, we were masked—and that helped. Not being recognized felt right. I'd later be recognized, even while masked, when I ran into a CVS for mascara. That experience unsettled me. I wasn't used to that kind of attention, and though the clerk was very friendly and helpful, it still felt odd to be recognized. I'd be spotted again a few times over the course of the next few months. Once, at a filling station/ small general store in Oklahoma, I spoke with two men from Florida who were on their way to Idaho to fly-fish. We talked about being really careful, as the virus was moving north. They thanked me and our messaging for having convinced them to drive rather than fly.

After our meetings were concluded in Texas, we set out for New Mexico. I wanted to go there specifically because it *wasn't* in crisis, yet two of the states it bordered, Texas and Arizona, were. It was important to see what New Mexico was doing differently. The state had far fewer per capita infections than the other two, and though I'm normally not keen on averages, that did say something big-picture about how it was responding to the crisis. From the governors' calls, we already knew that Governor Lujan Grisham had been very assertive. She had a strong belief in the public health value of testing and mitigation. (She also enjoyed being a Democrat in a state with a Democratic legislature.) In fact, Irum and I wouldn't be allowed to meet with the governor personally unless we tested negative for the virus. In spite of how lax the precautions had been in the hotel during our overnight stay in Amarillo, our subsequent tests both came back negative, which was reassuring but too early to mean anything.

Governor Lujan Grisham, her staff, and the state's health officials greeted us warmly. In our discussions, we got into some of the nuts and bolts of how the governor had advocated for and put in place restrictions on indoor dining and had tried to keep outdoor dining options available. Where other states had not, she had been able to keep retail businesses open with aggressive distancing and masking. She had provided options to the citizens of her state, and they had responded well. I made note of New Mexico's success, particularly with its tribal nations. Core to Lujan Grisham's approach was the constant expansion of testing. She had been the first governor to call me to ask for more tests back in March, and those requests continued as she used the tests to find the cases early to stop community spread. New Mexico's population is diverse, and its health officials had paid close attention to the state's unique needs and had supported all groups. Though she wasn't aware of the UPenn/CHOP model mitigations by name, Lujan Grisham had employed them.

Happy to have an example of how they worked outside a computer-generated model, we got on the road for Phoenix.

While the people we encountered over those first two days were nearly without exception friendly and courteous—even the Texas policeman who issued Irum a warning and not a ticket for her spirited driving—this didn't mean some people weren't angry. After completing early meetings in Phoenix, Arizona, I had a scheduled call with California governor Gavin Newsom at 5:30 p.m. I stepped out of the car into a blast furnace of heat. The governor wasn't able to get on the line, but his chief of staff and senior health advisor harangued me for the first ten minutes of an hour-long call, vehemently complaining about the lack of leadership from the White House. In that call and through future state and local meetings, I would hear multiple examples of Trump's faults and the damage they had done. My talk with Governor Newsom's people was merely a preview:

"If only the president would just talk about masks."

"If only the president would talk about the importance of widespread testing."

"If only the president would just—"

I'm sure my responses did little to assuage them.

"I understand."

"I know."

"I, and others, are trying."

I sensed that they needed to vent their (or the governor's) frustrations.

Once California was in a better place, they would be more receptive to the message I was formulating.

As painful as it was to feel the heat of the day and other people's displeasure, it helped solidify my vision: We had to travel to literal and figurative hot spots, where the weather and the virus were heating up. It was summer, and Arizona and the Southeast, where we would go once this three-state swing was over, were in the Goldilocks zone: too hot. With the extreme heat, people were being driven indoors, gathering together into air-conditioned spaces—optimal conditions for viral transmission.

The day after the call with Governor Newsom's staff, Irum and I walked into an office building in Phoenix to present our three-pronged mediation approach to safely reopening and staying open to Arizona governor Doug Ducey. He was open to a discussion of the UPenn/CHOP model.

"That sounds good in theory. But I can't see how it can apply here. Arizona's unique. We've got an aging population with lots of retirees. We have tribal nations at great risk. We've got a couple of major metropolitan areas. We've got people living in small pockets out in the desert and in the mountains. 'One size fits all' isn't going to work here. We're too diverse in our geography and in our politics."

He wasn't alone in this perception. Fortunately, Governor Ducey and his health team were willing to listen to our proposal. They reviewed the charts and graphs. Dr. Cara Christ, director of the state's Department of Health Services, was bright and hardworking, and she understood what we were trying to do. We learned a lot from Arizona's team, and Dr. Christ would become a critical partner in the months to follow.

We also met with Arizona's county health officers and community groups along with the governor. They had already started outreach in key Black and brown communities. The county health officers had a clear-eyed understanding of their communities and had developed a meaningful partnership with them.

One consistent claim—in Arizona, Texas, and California, and from many mayors in those states and elsewhere—surprised me. Many told us that outsiders had imported the virus to their locale. This wasn't a case of xenophobia, or of the despicable anti-Asian sentiment we had seen while watching media coverage. Instead, it was a variation on the "over there" problem. Leaders felt they were being responsible in their county or state, but that the state next door, the county next door, wasn't being careful, and now their area was failing due to others' carelessness. *People crossing the border are the ones responsible for the infestation of the virus, not us. It wasn't here. Someone brought it here.* In Southern California, someone said to me the problem was people from Arizona coming to the state to go to the beach. Texans told me it was the people from Louisiana, or Texans who had traveled to Louisiana (especially, to New Orleans), who had brought the virus back to Texas with them. Somehow having the virus in their state seemed more palatable if it was someone else's fault. *My community is good. Your community is bad. And now you've made our county sick. It was brought here.*

In Arizona, the governor and his public health officials were willing to implement the full mitigation measures that the UPenn/CHOP model showed would have an impact in two weeks. We made this point with them: We weren't imposing a strict plan, but rather an easily modifiable one that the governors and public health officials could revise and make their own. Having heard from Vice President Pence about how the governors thought, operated, and succeeded, this "You craft this, you own this, you have responsibility for this, and we will support you" approach proved effective in selling this message to the states. Conversely, we also gave the governors an out. If the measures we urged them to take failed, they could always blame the federal government for the ideas we'd brought to their state.

Ducey's actions were data-driven, as were those of many other state leaders. Governors who had come to public office after success in the business world immediately understood the graphics and the need to use data to make decisions. They had been doing that their entire business careers. I saw this data-driven approach from Governor Jim Justice in West Virginia, Governor Tom Wolf of Pennsylvania, and Governors Burgum and Ducey.

Governor Ducey reviewed the model and the data with his health advisors. Given the demographics he'd cited, the mitigations we were championing were a good fit for his state, especially with the large numbers of vulnerable Americans living there—older Americans and those living communally, Native Americans on tribal lands, and those living in long-term-care facilities. Ducey's progressive, analytic approach allowed Arizona to thread the needle between public health and economic interests, creating a road map we could offer other governors. Using that map, they could select the route they wanted to take to control the viral spread in their state.

In Governor Ducey's "We're different from other states" remark, I saw another objection I had to counter. It is always hard getting people to understand that two somewhat contradictory things can be true simultaneously. Yes, a state could be unique, but no, it wasn't the only state going through this. On this first trip and later, when in a given state, I found it helpful to point out that what they were experiencing was what other states were also going through or had already gone through. We'd then show that state's leader what other governors were doing and how it was working. This message of "You're not alone" helped offset the sense of failure and potential resignation governors might have felt in having hot spots in their state. This was a bit tricky at the beginning. Because the Northeast corridor had been the site of the first and most severe outbreak, the March–May 2020 surge, the virus was perceived as a big-city, "over there" problem. The places we were visiting weren't packed metropolitan areas, and the thinking was *It can't happen here.*

This wasn't true. Covid-19 was an urban, suburban, and rural problem—and very real. But we had to prove this.

Because they hadn't seen their own March/April surge, too many communities believed they were immune. This was a consistent theme in our conversations. All we could say in response was that, regardless of how the virus had gotten there, it was now time to concentrate on limiting and eventually ending its spread. In the end, this brief, busy, and nearly sleepless first trip accomplished what we had hoped it would. We remained concerned about those who were truly on the front lines, not those in power who had chosen to place themselves on the sidelines. If too many people put their heads in the sands of denial or anger or bargaining, instead of focusing on what we could do next, we'd all be grieving over what might have been. That was a place I never wanted to travel to.

On the flight back to Washington, I wrote to Jared Kushner, pulling no punches as I spelled out what the word was out on the streets about the White House leadership: The governors appreciated the supply support, but all of them wanted consistent federal messaging on masking, testing, and gatherings. This wasn't coming only from me. It wasn't coming only from Democrats or those on the left. It was coming from Republicans and those on the right. Left, right, and center—the persistent chorus contained voices that needed to be listened to.

In community meetings during that trip, I had also seen—along with frustration over messaging and, at times, anger—real concern. This was especially true for the women who gathered to listen to us and share. When I mentioned how important masks were, the women present nodded their heads. I saw them glancing at one another and then at their husbands. The first look was one of satisfaction at hearing that their concerns were legitimate. The second was a "See? Why won't you listen?" look. Across the board, moms, grandmothers, daughters, and sisters were concerned and anxious. Like me, they had vulnerable family members—sometimes a spouse, sometimes a child, someone with Down syndrome or severe asthma, sometimes an older parent. Although many were silent, their eyes held this worry.

I knew the feeling. It's hard to know that you're right but that it doesn't matter.

The entire executive branch needed to understand what was needed. We needed to listen, and to respond with support. Guidance and words on websites weren't enough. To truly support implementation, we had to translate those words into actions in deep partnership with state and local leaders. We couldn't simply issue guidance to the states and then take a hands-off approach. This trip showed me that we could provide the states with clear examples of what was working on the ground with specific populations. We needed to gather more of this kind of "data," and we needed to share those insights and successes during our weekly governors' call and the written governor's reports, and that's exactly what we did.

Irum and I were energized by the visits. At the state and local level, in counties and on tribal lands, solutions existed. You could see and learn from them not from DC or Atlanta, but only when you were out in the communities where good people struggled to follow complex federal guidance. You couldn't just issue a statement and expect it to be understood and followed. In order to adapt your communications approach to the people to whom you'd targeted your message, you had to meet those people where they were, figuratively and literally. That was the only way to know if you'd made a bull's-eye or were off the board entirely. This strategy, which had served me well around the globe, was important here in America, too.

CHAPTER 12

Battling the Herd Mentality

A few days post–Independence Day, shortly after Irum and I got back to DC, my assistant, Tyler Ann McGuffee, received an email that Jared Kushner had forwarded without comment. The original message was from John Rader, who had once been a part of the president's transition team, touting the work that someone named Dr. Scott Atlas was doing in support of the administration's pandemic efforts generally and in re-opening the economy specifically. Atlas had been on Tucker Carlson's show on Fox News and had written several op-ed pieces, all very much in support of the administration's current pandemic response positions, especially around testing and the "low risk" of Covid-19 disease.

I mentally filed Dr. Atlas's name under the category of: just another influencer coming from who knew where. The email from Rader was one of many FYIs I'd received and intended to get around to eventually. In the meantime, Irum and I were looking to build on our success in Texas and Arizona and our lessons learned from New Mexico with another trip—this one to the Southeast.

I can see now that I should have paid closer attention to that seemingly innocuous July 8 email. At first glance, I lumped Scott Atlas into that group of proponents of one message or another who had caught the attention of someone (frequently President Trump) in the White House. Their ideas were often at odds with conventional wisdom backed by substantiated data. Because I often had to speak with these individuals,

I believed I should find out a bit about them in advance of a conversation. Immediately after I got the email, I asked Irum if she had heard of Atlas; she hadn't. I had Tyler Ann do a bit of digging on my behalf.

Tyler Ann sent out feelers to other administrative staff, including Jared's assistant, with whom she had a good relationship. She reported back to me that Atlas wasn't a random, one-off contact with the president. He'd been in regular communication with the White House senior advisors and, potentially, with President Trump since early spring. Several White House officials, including Rader and Paul Alexander, at HHS, had been in touch with Atlas. At several points, Atlas had appeared on Fox News, delivering a message that aligned with the president's vision of what he hoped was true about the virus. Collectively, these people had encouraged Atlas to continue his good work in spreading a message that supported the president's position that Covid-19 was of low risk and more like the flu. Testing wasn't needed, Atlas claimed, unless you were really sick.

The headline for one of Atlas's op-eds reinforced my initial perception that he was out of his depth: "The data is in. Stop the panic—and end the total isolation." In this and other pieces he had written, Atlas claimed there was no need to flatten the curve with aggressive mitigation in the spring to preserve the hospitals across the country or enact any measures that would negatively impact the economy. His rationale read very much like what proponents of so-called herd immunity believed, and he advocated for a very limited (in some cases, nonexistent) set of mitigation steps focused solely on the most vulnerable Americans—read those in nursing homes.

Atlas believed it was possible to fence off the vulnerable Americans with significant medical issues and elderly Americans over seventy and let the virus move undetected and unmitigated among the full population. That was his theory, and if that were possible, it would be a powerfully effective tool. The problem was that it wasn't implementable. It simply wasn't possible to protect everyone in the community at significant risk and let the virus spread wildly without protective vaccines. There were more than 50 million Americans vulnerable to severe disease, 42 million over 70, another 7 to 8 million with immunosuppression, and another unknown number of younger Americans with significant underlying comorbidities. All of these Americans lived in our towns and cities across America, in addition to the 1.5 million Americans at the greatest risk in long-term-care facilities. Atlas's thinking was on the edge of the scientific community.

In every pandemic, there are brilliant scientists who offer different approaches at the boundaries of conventional science. During the early days of AIDS investigations, some researchers were convinced the syndrome was caused by an unusual fungal infection and not HIV, the human immunodeficiency virus. Unfortunately, as with those scientists, Scott Atlas didn't always reside at the edge.

Trained in nuclear radiology at the University of Chicago, Atlas had taught at Stanford University's Medical Center for fourteen years. He was also a senior fellow at the conservative think tank the Hoover Institution, specializing in health policy. Crucially, it was what he was *not* that most concerned me: He was *not* an epidemiologist, nor one with on-the-ground experience in the community in implementing a response to a deadly pandemic. He was also *not* an infectious disease expert. He was a respected health care policy wonk on the right.

Mistakenly, I at first assumed that, given his lack of direct experience or background in pandemic response, Atlas could be easily dismissed. Clearly, he was a contrarian, and I wondered if, like others of this type, he was more interested in being viewed as a contrarian than in actually advocating for his positions—which, to me, were untenable. Perhaps he wanted to initiate an academic debate of views but not change policy. But, he was not in an academic debate; he preached to a converted segment of the population, and within the administration, the benefits of let-it-rip "herd immunity." Regardless of the name it goes by, this approach, this set of beliefs—that masks are likely ineffective, that children cannot pass on the virus (and thus, by extension, that schools should remain open with students and teachers unmasked), that the role of government is not to squash the virus but to let it spread while protecting only the most vulnerable Americans—was dangerous. He also advocated for civil disobedience against state guidelines designed to prevent the virus's spread.

That Scott Atlas had found a receptive listener in Paul Alexander at HHS was not surprising. Michael Caputo, who was appointed assistant secretary for public affairs in early spring 2020, had hired Alexander as his scientific advisor, poaching him from his position with the Infectious Disease Society of America in DC. Alexander had engaged in a campaign to wrest control of the public messaging from the scientists and public health officials. Centering many of his efforts on the CDC and the NIH, he tried to exert influence over Bob Redfield and Tony Fauci. Eventually, the House Select Subcommittee on the Coronavirus Crisis would see some of the emails Alexander wrote to both doctors, accusing them of fear mongering and stating that the CDC's Dr. Anne Schuchat was out to embarrass the president.

Alexander would leave his position in September 2020, but not before Assistant Secretary Caputo showed his support for various conspiracy theories (among them, that the CDC had created a resistance unit to oppose the president and his coronavirus positions) and not before he accused various scientists of sedition. As bad as the pandemic had gotten, Paul Alexander openly expressed his support for natural/herd immunity, stating of young people and those at lower risk, "We want them infected."

For his part, Paul Alexander was a credentialed public health policy professional, but though he had a PhD and had graduated with a master's degree from Oxford, he was an unpaid, part-time professor at McMaster University. Scott Atlas shared the same set of beliefs, but had better credentials, affiliations with Stanford and the Hoover Institution. Whether Paul Alexander realized this or not, along with John Rader, Michael Caputo, and Fox News, he had managed to put Scott Atlas more prominently in front of the president and his senior advisors.

When I first read about Atlas and his views, I felt a sense of déjà vu: his views sounded familiar. I suspected that it was Atlas who had been urging the president, or someone close to him, to take his "never again" position on shutdowns. And then painting any level of mitigation agaist the virus as a "lockdown." I'd always wondered who had gotten to the president to make him shift his position so abruptly. It didn't seem to matter to Trump or his advisors that Atlas was not an epidemiologist and had no pandemic experience. Furthermore, it didn't seem to matter that he backed a flawed theory that couldn't be translated into an effective response, since there was no way to protect all the vulnerable without an effective vaccine. In this White House, appearances mattered most. As long as a person had graduated from the right schools and espoused the "right" thinking, they'd be a welcome addition to the team. Tony, Bob, and I weren't saying or doing what they wanted. Time to get someone who would!

Scott Atlas was certainly entitled to his ideas and opinions, and like any American, he could express those opinions freely and widely without censure. But his opinions—which flew in the face of hard scientific evidence to the contrary—took hold and influenced the president of the United States and others in his administration ultimately cost us thousands more lives and added greatly to the unfolding tragedy.

From the outset, I knew that Atlas's message and the beliefs that underlay them were dangerous. If the approach he advocated for were turned into action, many more lives would be lost.

Sadly, despite everything I and others said, despite the numerous warnings I issued, Scott Atlas's early influence and later continued advice to the president and others in his administration instigated an ongoing battle between two very different points of view. Atlas's hands-off approach was diametrically opposed to the proactive community-based mitigation efforts to stop community spread for which I and the rest of the doctors on the task force (not to mention most in the scientific/medical community) were proponents.

Like all things related to the pandemic, herd immunity was a complex subject often presented in simple terms. A "let-it-rip" approach among the "healthy"—that is, allowing a virus to take out a certain percentage of the population vulnerable to its worst effects—may be a viable option with livestock. If, for example, you don't mitigate against infection at all and 100 percent of the animals get infected, 10 percent of the animals might die, but the remaining 90 percent of the herd are then immune to that disease for an unknown period of time. It's a somewhat costly approach in animal husbandry, but in terms of scale and profit—weighing the costs associated with fighting the infection and treating the diseased animals against the cost of replacing the animals that die off—it has its appeal. Once developed in an animal population, if the developed herd immunity is durable it does decrease the rate of transmission, protecting the whole herd and reducing the chances animals in the herd will be infected by other animals carrying the virus.

There are two ways to achieve herd immunity: through natural infection or effective vaccination. But herd immunity is only possible if either natural infection or vaccination results in long-lasting immunity against both infection and disease. Many Baby Boomers were naturally infected with measles or mumps in childhood and that single infection resulted in long-lived immunity and therefore long-term protection from both infection and disease. The measles and mumps childhood vaccine (MMR) that was developed in 1971 and mimics the long duration immunity and protection of natural infection can create herd immunity in the population. Establishing herd immunity through natural immunity or the "let-it-rip" approach is dependent on natural infection or vaccination producing *long-lived protection from both infection and disease* that results in no reinfections. That is what Operation Warp Speed potentially offered us. Though even then we were concerned the vaccine would not produce protection from all infection but primarily disease. What is difficult for some to understand is that the calculus for developing herd immunity depends on achieving a threshold relative to the infectiousness of the virus. If the virus is highly infectious, you need close to 85 to 90 percent of the population vaccinated or naturally infected and, critically, you need that protection to be durable—not for months but for years. Get to what many believe is the magic threshold number with durable immunity, and you can achieve herd immunity.

What is absolutely essential to remember about herd immunity is the notion of it being long-lived, regardless if it is induced naturally or via a vaccine. Getting to that point, having durable immunity, or long-lived immunity, is difficult to achieve. As a consequence, each viral surge results in reinfections. You can also develop immunity to severe disease without durable immunity to infection. This is possible because vaccination or prior infection produces what we call a memory immune response. So, when you encounter the virus again and you get infected, your body immediately recognizes the virus and rapidly produces an effective immune response that clears the virus before it can spread throughout your body and cause severe disease.

It is particularly difficult to determine the threshold number when you are dealing with a novel virus, both as variants change the infectious nature of the virus and if immunity to infection wanes. As the Covid-19 pandemic wore on, many may have lost sight of the fact that this was a new virus. Yes, we shorthanded the name and dropped the "novel." Yes, the virus had an antecedent in the SARS family of viruses. But SARS-CoV-2 was still a new, unique virus, one exhibiting exclusive characteristics in terms of how it was transmitted; how severe a disease it produced in infected people; and how that disease, Covid-19, responded to various interventions. And because it was a new virus, and we didn't yet know how it behaved, those who advocated for the herd immunity approach early on couldn't have known what the threshold figure was going to be. We just didn't have enough data then. They also couldn't know if natural infection led to long-lived protection and immunity to any reinfection. Eventually, when we learned that SARS-CoV-2 was highly transmissible, this drove the threshold number up. Likewise, when we saw that, like all viruses, this one was going to adapt and that new variants would emerge, the threshold number was driven up again. Later we learned that natural infection did not result in long-lived protection from reinfection. Pushing for herd immunity or allowing the virus to move unmitigated through communities without making clear these limitations was ill-advised.

As we've recently seen from South Africa, where waves of viral surges have occurred and serial variants emerged, natural immunity wanes significantly between each surge. This inevitable decrease in immunity demonstrates the limitations of the approach Scott Atlas and others advocated. Using broad (essentially unmitigated) infection to achieve herd immunity, without vaccines to prevent severe disease in millions of vulnerable Americans, propagates new variants and leads to increased hospitalizations and deaths. Even within a vaccinated population, if that immunity wanes, the same series of events can happen: variants will develop that may evade our previously developed immune response.

Morally, as well as scientifically, I was never comfortable (and never will be) with the herd immunity approach. I also couldn't support "letting it rip" in certain age groups in an attempt to protect others. This doesn't work. When you have viral community spread, you can never isolate and protect one vulnerable group completely. You cannot ring a fence around all vulnerable Americans, as the majority are in the community, part of multigenerational households, or have care providers who routinely come to the home, potentially bringing the virus with them. This was the very problem with the hypothesis Scott Atlas and others espoused. With each surge, despite high levels of precaution and testing in nursing homes, we have seen silent invasion continue to result in high fatalities among our parents and grandparents.

Downstream from the idea of herd immunity was something even more troubling. If you don't want to mitigate, and if you believe you can build an impenetrable wall around the vulnerable, then testing for the presence of the virus becomes inessential. This approach ran counter to my belief that testing was one of the main pillars of an effective public health response. This belief was being borne out by reality.

We were seeing that significant closures of schools, offices, and other public places weren't necessary if we tested widely and wisely. Effective sentinel testing in certain sectors—in the film industry, in the medical community, at universities, and on sport teams, among other businesses where routine testing was taking place—dramatically decreased community spread and resulted in fewer limitations and closures, not more.

As an advocate of herd immunity, Atlas and his colleagues believed that no form of closures or widespread testing was necessary. Atlas believed that testing a young person (who he believed would not suffer any ill effects from the virus) and asking them to isolate, if positive, for ten days was tantamount to putting them in lockdown, a violation of their personal freedom. In the extreme, his preferred approach was the equivalent of telling people with asymptomatic HIV not to practice safe sex and to infect others as a way to fully exercise their personal freedom. During my visits to the states, people wanted to know if they were contagious and were more than willing to isolate for ten days to protect vulnerable friends and family. What they *didn't* want to do was repetitively quarantine due to exposure—but we had an answer for that: regular testing. I have found that everywhere I have been around the globe, people are responsible and want to protect others from infectious diseases that they may have. It didn't matter if it was HIV or TB or now Covid-19—people are protective of others.

It had been months since we talked about shutdowns or lockdowns in our task force meetings. Instead, we had pivoted to an effective strategy, one that was working in real-world communities: testing, masking, limiting indoor gatherings, especially when community test positivity rates rose. With this virus, we had learned well the sequence of events, how to see the virus early and how to keep retail spaces, schools, and workplaces open safely, and we had spelled all this out in the governor's reports.

Controlling infection using these available commonsense mitigation tools, and vaccines, to decrease viral spread to the millions of vulnerable Americans was, in my own view and from a public health standpoint, the only way to get to that elusive threshold number. Anything less would be taking too great a risk with too many lives, especially with vaccines on the horizon.

I understood that many in the White House and around the world had to consider the costs associated with the pandemic. As in military operational and tactical planning, some level of casualties was expected in conflict, but through training and aggressive mitigation, everything possible was done to decrease battlefield fatalities. As someone who had spent her entire professional life working to reduce disease-related fatalities, I couldn't perform the mental or moral calculations to derive an acceptable number of Covid-19 deaths if, through simple commonsense measures, they were preventable. It's not that I believed those kinds of decisions were better left to people above my pay grade. It's that I simply could not tolerate the notion of having 10 percent, 1 percent, or even 0.1 percent of Americans die a preventable death when we had the tools to protect them and ensure early access to lifesaving treatments.

SCOTT ATLAS DIDN'T OCCUPY all my time in July. Irum and I had continued to track hot spots around the country, and in mid-July, the Southeast was highest on our list of regions to visit. I had gone to Florida just before the Fourth of July. Then over a four-day period, we went to Louisiana, Mississippi, Alabama, and Georgia. I had gone to Florida just before the Fourth of July. As was true for our first road trip, we were pleased to participate in a productive exchange of ideas. Governors John Bel Edwards (Louisiana), Governor Tate Reeves (Mississippi), and Governor Kay Ivey (Alabama) were all engaged in active data analysis to see the virus's impact in urban centers and rural outposts. Even though they were a step behind the virus, they all understood how the pandemic was behaving in their state and were implementing effective mitigation interventions to save lives and pave the way for full school reopening in the fall. These would bear fruit over the next couple of weeks and blunt the ongoing community spread.

Again, the majority of vulnerable Americans being admitted to the hospital with severe Covid-19 disease often were being infected from household and community gatherings. With the mitigation efforts, workplaces and retail and public spaces were becoming safer, with the primary spread shifting into households. Still, for all the positive work across the Sun Belt, from California to Florida, one state stood out.

I was especially dismayed when, on July 16, I met in person with Governor Brian Kemp of Georgia. Sometimes leaders don't practice what they preach. We were jammed shoulder to shoulder into a very small conference room, with no space allowed for physical distancing, even' though larger rooms were available. At least we were masked. But what message were we sending to the public health officials and business leaders around that table?

The governor didn't engage directly with me, but spoke primarily to Bob Redfield, who was also at the meeting. I suppose this was no surprise. I was a part of the White House; Bob was not. Governor Kemp had been understandably upset when President Trump called him out publicly in the spring for opening his state too quickly. Kemp was also among the last of the governors to issue a stay-at-home order, finally doing so on April 2. The day before our meeting, he banned cities and counties from mandating masks. Though he didn't want the federal government to impose regulations on his state, he appeared to have no qualms about imposing a rule from above on Georgia's mayors and other local officials. We wanted him to allow local communities to make their own decisions based on the level of community spread—as Governor Ron DeSantis of Florida was doing at that moment.

Our time in Georgia was a teachable moment for Irum and me. With many homes lacking broadband internet, laptops, and other devices, the state needed to keep its schools open. Latino/Latina residents hadn't been fully engaged with effective, nuanced, culturally focused translations of health guidance. Also, for myriad reasons, testing was highly problematic. Many outdoor testing sites were located in parking lots whose blacktop became so superheated that it was difficult for anyone to wait in line. For those able to tolerate the hot sun and melting tar to get swabbed, test results were woefully delayed because of a lack of coordination among the Public Health Labs, university research sites, commercial labs, and the community. Pending anti-immigration legislation in the state had spread fear among Latino/Latina residents, making them hesitant even to get tested. While nongovernmental organizations were active in the state to try to ameliorate many of these issues, they weren't able to fully coordinate with state agencies. Also, they lacked PPE to protect their own people while out in the field, and tests and testing supplies were limited.

We heard a great deal of skepticism among the Black population about anything the federal or state government had to say or do about the pandemic. A very long history of abuse and distrust was responsible for many Black communities discounting federal messaging. We had to come up with a work-around for the usual methods. We knew from HIV-prevention work that the hubs for the Black community were beauty shops, barber shops, community centers, and churches. If we could get the trusted, culturally aware influencers in these hubs to spread the message, we'd have a chance of dispelling many of the untruths that were spreading nearly as rampantly as the virus. This was a tall order, particularly in the rural areas, where distrust of the federal government ran deepest. In fact, it was on that trip to Georgia that we first heard someone say, "I'm not going to get that Bill Gates vaccine." There was always going to be resistance to vaccines, and the persistent rumor that the Microsoft founder was using the vaccine to, of all things, implant microchips in people was one of the many confounding aspects of this.

I discovered another source of distrust. When I spoke with Georgia's public health officials, it became clear that Dr. Kathleen Toomey, the public health commissioner, was supportive of the governor's policies and actions. She noted that the state was improving—even though Irum and I were there expressly because it was not; the data demonstrated this. As was too frequently true, Dr. Toomey and her people were relying on state averages rather than the community-level data that showed precisely where cases were lower and higher. I saw a lot of raised eyebrows among the participants at the meeting when the disconnect between what state officials were seeing and what was actually happening in a particular locale became clear.

After the meeting was over, I shared a governor's report with one state official. I asked if he had seen one before. He hadn't. I couldn't help but wonder if valuable information was being withheld. Distribution appeared to stop at the governor's office or was perhaps limited to a select few at the top. I handed the official a few more copies and urged him to share them with his colleagues.

The governor's reports were a key component of our direct communication of the mitigation approach being executed across the Sun Belt. If that information wasn't being widely disseminated through the state systems in Georgia—whose governor clearly didn't believe the federal government offered any kind of useful guidance—then what were health officials there relying on? Who or what, I wondered, was influencing the incumbent governor?

It was absolutely essential that the public across the country receive the information contained in those reports. The message, those three key moves that could make the difference between a full-on surge and isolated, manageable outbreaks—masking, closing bars and restricting indoor dining occupancy, and limiting social gathering indoors—had to be delivered. In Georgia, it wasn't. Worse, Governor Kemp (and, later, Governor DeSantis of Florida and many heartland governors) wrote to Marc Short complaining about the very existence of the documents my team was sending out. If the state- and county-specific pandemic summary and recommendations had been widely distributed, the public would have seen the depth and breadth of the pandemic in their particular state. This would lead to questions regarding why our recommendations weren't being implemented or even considered. Information is empowering, and information provided in an objective and culturally sensitive manner allows members of a community to make informed decisions about their health and how to protect their friends and family. Several governors even said they were worried that the Freedom of Information Act would give their constituents ammunition to use against them. Better that the public not know, the thinking went, so as not to judge. This was crazy, but it was happening.

I appreciated that some states—Kentucky, for instance—posted the weekly governor's report to their public internet sites. Still, the kind of pushback Marc Short was receiving, and that he was reiterating to me and sharing widely within the administration, continued throughout the summer and intensified in the fall, as the pandemic became far, far worse.

The message from some governors was *We don't want to know what you are seeing in our counties, our nursing homes, and our hospitals. Let us go our own way.* This resistance always made me wonder, *If they're not willing to listen to us, then whom are they listening to?* The potential answers to this question unnerved me.

AFTER RETURNING TO WASHINGTON on July 18, I received the text of a speech being prepared for the president, along with attached graphs and data I'd never included in my daily reports. The information in the speech was inconsistent with what our team had been integrating from across the country. This was further evidence that the president was receiving parallel data and analyses to the exclusion of mine. I went up the chain of command to get an explanation for why these figures were being used instead of what we knew to be real numbers. My not knowing this already, though I was the coronavirus response coordinator, was additional evidence that I had been pushed even farther outside any sphere of influence. At least with the infamous CEA memo, with its optimistic projection of only 26,000 deaths, I was fully informed about who and what underlay that prediction. This was a speech, not a formal report. Still, it bolstered my belief that other teams or individuals remained at work providing the president with data, that others had a different level of urgency than mine.

Because I had long suspected that something of this nature was going on from, very likely, the onset of the pandemic, the governor's reports and my daily analyses were critical to counteracting these other numbers and the false impression they created. With those reports, the governors and the White House could never say they didn't know what was happening across the country. Fortunately, the action teams throughout the U.S. government, including the task force, were using my team's daily data to move supplies and personnel to the states and counties. And the vice president continued to use our data to alert the governors of their state's status and to direct action within the task force. The actions remained driven by my team's data. The president's perceptions and rhetoric seemed to be driven by another set of numbers.

With this speech, as was too often the case, the words coming out of the West Wing were at odds with the actual response. Surface and substance frequently collided, making the inconsistent messaging even more problematic. I had seen this kind of rhetoric-versus-reality distortion play out many times in my career all around the world. I've made the point before that all pandemics are political. I'd seen it in Africa and Asia, and it was now playing out in the United States to varying degrees. It is tempting to say that the president's speech was just words on a page or vibrations being transmitted. Yet, not only did it produce a distorted view of reality, but it eroded confidence in both the man delivering those words and the agencies working hard to ease the collective suffering and save lives. In developing a pandemic response, it is important to realize both how much words matter and the consequences any utterance can produce. Words and numbers are both part of our language. When one or the other is used imprecisely, the degree of shared meaning is reduced. When that's done repeatedly, no agreement is possible. People tune out. They lose trust. I was very fearful that this erosion of trust in our leaders and institutions was very real and would damage our ability to communicate effectively to motivate people to do what was needed to protect themselves, their families, and their communities.

That same day, I received another unwelcome reminder of how words could be manipulated in the form of a *New York Times* article in which I was characterized as "the chief evangelist in the West Wing for the idea that infections had peaked and the virus was fading quickly." The article also stated that I had been roaming the halls in April stating that we had hit our peak and that, in morning meetings in Mark Meadows's office, I declared that "all the metros are stabilizing." I was even "quoted" as saying, "We're behind the worst of it."

I wrote to Marc Short and his communications staff immediately, attaching the internal report cited in the *Times* piece, from April 11. It was true that I had written in that report that New York City and the New York metropolitan area (which encompasses the state's Mid- and Lower Hudson Valley, Long Island, and parts of New Jersey and Connecticut) were stabilizing and reaching peak. Yet, in the very next bullet point, I had also written that Chicago and Boston were still worsening, Houston was in full logarithmic phase (aka the period when cases rise exponentially), and new hot spots were developing in the DC, Baltimore, Philadelphia, St. Louis, and New Haven, Connecticut, metro areas.

There was no evidence that I had sugarcoated anything in my April 11 internal report. Instead, I'd stated the facts clearly. Perhaps the administration had cherry-picked my first bullet point referring to New York City and its surrounding counties improving and had ignored the rest. I have no idea why this story was published then or for what purpose. I wondered if someone in the administration was trying to discredit me and what I was currently saying about the severity of the southern surge—which was serious. With that surge, even with our current level of action, I predicted that another one hundred thousand Americans would succumb to the virus that summer.

I suspected that someone within the vice president's communications team was responsible for leaking the false story. I was livid. Clearing up this misinformation, in addition to taking me away from other valuable work, forced me to go to the vice president.

"Mr. Vice President, I believe this leak came from within your office."

He shook his head immediately. "No one on my staff would do that. What proof do you have? That's a serious accusation. I'm sorry, but you're going to need something more specific than a suspicion," the vice president said. "I'm sure this is all upsetting, but these things happen."

This was true, but it was no consolation. Still, it was clear to me that the vice president wasn't involved in any way in the creation of the "fake news." While I was out there trying to build rapport, trust, and influence with the governors from both parties, someone had been at work inside the White House to discredit me. The suggestion that I was out of touch or talking out of both sides of my mouth could potentially hamper those efforts.

Next, I went to Marc Short's office.

Short, of course, denied being complicit or having any knowledge of who the leaker was. "Wasn't me," he said. "I'm not saying that it wasn't someone in the West Wing, but I don't know anything about it, really." His words said one thing; his slight nod toward Mark Meadows's office said another.

I wanted to make certain that this didn't happen again, but without knowing who was responsible, and considering that the leak could have come from so many places, this would prove very difficult. Whether it was because I had been raising the alert about the South for weeks, because of what I was saying now, or because of what was in the governor's reports—whatever the cause, I knew in that moment that the White House was creating its own fake news. This wasn't a case of leaking, but of deliberately lying to mislead the press.

In that moment, I realized that the White House had learned to fight fire with fire. They had continually accused the press of creating "fake news," but they were, in this instance, creating their own fake news to deflect attention away from the administration's handling of the response. And apparently, they wanted to make the public believe that I was the one misleading the administration, telling them that all was well across the country, suggesting it would then be okay to stop all mitigation and return to engaging in our prepandemic behaviors.

When you had an administration with a long record of using leaks to damage or discredit anyone who disrupted the narrative the White House was crafting, setting the record straight would never be enough. The July 18 *Times* piece wasn't the first instance of this; it was just one of the most blatant and egregious. When I told Steve Hahn weeks before that we task force doctors would be lucky to survive working in that White House environment, I'd anticipated just this situation. Back then, I didn't know what *form* the backstabbing would take, but back-stabbing was inevitable. It began to be more open with this leak, and variations of it would continue for many, many months to come.

This was how this administration operated. But they didn't yet fully know how *I* operated. My reputation mattered to me, of course, but it mattered less than saving as many lives as possible. I was hurt by this false depiction of me. I'm human, after all, and not impervious to personal attacks. But I never underplayed the seriousness of the pandemic or the extent of the damage it was doing. Maybe those who had characterized me as overly optimistic for the *Times* had misinterpreted my belief that, in the end, we would prevail. For me, this had meant we would do the right thing, and we would do it over and over. Connive, leak, and distort all they liked—that was their reality, but it wasn't mine. I may have lived in it, may have adapted to it, but I wasn't going to adopt it. I wasn't going to play their game.

I was never an anonymous leaker to the press. I never went on back-ground or off the record unless specifically requested to by the White House. But I couldn't allow these moments to distract me from the mis-sion, and I never did—until a leak criticized my family and brought them into the news cycle fray. Even then, I vowed to keep pressing on. I'd already lost credibility within the administration. Now, with this leak, *Debbi Birx is one of them!* might as well have been written across the skies.

It was around this time that Scott Atlas's name began to come up more. After being informed of his presence via Jared's email, I was asked by Jared's team to meet with the man personally. He was on my calendar for the end of the month, but I suspected the name "Scott Atlas" had been on others' calendars since long before I got that memo.

With this recent leak, senior leaders in the White House, as I had long suspected, appeared to be engaging in a common strategy for get-ting rid of people in an organization who are no longer wanted. It felt like the *New York Times* piece (for which I was never contacted for a response or confirmation) was just another step in the process of dele-gitimizing me, deescalating the intensity of the pandemic alarms I was sounding, and paving the way for the installation of a well-credentialed physician whose flawed views were more fully aligned with theirs.

Misery loves company, and in the process of making life miserable enough to get *me* to quit, they appeared to be doing the same to *Tony*.

While I was on the road, the administration seemed to have launched into a campaign of *While they're divided, let's conquer.* Peter Navarro, Trump's top trade advisor and coordinator of the Defense Production Act, wrote an op-ed that was published in *USA Today.* In it, he said Tony Fauci was "wrong about everything." (To their credit, the editors at *USA Today* later issued a statement saying that the piece "did not meet *USA Today*'s fact checking standards.") The media also got ahold of a cartoon mocking Tony, and the *New York Times* was in possession of other internal White House documents with research "proving" that Tony had made mistakes.

Of course, the president and Mark Meadows tried to distance themselves from what Peter Navarro had written about Dr. Fauci. But they remained curiously silent when it came to the attacks on me. As some sort of twisted non-apology, I received word on July 20 that my daily report would now be used across all White House communications teams for the sake of consistency.

After I'd been dismissed from Mark Meadows's daily chief of staff meeting, Jared Kushner, seeing the numbers rising across the South, convened a new group called the Covid Huddle that would meet until the administration left office. From mid-March to mid-May, during the spring surge, the vice president had run the Covid Operations Group, which was active in coordinating efforts across the White House and the agencies, ensuring states had the supplies they needed. It became clear again in early July that they needed this operational element restored. The name "Covid Operations Group" better described the Covid Huddle's function: its members were the tactical worker bees who got things done at the implementation level. When the group was reinstated as "the Covid Huddle," Jared Kushner, Brad Smith, Chris Liddell, Brian Morgenstern, Hope Hicks, Adam Boheler, HHS's Paul Mango, Morgan Ortagus from the State Department, and I continued this operational-level work while also developing strategy until the Biden administration took over.

In late July 2020, the Covid Huddle team was in the middle of formulating a forty-five-day strategy with an emphasis on revised guidance on testing, therapeutics, and vaccines. Included in this was our "Embers strategy," a geographically focused campaign devised to warn the public at the very first increase in test positivity in their area. Through local media appearances, we would emphatically remind people to aggressively mitigate when a surge was about to hit their region.

The president resumed his press briefings on July 21, and I noted that he had accepted many of my edits to his prepared remarks and stuck closely to the script that day. In the weeks ahead, I'd sometimes get the opportunity to edit his talking points. I tried to calculate the number of days since he and I had spoken in person. It was sometime in April, and now August was just around the corner. But the administration gave with one hand while taking away with the other. On July 23, I received another "no" from Marc Short with regard to my making a public appearance.

I had Jared to thank for the president's sticking closer to the script. I'd been sending Kushner email after email and had spoken on the phone with him numerous times, letting him know the importance of clear, accurate communication. He was an important conduit to the president, and I needed him on my side. Jared was concerned about the surge in the South and, behind the scenes, was able to continue his work regarding testing, communications, therapeutics, and vaccines. We didn't have time to talk about what hadn't happened in the past. Any conversations we had were about moving forward. Along with the vice president, Jared remained my key go-to person throughout the summer, fall, and winter, until January 2021.

I wrote to Tony and the rest of the docs on the task force, once again urging them to throw their full weight behind the message and the new forty-five-day plan and guidance we needed the White House to agree to. It was a variation on the UPenn/CHOP model:

Masking works and you need to do it anytime you are in public indoor spaces. Practice good hygiene, especially in multigenerational households with residents over age sixty-five. States in the red and yellow zones should restrict indoor dining capacity, close early enough to reduce alcohol-related crowding (a lesson we had learned from South Carolina's governor Henry McMaster), and offer more outdoor dining options. When viral spread exists in your community, don't gather in indoor groups larger than ten. Schools in areas with fewer than fifty cases per one hundred thousand should open. Schools in areas with greater than fifty cases per one hundred thousand could open with testing and masking to prevent the already present community spread from invading the classrooms. (The latter represented the carrot and the stick. If you want your schools to open, a major priority, then you may have to give up your socializing and your drinks when cases are rising.

I learned much of this from West Virginia's governor Jim Justice, whose state officials used a color-coded system to communicate easily and directly to residents in all counties when an area was seeing increased cases and community spread due to increased social interactions.)

The doctors were on board, unified in their agreement. Time would tell how the administration would respond. With this administration, it was always difficult to predict how things would be received. Regardless, we were going to take this message directly to the states ourselves.

I CONTINUED TO TRAVEL to the states up through Appalachia, reaching Virginia by July 27, for a meeting with Governor Ralph Northam and others. I did a number of press events there, pointing out that the state should consider what had been shown to be effective in other places: testing, masking indoors, reducing size of indoor gatherings.

In a presser in support of the governor and his plans to increase mitigation in the hot zone counties, I said, "What always worries me is that there's people that have gone to the Virginia Beach area or the Portsmouth area or the Hampton area, and unknowingly bring that virus back. Even if it's a neighbor down the street, you don't know what their vacation and travel history has been, and we are seeing significant outbreaks occurring from birthday parties, graduation parties, family reunions. I know no one intends on bringing the virus into those situations, but we have to remember that every Virginian who is under thirty-five—most of you—will not have significant symptoms, and you might not know you're infected."

That night, I received a text from Marc Short. He was very angry. Members of his family frequented bars in Virginia Beach. He accused me of taking away the livelihoods of people his family knew and cared about. Whether it was the schools having to switch from in-person to virtual or bars restricting hours or closing altogether, Marc Short always seemed to take the ramifications of the pandemic mitigations personally. I gritted my teeth and took a deep breath, wondering why he was holding me accountable for doing damage when what we were doing was encouraging governors like his own to save lives.

Stunned, I was still fuming when the vice president's communications team followed up on my Virginia media spots by claiming I had misrepresented the state's test positivity rates. I had to wonder again if this was another case of the administration presenting an alternative reality to the vice president and the president. But I was using the same data streams we had worked to develop in partnership with state officials—data that Marc Short and Mark Meadows (essentially, everyone in the offices of the president and vice president) saw every day, data from sources anyone could have verified. Yet, I was "misrepresenting" it? How was that possible? Where and who was doing this counter-analysis? Not for the first time, the name "Scott Atlas" crossed my mind.

There was a precedent within the Trump White House that no White House advisors would testify before Congress. As a result, whenever Congress wanted an update, Bob, Tony, Steve Hahn, Jerome Adams, or Brett Giroir, all of whom worked for public health agencies but not for the White House itself (as I did), would be called to testify. I could not; that was the reality. No exception was going to be made for me. So, I did what I needed to do and wrote extensive reports and provided these men with the numbers and analyses they needed to communicate effectively to our representatives and to the American people.

I wanted to speak directly to Congress; I wanted to talk to the American people—but with a few exceptions, I was still largely limited to doing local media. Not having that broader national platform, that broader audience, hindered my getting the accurate national data and solutions out to the public. I wanted the American people to understand that the changes they were implementing in their lives were making a difference. This message would be critical for the coming fall, and I wanted them to feel empowered to confront the virus and change the course of community spread.

I also wanted people across the country to understand what we were seeing, what we had learned, and how our new insights and approaches were working in various states. I wanted the American people to see the work at the state level, the dedication of their state teams, the innovations that were being implemented across the country, like the amazing work of mayors to create more outdoor dining spaces using the road lanes and sidewalks; I saw this taken to new levels in Philadelphia. I wanted them to see the work of the private sector to save lives and how frontline hospital workers continued to innovate clinical care. People needed to know that with new treatments, fatality rates for those over seventy had dropped by nearly 70 percent in three months.

Behind the scenes, Marc Short continued to question every recom-

mendation I made. In those weekly governor's reports and elsewhere
ran the same theme: Masks work and should be mandated in public
during surges. When test positivity and cases rise, close indoor standing bars where people can't socially distance safely. Reduce indoor dining capacity. I was like the regular at the local diner who always ordered
"the usual." At one point (at many points, in fact), Short wrote to me
asking, "Where do you get your bar and restaurant data from?"

I wrote back, "I got it from the UPenn/CHOP expert whose model is
based on people being in that space for 60 to 90 minutes in close prox-
imity to others and using the county infection rate to determine the
number of diners likely to be infected based on the typical ventilation
standards for a space of various size establishments."

Later, he would respond that I was wrong. The model I was putting
forth was a worst-case scenario, he said. It was exaggerating the negative effects of indoor dining.

I asked for the source of his data. He didn't respond.

Throughout my time on the task force, I'd relied heavily on big data
sets, with extensive charts and graphs to support my position. At times,
to counter my projections and presentations, one member of the administration or another, including the president, would show me *their*
visuals. I'd wondered at the time who was helping them with the data
and the demonstration materials. I now suspected that it was Scott Atlas or other outside teams. Marc Short's refutation of population-based
mitigation (mitigation tailored to a specific demographic), which had
been tested and was working effectively on the ground in real life, fit
too neatly within the parameters of a Scott Atlas–based response.

Fortunately—though, ultimately, it had little effect on the perceptions of Marc Short and many others in the White House; they had already made up their minds—enough time had lapsed since we'd been to
Arizona and since Governor Ducey had enacted the UPenn/CHOP measures for us to see results. I contacted Dr. Christ, the director of the Arizona Department of Health Services. She and her tiny team had done a
fabulous job in the trenches, constantly adjusting their approach to the
evolving needs of their communities. At the very end of June, Governor
Ducey had issued his mandates. Case numbers in the state reached their
peak within the next two weeks and then rapidly declined. The simple
measures the state had applied—the triple platter of closing bars, reducing indoor dining capacity, and instituting a mask mandate—were

having the desired effect! Arizona's case numbers would continue to decrease throughout the summer.

I asked Dr. Christ to write up a report on this success story as part of the CDC's *Morbidity and Mortality Weekly Report* (MMWR), an epidemiological digest of health information and recommendations sent to the CDC by state public health departments. This way, public health officials across the country could see the positive impact mitigation had had in Arizona. Despite everything on her plate, Christ and her compatriots achieved this herculean task in partnership with the CDC, and the MMWR containing the Arizona figures was published in early August. I shared it with the task force, Marc Short, Jared Kushner, and Mark Meadows, pointing out that what had begun as a theoretical model had brought about actual, real-world success without "shutting everything down." Population-based impact data. Data that mattered. Not small, biased samples. Not theory and perception. But population-based impact. *Here's the proof. If we do this early, at the very first rise in test positivity in the fall, we can blunt the community spread, not just flatten the curve.*

What I often got in return was an unsatisfying, hybrid response that combined the worst elements of a patronizing *We hear you* and a dismissive *That's nice.* They could barely be bothered to pay attention to anything I put in front of them. The additional message was *We don't care. We've got other, more important things to focus on.* I had gone from banging a drum, announcing that we needed to be vigilant and proactive, to banging my head against the wall in frustration.

We had an evidence-based solution. To prevent the cascade of cases that ultimately results in hospitalizations and deaths, we had to blunt the first viral spread, the silent invasion. We could identify that first incursion through aggressive testing and stop it there. Later, I went public with a statement that we could have prevented 30 to 40 percent of the deaths due to the surges if we had acted together at the first evidence of spread.

Our government-led efforts didn't do this in the spring of 2020. Without tests we couldn't see the virus. But now we were learning and could clearly see the early warning signs and needed to heed them. Yet with the summer, fall, and winter surges of that year we failed to act at the earliest evidence of community spread, a pattern that continued throughout 2021. Collectively, we kept repeating the same mistakes, including not sending a clear warning to the American public until hospitals filled up, far too late. I hope this will change in 2022 and beyond when we reevaluate pandemic preparedness and what it takes at the local level with data and actions. Follow the data, react early, protect the vulnerable, test to keep schools open, and test to visit the vulnerable family members.

In preparation for the fall surge I saw coming, I was offering a set of measures to the states that were easy to enact to reduce its severity. In addition to distributing that information widely to all states through our reports, holding press conferences with the full weight and imprimatur of the federal government would have encouraged more states to stay the course throughout the summer. More critically, the measures provided a road map for effective mitigation for the upcoming fall and winter. Tragically, in a hurry to get to their destination, the Trump administration raced ahead like a driver stubbornly following erroneous GPS instructions, heedless to the passenger in the seat next to him waving a paper road map and shouting that he's going the wrong way.

But for all my frustration with some, the Covid Huddle and the vice president were helping. Day by day, they quietly labored behind the scenes, shepherding through the new communications strategy directly to states and ensuring that critical supplies got where they were needed, all without fanfare.

At the very end of July, I saw warnings that the administration's visible coronavirus response was about to run off the road and plummet into the abyss. And it wasn't just because of the numbers. It was because I finally got to meet Scott Atlas in person.

When Atlas came into my office on July 31, we were both cordial. Up to that point, we had no reason to be otherwise. I had, however, done additional research on him and had read each of the scientific papers he'd written. Atlas was a credible nuclear radiologist. He'd edited a textbook on magnetic resonance imaging of the brain. He was on the board of journals and professional associations. He'd written more widely on public health policy, penning four books on it since 2005. He was also a critic of the Affordable Care Act.

I read, and I scanned—but I couldn't find anything he'd written that indicated he had any expertise in the field of epidemiology or pandemic response. He had gained notoriety and credibility within the White House and with the president because, for years, he had backed the prevailing views of the Republican Party on health care reform. Now his views on the pandemic were aligned with those of this administration.

With his lack of experience in the field, I came to the conclusion that, as far as his views on the pandemic response were concerned, there wasn't much substance there. He was a health policy expert—but this alone wasn't enough for him to possess a substantial or nuanced understanding of how SARS-CoV-2 was uniquely different from other pathogens or how its characteristics should shape the response and guide public health actions.

For more than an hour, I recapped the current state of affairs across the country, providing Atlas with an in-person version of the morning summary I had distributed. People are sometimes misinformed. Sometimes they develop beliefs based on incomplete data or in the absence of data. But as was typical for me when faced with the misinformed, I believed that if I presented Atlas with a reasoned argument backed by clear graphs and data, he would agree, disagree, or attempt to refute what I'd said or at least ask for further elaboration.

"That's interesting," he said, when I'd finished speaking. His voice was flat, his expression neutral.

I went through all the data streams with him, then moved on to one of the most salient points about the data—not just its current status but what it had been able to predict. "Once we were able to expand testing," I said, "we can see that rising test positivity is the first incidence of community spread. There is a very predictable and repeatable pattern that reveals itself in a rise in cases and then hospitalizations fifteen to twenty-one days later."

"That's interesting." This time, he pursed his lips as if he were going to say more. I waited. Again, he didn't comment or question.

Knowing his belief that testing was not a very useful mitigation tool, I stressed again how important it was to expand testing, but to do so strategically.

"That's interesting." He brought his fist up to his mouth and softly coughed.

I knew it was too much to expect him to wear a mask, especially given his position on them, but I did wonder if he had been tested for the virus. Maybe his affectless "that's interesting"s were the product of his feeling ill (though they more likely indicated *dis*interest). Now that he had ready access to the White House and, presumably, the West Wing—which had a checkered history in requiring employees to regularly test or wear masks—my concerns were legitimate. More to the point, it wasn't so much about whether he himself had been tested, but whether he had been influencing the president's views on testing. It was one thing for Scott Atlas to personally not want to be tested, but quite another for him not to want to test others. If he was influencing the president's decision to advocate for less testing—the president's anti-testing statements up to now seemed to paint this picture—then he needed to be stopped. Not later. Now. But I wasn't going to show my hand. That would give him a reason to double down on his anti-testing message with the president.

In sharing the data with him, I tried not to assume he didn't fully understand the points I was raising. I went through all the science, the data, the solutions supported by the science, and the evidence base that showed those solutions were working. I asked him if he had any questions.

He shook his head.

Next, I highlighted for him the successful intervention in Arizona, which had been driven by the UPenn/CHOP model. "So, within two weeks of mask mandate, decreasing indoor dining, closing bars and gyms, the case numbers peaked and then immediately came down. So, now we have a path forward that other states can follow. They can keep retail open. I think that this is exciting. This is the approach everyone is looking for—preserving as many portions of the economy as possible and saving lives."

Atlas sat regarding me for a moment, nodding. His eyes revealed very little. "That's interesting," he finally said, again.

I was surprised by his lack of pushback on any of the points I was making. He made no mention of the charts and graphs and data. I knew we had definite points of disagreement; he'd shared his widely before in the media. Prominent among them was his belief that the rise in infection rates among younger people, the least vulnerable, was, in the long run, a good thing. I saw infection in anyone as a bad thing without a protective vaccine, as another link in the chain leading to the most vulnerable. I would have liked him to challenge me on those points of disagreement now, in person. I was willing to believe there were legitimate reasons for the views he held and had expressed so openly before. And I believed I could refute them.

When the meeting was scheduled, I didn't think I would be doing a ninety-minute solo presentation. I was expecting, and looking forward to, an exchange of ideas, dialogue. That's what I'd always encouraged within my small data team; that was the ethos Vice President Pence brought to the task force; and that's how Bob, Tony, Steve, and I conducted ourselves within the formal task force meetings and in our smaller, doctors' sessions. But with Atlas, even though I stopped frequently to ask if he had any thoughts, he expressed no interest, asked for no elaboration; nor did he rebut any of my points.

I went through all the evidence that supported each element of our approach and the data-driven recommendations I was including in the governor's reports. There was no dialogue, and he presented no evidence that supported his concept or perceptions. There was none then, and none was ever presented to the task force.

When he left my office, I felt I knew him no better than I had when I first came across his name. I knew what he believed, but he hadn't expressed any of those beliefs in the meeting; so, we'd been unable to talk through his ideas. I also had no clearer sense of what his current or future role would be. At this point, he was simply a White House visitor. Still, my antennae were up. It wasn't as if any of the positions I'd shared with him were secrets—they were right there, in black and white, in every one of my reports—but as I turned back to my desk, I had the nagging feeling that in showing him the data, I had taken a risk.

I was right. The backstabbing began immediately.

Once Atlas formally arrived in the White House, on August 10, an unpaid—I assumed he continued to receive his salary from Stanford or the Hoover Institution—senior advisor to the president, I had added his name to the distribution list for my morning report. I had issued this report every day since arriving in March, sending it via email to at least forty people—the task force doctors, cabinet secretaries, and anyone who had even a modicum of influence on decision making in the White House. Atlas began his new job by refuting the analysis in my reports. I wasn't aware of his open denunciations because they weren't truly open. He didn't include me in his emailed critiques, hitting Reply All, but deleting my email address before hitting Send. It was a deliberate but also highly disrespectful form of subterfuge, especially for someone from academia and think tanks, where dialogue and discussions are applauded. I was fine with someone having a dissenting opinion—it's something one sees in a collegial environment, especially one where the free and open exchange of varying opinions and perspectives is encouraged. But I wasn't fine with someone going behind my back in this way. Not only was Atlas dismissive of the data and analysis of the pandemic, and offering dangerous counterfactuals, but he was dismissing me and denying me the opportunity to rebut his claims.

It was part of a larger pattern. I'd been cut out of the loop. I was no longer attending Mark Meadows's chief of staff meeting, and those within the White House knew I'd been excluded. I was no longer visible in the national media, and the rest of America presumed I'd been dismissed. And now, by not including me in his responses to my reports and analyses, Scott Atlas had essentially declared me a nonentity.

I was angry. I was frustrated. What I wasn't was surprised. His was a move you pulled when you had little else to rely on but antagonism. It was what you did when you'd cast your lot with the bullies and the bluffers. Atlas had wheeled out a rhetorical straw man and skulked away.

Atlas misrepresented my position. He called any form of mitigation a "lockdown" to immediately tap into the nightmare that this word represented to many in the White House. That was what caused all the economic damage. The use of the word *lockdown* captured the perception he hoped to create. In his view, I would limit freedoms and damage the economy and people's livelihoods by recommending measures that would, in the end, save no more lives than if we did nothing and allowed the herd to be culled. Of course, Atlas was smart enough to claim that his position wasn't "herd immunity." He knew that in order to have some scientific credibility, he wouldn't be able to use those exact words. In essence, he was making a case for herd immunity, but wrapping that suspect product in different packaging. What he was advocating was an untested approach built on assumptions and perceptions—when we *had* a tested approach that was working!

When I found out about Atlas's cutting me out of his emails, I went ballistic. I told Mark Meadows in no uncertain terms that Atlas's egregious stunt was not something I would tolerate. I didn't want Atlas around, and I didn't understand why anyone should have to tolerate his blatant maneuvering and manipulation. Meadows's response? The equivalent of "That's interesting." Despite Meadows's somewhat ambiguous response, Atlas did start to include me on his distribution list thereafter.

Marc Short took stronger action. Heeding the long-standing advice to "keep your enemies closer," he invited Atlas to join the task force. Short believed it was better to have him where we could keep an eye and ear on him. I wasn't convinced this was a wise move. Allowing Scott Atlas behind the proverbial curtain made me uncomfortable. Throughout my time on the task force, I had been keenly aware of how easily, how frequently, what I saw as objective information was distorted through subjective interpretation. It would be better if Atlas remained a senior advisor to the president without any of the trappings of credibility membership on the task force would give him. It was too soon for me to make the demand of "either he goes, or I will." At that point, there wasn't enough evidence to demonstrate just how dangerous a presence he was.

There were a number of medical professionals on the task force—Bob, Tony, Steve, and me among them—and it was clear that the views of Scott Atlas, another medical professional, would not align with ours. We four sensed that Atlas was being brought in not so much by Marc Short, but by others in the administration, including the president, who perceived him as a golden child. I was never certain how faithfully or accurately the task force's discussions, operations, and management were being communicated to the Oval Office. I could only gauge the president's response based, primarily, on what was also visible to the American public. Certainly, informal and formal communications were transpiring within the halls of the White House and the Eisenhower Executive Office Building. With Atlas as a favored advisor, one with far greater access to the president than any one of us doctors, his views took on more heft than ours—heft that eventually proved greater than all of ours combined.

If I was guilty of one thing during my tenure at the White House, it is that, for too long, I clung to the notion that reasonable, intelligent people would eventually see the light. I trusted that data, that logic, that a substantially formulated, critically reasoned approach would win out over suspect science and entrenched and wishful thinking. I had been right about the depth and breadth of the surges and about the

solutions that could be deployed to limit community spread. At least
with Scott Atlas, I knew where he stood and what he stood for. It was
the latter that was most problematic.

In the end, I suppose, it didn't matter that Marc Short had invited
him onto the task force, or that Jared Kushner had asked him to be a
part of the Covid Huddle. Scott Atlas was already an insidious presence
in the White House, and I suspected he had been so for quite some time.
He had made himself comfortable, and now he seemed poised to make
me and many others very *un*comfortable.

ON AUGUST 1, HEADLINES around the country announced a hard reality: In July, the United States had more than 1.9 million infections. That
figure represented 42 percent of the total 4.5 million cases we'd seen
since the pandemic began. I'd been trying to make it clear that because
the summer rates were so terrible, much worse than the spring rates
had been, unless we intervened strenuously, the fall was going to be
even worse.

Early that month, I was given the chance, at the last minute, to speak
on *State of the Union*, a Sunday show on CNN, with anchor Dana Bash.
I don't know how this opportunity came about, but I wanted to use
it to tell every American to be on alert for the coming fall surge and
that the spread was equally in rural counties and large metropolitan
areas. Unfortunately, on August 2, prior to my appearance, I came under attack—this time from an unlikely source, Speaker of the House
Nancy Pelosi. Publicly, she accused President Trump of spreading disinformation and accused me, as his appointee, of doing the same. It was
guilt by association. When asked specifically about my effectiveness in
my role, Pelosi said she didn't "have confidence there, no."

I had too much respect for Nancy Pelosi, and all the work she had
done over the years to help support HIV/AIDS initiatives, to engage
in any kind of back-and-forth. I wondered if she was basing her comments on that July 18 article in the *New York Times*, which had mis-
characterized me as overly optimistic about the pandemic. The media
company Politico had reported that, in a meeting with Mark Meadows
and Steven Mnuchin, Pelosi had said, "Deborah Birx is the worst. Wow,
what horrible hands you're in."

When I appeared on *State of the Union* on Sunday, August 2, of
course, the subject came up. I began by saying that I had nothing but

respect for Ms. Pelosi. I explained that I had not had any direct conversations with her or her staff since coming to the White House in March, but that I had known her and her staff for decades, from my HIV/AIDS and global health work. In responding to Pelosi's comments, I went on record as saying of my views on the pandemic that they were "not Pollyannaish, or nonscientific, or non-data-driven. I will stake my forty-year career on those fundamental principles of using data to implement better programs and save lives." I went on to add that the United States had entered a "new phase" in the pandemic and that Covid-19 disease was "extraordinarily widespread." How our views could have been construed as Pollyannaish, how our state-by-state visits to hot spots—carrying a consistent message about increasing vigilance and enacting more stringent mitigation methods—could have been seen as overly optimistic is difficult for me to understand to this day.

In speaking with Dana Bash, I seized on this rare opportunity to address the American people directly, to set the record straight: "I want to be very clear: What we are seeing today is different from March and April. [The virus] is extraordinarily widespread. It's into the rural as [well as] urban areas. To everybody who lives in a rural area, you are not immune or protected from this virus."

It was a realistic assessment, and stating it on national television was a risk worth taking. I knew what the consequences were likely to be, but I didn't care. This might have been my last shot at national media. It might even have been my last public statement as response coordinator, but people needed to hear my warning.

The next morning, August 3, I was reviewing data and monitoring supplies—as bad as things were, they were certain to get worse and the last place we wanted to be was in the position we were in the spring—when my cell phone rang.

"Hold for the president."

"Mr. President, this is Debbi," I said.

A moment passed before I heard that familiar, pained tone: "What do you think you're doing? Don't you ever do that again. 'Extraordinarily widespread'! Do you know how those words scare Americans?"

I took a deep breath, shut my eyes for a moment, and then opened them. I flashed back to all the times I'd asked someone in the vice president's communications office why I wasn't doing any national shows, just local spots. It seemed clear they didn't want me to warn Americans about what was ahead. I tried to summarize for the president why I was saying what I had and what the data was showing. I was assuming he hadn't been briefed on the full extent of the summer surge and the coming fall threat. I never was able to get through much of the data on that call.

I didn't want to frighten people; I wanted to *inform* them, empower them, provide them with a rationale for changing their behaviors. I wanted them to elect to do the right things for themselves, their families, friends, and colleagues. I wasn't running for office. I was running as fast as I could, on so many fronts—all in the service of the American people, not my own interests. That's what a life in public service is all about.

The president diverted to a different tangent. "Who did this? Who in Comms let you do this?"

For everything the public and press thought they knew about the Trump White House, there were people there who worked with me every day to get the right information out to the American people. I had allies. Week after week, those allies in the White House set up local news hits for me; Comms Office people went into the field with me to facilitate press appearances. Every one of them believed in what I was saying and helped me get the message out. They needed protection.

The president wanted retribution. The optics of firing me would be too bad, so he'd do the next best thing. But even if I had known who scheduled me for Dana Bash's show—I'd received an email from only a support staff person—I wouldn't have given him their names.

My legitimate claims of ignorance infuriated him further. After minutes of asking, "Who was it?" and "Who did this?" he resumed railing against my statement and me. "You should be paying more time and attention listening to what we know instead of spreading reckless lies. These people live for fake news, and all you've done is feed them." He paused briefly.

"Mr. President, the virus is widespread—"

"That's it! Do you understand me? Never again! The virus is under control."

This conversation had only hardened my resolve not to back down. I wasn't going to change my approach, my concern, or my focus on the states. The president was wrong. We *didn't* have the virus under control. Though we were making progress, it was *despite* Scott Atlas and some in Trump's inner circle, not because of them.

I stood my ground: "Mr. President. We are in a very, very, very serious situation right now. It is only going to get worse. If we don't take action now, then the rest of the summer, the fall, and the winter are going to be far worse than what we saw in the spring. The American people need to know this. We have an effective solution that will save lives and not do undue damage to the economy."

I felt as if I were speaking into a dead line. I was.

That same day, the president went on the offensive, using the kind of divisive rhetoric and tactics that had marked his entire presidency. I hadn't given him the names he wanted, so he started name-calling, tweeting, "So crazy Nancy Pelosi said horrible things about Dr. Deborah Birx, going after her because she was too positive on the very good job we are doing on combatting the China Virus, including Vaccines and Therapeutics. In order to counter Nancy, Deborah took the bait & hit us. Pathetic!"

He didn't want to listen to me. Instead, he was listening to a radiologist who believed the pandemic was being "overhyped" and "overmitigated." Scott Atlas was a false prophet with a ready group of followers, all of them singing lyrics to the same tune that many in the White House had been humming for the past six months.

They had their solution. Unless we made certain the Atlas approach was never executed, it would cost thousands of American lives. He might be a talking head, but we had to make sure his words were never translated into actions.

CHAPTER 13

Scott Atlas Shrugs

On August 8, a few days after President Trump tore into me on that phone call, I walked into the Oval Office for a major vaccine briefing with him. Since April 21, I'd been to the Oval only a handful of times, either to serve as a prop for a governors' meeting or for a brief huddle prior to a press conference on vaccines. I saw this as my first opportunity to engage in discussion of real, substantive policy with the president in the Oval in 109 days. Unfortunately, it only confirmed a fear that had been steadily building in me: not only were Scott Atlas and his dangerous ideas now controlling Trump's message, and could impact policy on the pandemic, but in Atlas, Trump had found a scientific messenger willing to tout his views, regardless of how many lives they cost. This meeting made it disturbingly apparent that when it came to the pandemic response, the Oval Office had become an echo chamber.

The White House was on its back foot at this time, trying to recover from a revelatory interview. On August 3, President Trump had appeared on HBO's *Axios* series. Journalist Jonathan Swan conducted an interview in which the president floundered. He repeated his claim that there were those who said (including him) you can test too much. When pressed for who these people were, Trump claimed there were "manuals" and "books" where this point was substantiated. When pressed again, he couldn't name any of them. The president also again said that we had higher cases of the coronavirus "because of the testing." This didn't get much attention given that this was the same interview in which he claimed that he had done more for the Black community than any president, with the possible exception of Abraham Lincoln. As President Trump made his way through a series of questions, he held up "supporting" graphics (which I had not made) to defend his position. He handed one over to Jonathan Swan, believing that it supported his contention that the United States had the lowest rates in the world in several categories.

Though his base may have seen the interview as *the left-leaning media attacking the president*, with it, President Trump once again laid bare the issues that existed in the White House. Who had made the charts and graphs? Who was briefing the president and what were they telling him? It wasn't me. It wasn't the task force. The figures he was citing weren't from the data sources we'd vetted as accurate.

In the same interview, the president said of people dying of the virus, "And it is what it is." Of the lag time between tests and results: "There's nothing you can do about that." Worse, he continued to make the claim that the virus was now under control.

I'd made clear dozens of times that the virus wasn't under control, that we could do better, that we had solutions to the testing issues he'd said were not solvable. He wasn't listening to the task force. He wasn't listening to Bob Redfield or Tony Fauci. He was listening to someone, or several someones, people were manipulating data and giving it to the president without review or discussion. I was reminded of the disinfectant moment at that press briefing.

The next day, Stephen Miller told me that President Trump needed to know exactly where things were improving. I wasn't sure why he needed to know that now, but I suspected he was looking to support the claims he'd made in the *Axios* interview. Miller said that he and the other speechwriters believed the "vigilance" message would be more effective if they leavened it with a dash of hope. I'd chafed against this kind of snapshot thinking before. Yes, there were counties where case and fatality rates were falling—parts of Texas and Florida came to mind—but cases were far worse in Oklahoma, Missouri, and Tennessee. The virus was relentlessly on the move through the heartland and would reach the Northern Plains just as cooler weather settled in. Highlighting improvement in one place would give a falsely positive picture of the scope of the pandemic. It wouldn't account for rising and falling patterns in regions, for how it improved in one spot and deteriorated in another. Instead of supplying the rosy picture Miller wanted—presumably, for campaign rally speeches—I sent him the now-familiar U.S. map showing the status of every state. The map was overwhelmingly red and yellow, with few pockets of green. That was the state of things as the summer surge ebbed slightly and the fall surge approached. Knowing what a trusted advisor Miller was, I also sent him the Arizona data, to show how effective proper mitigation could be.

Whenever anyone—Mark Meadows, Marc Short, and Stephen Miller being the most prominent—came to me with a request for good news, I complied with specific, honest, detailed figures. I always pointed out that such snapshots were distortions. *Yes, these specific instances of improvement are occurring, but each state was in a different place in the cascade. Even if smaller areas are showing improvement, others are showing an early increase in positivity. Here's the projection that shows that those areas are going to worsen.* I knew the winter was coming and we needed to prepare, and part of that preparation was to make sure both the president and the public understood the risks.

The Trump reelection campaign was rolling along. That week, they claimed to have knocked on one million doors. How many of those canvassing workers and respondents socially distanced or wore masks is unknown, but if they did not, it showed a continued disregard for best public health practices. President Trump spoke to one hundred supporters in Cleveland, attended two fund-raisers in the Hamptons, and launched two campaign tours (one from Pennsylvania, the other from Florida), with buses traveling across the country until the general election, going to rally after rally.

Though most large White House events were moved outdoors, into the Rose Garden—a step in the right direction—indoors, people gathered unmasked, a failure that resulted in one superspreader event after another. There was always someone in their midst who was silently infected, sometimes within hours of a negative test, with the virus exploding from the cell factories in their noses and spreading to those around them invisibly in the air. No masks. No physical distancing. No mitigation. No worries. In the months to come, the West Wing became a hot zone over and over. My assistant, Tyler Ann, and I masked consistently. To avoid risk of exposure, other task force doctors called into work rather than coming into the White House. I didn't have that luxury. Mitigation works. Tyler Ann and I are uninfected proof of that truth. In the end, in this White House, it became easier to count who *was not* infected than who *was*.

THE OVAL OFFICE MEETING of August 8 was ostensibly to discuss Operation Warp Speed and the progress on vaccine development, and because of this, I walked in hopeful. Dr. Moncef Slaoui delivered the news that the vaccine trials were on track. They would be conducted according to established rigorous scientific protocols. I'd heard a number of times from Steve Hahn that the president had repeatedly questioned whether that process could be sped up. The president peppered Dr. Slaoui and others with questions. Dr. Slaoui and Steve endured. They pushed hard, but always with safety as the primary aim, resisting the president's urging for "sooner" and "faster" and his incessant "Are you sure?" regarding the projected approval time line. Unstated, but very clear, was the president's desire to have the vaccines available before the election. Moncef, Steve, and Alex Azar stayed the course. They'd cut no corners, but would very quickly deliver safe and effective vaccines—not in time for the election, but in record time nonetheless.

I was motivated by a different set of numbers, numbers that reflected what was happening in the immediate—test positivity cases, hospitalizations, and deaths. As a result, as the meeting was drawing to a close, I decided that, though I wasn't on the agenda, and despite months of the president's completely ignoring me, I would reintroduce him to the reality that, as vital as a vaccine was, we had other aspects of the pandemic response still to deal with.

"We are currently at just under fifty thousand cases a day," I said. "We are trending downward from the July peak, but I am very, very concerned about the fall. That could be an incredibly deadly period. The way to prevent that is to mitigate—expand testing, masking, and reduce indoor unmasked gatherings. We've got to fight the virus that's here now and coming to new communities in the fall."

I noted that the president didn't do what he had so frequently done in the past when I gave him unpalatable news. He used to turn his head to the side and put his hands up, as though shielding himself from me. This time, his reaction was more muted. He sat back and folded his arms, nodded, and then shook his head, as if he'd heard all this before and wished I would come up with something new. When I pressed on, he waved me off and turned toward Scott Atlas: "Is that how you see it?" He didn't so much *ask* the question as offer up a prepared introduction to what he knew was coming next.

"I couldn't disagree more strongly," Scott Atlas said, leaning forward, eyeing me without turning toward me—both of us were seated facing the president. "No matter what we do, the outcome is going to be the same. *The same!* You don't need to mitigate. You don't need to test. All we need to do, and we've already done this, is to protect the most vulnerable." He shrugged and held his hands out, palms up, as if it were all that simple and straightforward.

Ignoring the satisfied look on the president's face, I challenged this grossly inaccurate and inadequate statement: "We can't protect the vulnerable if the virus is already in the community. If we—"

"That's just wrong!" Atlas said, cutting me off.

For the next five to ten minutes, Scott Atlas and I argued. Back and forth we went: I'd make my case based on the data, and he'd cut me off, his rebuttals increasing in frequency and intensity, each one beginning with either "You're wrong!" or "That's just wrong!" It wasn't a case of differing data; it wasn't about diverging points of view. It was black and white: I was wrong, and he was right.

I refuted each of his dangerous assertions:

That schools could open everywhere without any precautions (neither masking nor testing), regardless of the status of the spread in the community.

That children did not transmit the virus.

That children didn't get ill.

That there was no risk to anyone young.

That long Covid-19 was being overplayed.

That heart-damage findings were incidental.

That comorbidities did not play a critical role in communities, especially among teachers.

That merely employing some physical distance overcame the virus's ill effects.

That masks were overrated and not needed.

That the Coronavirus Task Force had gotten the country into this situation by promoting testing.

That testing falsely increased case counts in the United States in comparison with other countries.

That targeted testing and isolation constituted a lockdown, plain and simple, and weren't needed.

Point by point, as I dismantled his case, Atlas grew angrier, cut me off more aggressively. To further emphasize his points, he'd push himself back in his chair and splay his arms in the air while nearly shouting his new mantra, *That's just wrong!* Using alternately defensive and offensive language to bolster his position, he became animated and boisterous, his thinking neither scientific nor logical.

I should have expected as much. He'd started his campaign against the facts (and, by extension, me) by editing or refuting my daily reports. For weeks since that inauspicious start, he had been discrediting me and the science through emails and in person during Covid Huddles and in task force meetings. He never offered clear evidence to support his position. He bullied instead of debated using facts.

Now, as I sat there, alternately eyeing him and then the president, Atlas rose up, leaned forward, and said, apparently out of desperation, "You're going against the president's policies!"

I flinched, surprised and concerned. To this point, the president hadn't articulated any policies beyond we will never shutdown this country again. All along, whether it was through the CDC, the White House, the task force, or other agencies, the federal government had consistently been developing *guidance*, not specific policy, guidance that was approved by the Staff Secretary and OMB. We'd been providing the states with options for how to implement effective public health recommendations. But those recommendations weren't policies. They weren't hard-and-fast *rules* or *regulations*—a synonym for *policy*. Nothing we had developed and distributed had used that word unless it went through the Staff Secretary. Nowhere had we stated that these were actionable, enforceable measures or that failure to abide by them would result in consequences. There would have been a firestorm of protest if we had. Most governors, and the American public, would have seen this as the federal government overstepping its authority. Now here was Atlas attempting to speak for the president and set policy from in front of the Resolute Desk and not through data-driven discussions within the task force.

Since he'd told me "never again" with regard to shutdowns, the president had given neither me nor any other person on the task force a policy objective. As for the rest—masks, testing, and other elements of the response—the president hadn't explicitly elucidated (his tweets notwithstanding) any measures or approaches of the kind Scott Atlas was now insinuating existed. If Atlas was overstating the approval the president had for his, Atlas's, opinions, that was one thing; if Atlas was speaking on *behalf* of the president, that was another, far more dangerous thing for the country.

If Scott Atlas was now the voice of President Trump's policy, it was worse than his saying I was wrong. The test of whether this was Scott Atlas overstepping his bounds or whether it indicated that he had the president's approval to speak on his behalf like this was right there in the room with us. In that moment, the president neither endorsed nor refuted Atlas's positions. He was silent. He didn't say "Scott, that's right. Those are my policies," and he didn't say they weren't. This left the field of play completely open, where it was before I entered the room.

I countered Atlas's points with the same message I'd delivered consistently. The summer figures were harbingers of more dangerous and deadly times ahead. I continued this drumbeat: the time to more proactively and aggressively mitigate was now. I'd been saying the same to Kellyanne Conway and Hope Hicks, to Alyssa Farah, Morgan Ortagus, and Kayleigh McEnany, and to every senior advisor in the West Wing. I delivered the same message now in front of the president: Scott Atlas's words weren't just irresponsible. They weren't backed by accompanying evidence to support them.

For every unsupported "That's just wrong," I offered a substantiated "Here is the proof." What he offered by way of a rejoinder was a reiteration of the herd immunity theory without directly calling it herd immunity.

I pointed out that now that we had an even more accurate and efficient data collection system, with 93 percent of hospitals reporting, there was even greater weight behind my assertions. We knew the majority of hospitalizations and deaths were from the community, not just nursing homes.

I used Arizona's turnaround to demonstrate the efficacy of the UPenn/CHOP model.

Atlas shook his head and entire torso while shouting, *"That's just wrong!"*

Meanwhile, President Trump sat expressionless, his arms folded across his chest. The posture said: *I don't want to hear this. I don't care what you're saying.*

As the confrontation drew to a close, a bemused but detached expression animated the president's face. It wasn't what we were saying that had grabbed his attention as much as the volume at which it was being said. Atlas was assertive, aggressive, perhaps hoping that his passion would carry the day. If he could speak loudly enough and long enough, perhaps he'd demonstrate that he was right—at least to someone who didn't understand the data and enough to earn the status of senior Covid-19 advisor. Bombast versus substance makes for interesting spectacle when there is no legitimate reality at stake. But in a room representing the highest level of American government, with a viral pandemic raging and countless lives on the line, the he-who-speaks-loudest and he-who-speaks-what-I-want-to-hear nonetheless seemed to carry the day.

It was tragic. It betrayed people's trust. It cost lives.

I knew that if I rose to match Scott Atlas's fever pitch, I'd be characterized as shrill or hysterical, the doom-and-gloom lady out of control. The adjectives used to describe women are often laced with hidden meanings that trivialize or undermine. The proverbial rock and hard place—a man's passionate stance versus a woman's letting her emotions get the better of her.

As with so many meetings and crisis points, this one ended with no resolution, no dramatic shifts. I knew that making a final push, imploring the president not to listen to Atlas or others, was going to fall on deaf ears. The president had enjoyed the show, and just as he'd done in the very first meeting I had with him, he tuned to another station.

"Very good. We're done."

I had a sense that the first two words were directed only at Scott Atlas and the last two were for me.

As I got up and walked out, the feeling was mutual. I was done. I was done with this type of spectacle in front of the president. I was done with Scott Atlas. I was done with beating my head against the most stubborn and unrelenting wall I'd ever come up against in my

professional career and my life. I was angry with Scott Atlas for his handling of science. I was furious with whoever had handed this man the
keys not just to the White House but to the vehicle that could steer the
response. Whoever that person was, they were accountable for whom
they brought on board, how they managed them, and what they allowed them to do.

Though the president didn't utter the catchphrase, "You're fired," I
very much had the sense that I had been.

I'd long suspected that the president was being influenced by someone like Scott Atlas. That was obvious. But to sit in a meeting and see
that person, to experience viscerally how the shift in influence was affecting the president, left me feeling lightheaded, disembodied. As the
meeting ended, I felt no sense of relief, only overwhelming sadness and
frustration. I knew that while continuing to fight the virus, I would still
be fighting Scott Atlas, too.

FOLLOWING THE OVAL OFFICE meeting, Scott Atlas did two things to
establish his authority and take over the direction of the federal response to the Covid-19 pandemic:

First, when rebutting anyone's point of view, he used the president's
name at every opportunity—not data or analysis; just the president's
name, to give himself credibility. I'd seen this move many times before:
someone with a tenuous position in an organization, or an untenable
position or approach, will attempt to firm up both by citing the name of
the person with the highest level of authority. Basically, Scott Atlas was
saying, *I know what the president wants. I have insider knowledge that
you lack. I have access to him you don't. Consequently, if you disagree with
me, you are disagreeing with the president.* This ploy had never worked
with me.

Second, Atlas kept repeating his "policy" comment. When, in task
force meetings, we discussed making some modification to guidance,
he'd frequently say, "Those are not the president's policies." But the
president hadn't defined any new policy positions, in the Oval meeting
or elsewhere. His sole policy position that we never shutdown the country again remained in place. Nothing had changed. Atlas may have been
using the president's name, but until the president or vice president
specifically told me about a new policy, what Atlas espoused would
be irrelevant. When it came to this president, and in military terms, I

wasn't violating a direct order. Perhaps I was going against the president's wishes, but wishes are not the same as orders or a stated policy.

To be clear, I was never told by anyone in the White House to stop issuing my daily report, the critical four-pager I created for the Covid Huddles, which clearly showed day by day where we were in the pandemic and that charted actions and local communications. I was never barred from taking my messages to the states and counties by phone or in person—indeed, the White House and Vice President Pence facilitated my trips. I was never prevented from doing local press to get the message out locally. Though the president might have been doing the opposite of what I was recommending and saying the opposite of what I was saying, his White House supported my actions, and I was never stopped.

Despite Atlas's saying that policies existed, they didn't. But because I didn't demand that such policies be formally spelled out and confirmed, I could continue to operate on the razor's edge of Scott Atlas's opinion and the president's failure to make clear what his policy was and issue direct orders to implement it. As a result, I could still go to state and local governments with clear science and data-driven recommendations to mitigate community spread without defying the commander in chief's orders. If he didn't specifically lay out what was permissible, then I was free to do and say what I knew was necessary.

EARLY ON THE MORNING of August 11, I wrote to Bob, Tony, and Seema, laying out in detail the seven points of contention I had with Atlas's position that I'd gleaned from my "debate" with him and from his previous statements and op-ed pieces. In the aftermath of my confrontation with Atlas in front of the president, I'd committed myself to working on two parallel but related tracks—continuing to coordinate all aspects of the pandemic response, and managing Atlas in the task force and ensuring he wasn't viewed within the White House as having any real influence or impact.

I pointed out that Atlas had created an alternate reality, a parallel universe where the virus could spread unmitigated among younger people without infecting any of the most vulnerable, the thirty-five million Americans over seventy, a group that had suffered and died at the greatest rate since the pandemic began. We had seen this in the spring and in the summer: there was no way to lock thirty-five million

Americans away in an impenetrable bubble. What Atlas had offered wasn't a plan to mitigate the spread; it was a plan to speed it up. What looked like something logical on paper had the fundamental flaw of being impossible to implement. There was no way to separate the vulnerable from the community in which they lived. We needed to expand testing, not decrease testing.

Normally, theories, models, and approaches in science are tested with the use of well-designed experiments or at a population level. In this case, Atlas's and my positions weren't mere laboratory experiments. They were being conducted in the real world, in the lived experience of the American people. From the beginning of the outbreak through the end of 2020, 180,000 of the 250,000 recorded Covid-19 deaths were vulnerable Americans aged seventy and over. And that was with inconsistent mitigation efforts in place. We can't calculate how many would have died if Atlas and other proponents of herd immunity had had their way, but we know that figure would have been so much higher.

Even with our mitigation plan in place, the virus was moving among younger people and reaching the vulnerable. Seema Verma worked around the clock with nursing homes and nursing home associations to protect their residents. Despite those best efforts, time after time, the virus penetrated our defenses, and by June 2021, at least 185,000 Covid-19 deaths were found to have occurred in nursing homes. It simply wasn't possible to protect the elderly without taking measures to stop community spread.

Even once the vaccines were available, we knew that not every vulnerable person would have a fully effective immune response to them. As we age, or for those with underlying immune deficiencies, or those who undergo treatments that suppress our immune system, our immune responses to vaccines are blunted. If SARS-CoV-2 was actively circulating in communities, vaccination alone would never be enough to protect everyone. Therefore testing and masking would need to be continued in the presence of vulnerable family members (the elderly and those immunosuppressed) whose immunity status was uncertain despite vaccination. Full stop.

Thanks to the UPenn/CHOP model, we knew which measures were effective: proactive testing, masking, and decreasing gatherings. What had begun as a theory, and then became a computer-generated model, had now been put to a real-world test. As Arizona and other states that followed the model had shown, we could mitigate against community spread; the numbers would decrease.

Scott Atlas didn't offer a plan or proof—but he did offer something: the distraction the White House needed and wanted in the moment. The administration wanted to campaign, to raise money, to hold indoor gatherings—and they wanted someone to tell them it was all going to be fine.

On August 13, Seema Verma shared with me an email she had received on March 21 from Scott Atlas. My warning emails to the task force had prompted her to recall this past message from him. In it, Atlas claimed that "the total lockdown is a *massive overreaction* [his emphasis] and super harmful to our entire society, destroying the economy, inciting irrational fear, and even diverting medical care away from sick people." At the time he wrote this, we were only five days into the initial Slow the Spread campaign. By no reasonable person's estimation were we in "total lockdown."

What astounded me about this email was what it revealed about Atlas's thinking, which had not evolved from the end of March to the beginning of August. Over those four months, we had learned a lot about the virus and had spent months working to refine all aspects of the response. Atlas believed that we needed to protect nursing home residents, 1.5 million people. But he didn't seem to understand, or didn't care, that there were far more elderly Americans who didn't live in long-term-care facilities but who had comorbidities that put them at great risk—35 million-plus of them. He didn't understand, or didn't care, that those who died in long-term-care facilities were infected by members of the community (nurses, orderlies, janitors, visiting family), not by other residents.

Atlas still believed that, despite the mitigation efforts we had put in place, this virus had a natural course to run. He attributed the success of the forty-five days of Slow the Spread not to all the measures governors had enacted, but to the virus doing what it was going to do. Even if we did nothing, his thinking went, we'd still see the same numbers of cases and deaths. I knew this wasn't true, as we could see direct, temporally related consequences of mitigation. We could see it in differences in the deaths-per-one-hundred-thousand between states with mask mandates and those without.

As much as I blamed Scott Atlas for his deeply flawed approach to pandemics and epidemiology, ultimately, the responsibility for this turn of events lay on the president's desk. Someone in his administration had brought Atlas in. They had allowed him to make reckless, scientifically unsound recommendations. I had serious doubts about whether the president, beyond watching him on Fox News, had ever actually had Atlas vetted. And I felt very strongly that our contentious Oval Office meeting represented the first time Atlas had ever been seriously challenged. I don't know whether it was a case of the administration lacking a nuanced understanding of what Atlas was advocating, or of their lacking the will to challenge someone who clearly had the ear of the president. It really didn't matter; the end result was the same: Atlas was inside the building. It was my job to make sure that was where his views and his influence were contained.

Of course, Atlas couldn't just leave things at that. He wanted and needed to substantiate his claims further, to bring more people within the administration and around the country over to his side. So, on August 13, shortly before I departed for another visit to the states, while the participants of a Covid Huddle were sitting around the conference table in the Roosevelt Room, he announced that he was convening a panel of infectious disease experts for a roundtable discussion. (He would later call it a Medical Experts Roundtable.) He would invite Dr. Martin Kulldorff, PhD, from Harvard; Dr. Joseph Ladapo from UCLA; Dr. Jay Bhattacharya from Stanford; and Dr. Cody Meissner from Tufts—essentially, the same group that had been championing herd immunity from the start. He wanted the president and vice president to attend, and he wanted the event to be open to the press. I believe Atlas hoped that the intellectual heft of the participants' academic positions would bend public opinion his way and give a White House public stamp of approval for his theories and his approach.

As Atlas outlined his plans for the event, I sat there living the nightmare of the moment while worrying about the one to come. The people he wanted to include in his roundtable were academics who lacked on-the-ground, commonsense experience in public health and infectious disease epidemiology. As he ran through his plans for the roundtable, Kellyanne Conway, who was seated near me, slid a sheet of paper over to me on which she'd written, "GOOD LORD!" I nodded, envisioning dark lords of misrule entertaining the president while the vice president sat there helpless. I hoped Trump would see the dog-and-pony show the event was sure to be as merely another example of unreality TV being captured inside the White House.

Once the Huddle broke up, I immediately contacted Kellyanne with this message: Scott Atlas was a dangerous man, and his roundtable would be even more dangerous. Allowing him access to the press with the president and vice president in attendance would give him and herd immunity the executive branch's seal of approval. We couldn't let that happen.

I did everything I could to block the planned roundtable event from happening. I went to the vice president's staff, the White House communications team, and anyone who would listen—letting them all know what a mistake it would be to validate Atlas's pseudoscience in public. The event shouldn't take place at all, but especially not at the White House, and not in front of the press. We couldn't be seen as endorsing this theory in public. I wrote to the other doctors with my concerns.

We were heading into the deadliest time with this virus. We needed to be talking about testing and mitigation, not about a recklessly damaging approach that the vast majority of scientists had rejected. I had to continue to try to prevent the Atlas event from taking place. Any public perception outside or within the White House that his views were impacting the direction of the response would further erode the public's already tenuous faith in the gains we were making.

Later, I could see how undermining Atlas's authority and the integrity of his plan took me places I'd hoped never to have to go. It wasn't enough to keep him and his cohorts out of the public eye, and it wasn't enough to keep his plan from receiving presidential approval. I also couldn't let him believe he could make headway with anyone on the task force, beginning with the doctors. If Atlas found a foothold there, who knew how high he'd climb or whom he'd get to accompany him.

Vice President Pence wanted all the doctors to sit down with Atlas to try to come to consensus. After a phone call with Tony in which I reiterated my position that Scott Atlas was a clear and present danger, Tony wrote to me and the other two docs. He agreed with me, shared my concerns. Like the vice president, he hoped that we could all sit down and have a "non-confrontative discussion to go over in detail the basis of his claims." But Tony hoped we could wait a week to do this, until he could resume speaking after a procedure that prevented him from doing so.

I was beyond working it out or reaching consensus. That wasn't going to work. My position and Atlas's were too far apart. Based on my experiences with him, I knew he wasn't going to listen to reason. He wouldn't engage with evidence. He was wedded to his beliefs and theories, and that was that.

I informed the docs that I was done with trying to reason with him. I just couldn't do it. I couldn't invest one more ounce of energy refuting his claims. I had told myself that I would have to fight both the virus and Atlas. His opinions needed to be contained within the White House. My priority was on mitigating the virus.

Fortunately, Tony reconsidered his position on a face-to-face meeting. He wrote to me: "I know what I'm going to do. I am going to keep saying what we have been saying all along, which contradicts each of his 7 points listed below. If the press asks me whether what I say differs from his, I will merely say that I respectfully disagree with him."

I responded to them all: "Perfect—will do the same."

I don't say this lightly: affording Atlas any professional respect after his actions in the White House would have been a mistake. Some doctors on the task force believed that because of Atlas's position as a medical doctor, he should be afforded professional respect. I had started from that place. I had sat down with him and gone through all the evidence. But he had chosen to say nothing and to delete me and refute my data behind my back. He didn't engage in dialogue; instead, he bullied. In the end, I believe all the doctors came to the same place, but it was a rough several weeks getting everyone there.

I understood from past experience that any pandemic response will be tied into politics. For decades in other countries I'd worked to get leaders to act against their own political self-interest in order to get needed public health measures enacted. But I hadn't understood that with this administration, the order of political magnitude was going to be so much higher. I also didn't anticipate that some physicians would use their MD to bring credibility to their deadly theories.

Over the days and months since that Oval Office meeting, with Scott Atlas's "You're wrong!" still ringing in my ears, I had to face this harsh reality: The president and some of his senior advisors wanted someone who could match their cynicism and their perceptions of the pandemic, which were driven by personal opinion and perceptions rather than data. Now more than ever, I knew I couldn't quit. I could still make progress—by coordinating the task force, by working with the vice president, by working through Jared's Huddle to ensure we were meeting the needs of the state and local governments, and by speaking with state and local officials. I would continue learning what was and wasn't working and get that information across the country through our reports and trips.

Scott Atlas could continue to believe he was speaking for the president. I couldn't change his mind, but I could make sure his influence didn't extend beyond the Oval Office.

THREE DAYS BEFORE I met with the president and confronted Scott Atlas, I had received a draft document titled, "Considerations for Covid Testing." As I suspected it would, it didn't talk about the clear importance of testing—the need to test both symptomatic and younger asymptomatic individuals to stop community spread. Proactive testing would (as I hoped and wanted) drive up the number of cases, alerting asymptomatic people to their infected status and preventing superspreader events from occurring. As was being shown by those routinely testing, doing so rapidly drove down spread, significantly blunting the outbreak. In a task force meeting, Atlas expressed agreement with the president on our needing new testing guidance posted to the CDC's website and said he would be the person to make this happen. I knew what he would advocate for. In his March email to Seema, he'd come right out and said it: "*Fear of exposure, or people who are without symptoms who are outside priority groups, do not need urgent testing and should not seek it* [his emphasis]."

As I read the draft of his proposed testing guidelines, I was dismayed to see they reflected a lack of strategic implementation that would be effective at preventing community spread. I added a bullet point about the need for people in high-transmission zones who were unmasked for the duration of any indoor gathering of more than ten people to consider getting tested. Such gatherings constituted superspreader events. Nurses I'd spoken with in hospital after hospital reported that the patients on ventilators had been infected at family gatherings—birthday parties, weddings, funerals. But Atlas's proposed guidance supported prioritizing testing only the symptomatic, not the asymptomatic, too. We were still getting hung up on this issue.

I received pushback on my bullet point addition from Brett Giroir. I respected Brett's position as testing czar, and up to that point, he'd agreed with me on the role of asymptomatic spread and the need to detect it early through widespread testing. Though the language I had used in my bullet point was intentionally mild, Brett objected to it. I had known that any strong signaling of the need to test asymptomatically wouldn't get past the staff at the CDC, given their continued resistance to widespread testing independent of symptoms when test positivity was rising in the community. But now it hadn't passed the sniff test with Brett Giroir, who'd been supportive of my position up until now. I suspected that Brett had either come under the influence of Scott Atlas, whose strong opposition to testing was so essential to his championing of herd immunity, or been instructed by HHS to find a middle ground. That would mean supporting Atlas while still trying to maintain as much testing as possible.

In response to my questioning about his shifting view, Brett wrote to me on August 8, saying that "people know that you can't fix this by public health guidance with testing everyone on the street." The language was right out of the Book of Atlas. This about-face by the country's testing czar couldn't have come at a worse time: testing needed to expand in preparation for the fall to blunt community spread and to ensure kids stayed in schools.

We discussed the new CDC testing guidelines in a task force meeting that same week. Bob Redfield and his people were the ones in charge of what was released as CDC-endorsed actions and guidance. And in task force meetings, it was Bob who usually brought up CDC-related matters. This time, though, it was Brett Giroir who went over the proposed guidance that had clearly been influenced by Scott Atlas. Essentially, it was another case of "Let's test less." In part, the revised guidelines said that—with the exception of people who had developed symptoms, were elderly, or had a medical condition that made them vulnerable— people who knew they'd been exposed to a test-positive infected person didn't need to be tested. These revised guidelines were an eerie echo of what Atlas had spent months calling for publicly.

If Atlas got the CDC to make a definitive statement like this, my job would be made much harder. The CDC and I continued to engage in back-and-forth on the degree to which silent spread among young adults contributed to community spread. This new guidance was exactly the opposite of what I'd been advocating, which was more testing of young asymptomatic individuals. Sports teams were using regular testing to find the infected early to prevent spread. If approved, the Atlas-influenced new guidance would not only reduce the number of tests being conducted, but would also make contact tracing impossible and prevent those without symptoms but engaged socially from being tested, effectively cutting two legs off the platform on which a pandemic response must be built.

I voiced my opposition. We had to stop discouraging people who thought they'd been exposed in indoor gatherings from getting tested. I believed that by proactively testing people who were in high-risk environments like bars and indoor gatherings, we could prevent spread to parents and grandparents. Tony wasn't present, but the other docs chimed in to support me.

So, on August 26, when Brett presented the task force with the final draft, I spoke up again, saying clearly, "I don't approve this. I can't." Scott Atlas stepped in and again went after me, saying that I was wrong about testing, wrong about asymptomatic spread. He concluded with the statement he'd made in our heated Oval Office exchange: that his views represented the *president's* position and policies.

Angry but under control, I said again, "I can't approve this."

The vice president stepped in to say, "I really want us to have consensus on this."

"I can't approve this," I said. "I can't keep the CDC from issuing this. I don't have oversight of them. I do have some oversight, though, over what gets put out from the White House. If the CDC wants to do this, that's one thing. But it can't go out from the White House as something the task force approved. It can't go out on the White House website as something the task force endorsed." There were limits to what I could do, but I had to strongly speak my mind no matter what.

Bob Redfield was the head of the CDC. If he disagreed with the document's points about asymptomatic spread—as I believed he did—then it was up to him to take on this fight. I looked to Bob. He wore the ex-

pression of a condemned man resigned to his fate, and it became clear
to me that this document, which undercut so much of my belief about
strategic testing, was being forced down his throat.

In the end, the revised testing guidance, which aligned so closely
with what Atlas had called for publicly, went out, despite its flaws. I
was flabbergasted when, in announcing the new guidelines through
the national media, Giroir said the White House Coronavirus Task Force
had approved them. I immediately contacted him, asking him how in
the world he could have done this. He offered the flimsiest of excuses,
saying that he thought I had said it could go out. But the point I'd made
hadn't been at all subtle; I reminded him of what I'd said: the CDC could
issue what it wanted to—I couldn't control that—but what came out of
the task force, and therefore the White House, was under my control.
And I'd make it clear that this document hadn't received White House
approval, either directly or tacitly. The task force unanimously had not
approved the CDC's recommended testing guidelines. Period.

Lowering testing levels was exactly what we'd been fighting against
for months. Initially testing rates had been low thanks to the CDC's
early errors in scaling up testing. We'd been gaining ground and the
new point of care/rapid antigen tests were helping us do that, but testing remained a long-standing problem. Then the president had made
the illogical (and public) connection between the role testing played
in the increase in cases. Now, with Scott Atlas having influenced the
guidance, he was putting into practice the push for lower testing President Trump had spent months calling for publicly. In the history of the
president's rhetorical assaults on testing, this was the first time there
had actually been a guidance shift from the CDC. I didn't know if he
had directed Atlas to do this or if Atlas was taking it upon himself to
influence one of the president's "policies." Either way, Atlas was now
the de facto testing czar.

That all this was happening right as the political campaigning was
shifting into high gear was hard to ignore. At a crucial time in the pandemic response and the election cycle, testing and campaigning were
now intertwined. The administration knew the two key issues most
prominent in Americans' minds: the economy and the pandemic. The
two were inextricable, of course, but in the world of politics, simplification works best when it distorts most. Even if there was a false sense
that Covid-19 cases were decreasing, the Trump people could tip the

scales in their favor. Scott Atlas and his approach to testing was the now-visible thumb on that scale. If we didn't test as much, the number of cases being reported would decrease, and it would appear as though this administration's handling of the pandemic was producing results. The reduced case numbers could be used to prove a false claim: that the United States was triumphing over the virus. We weren't.

I was surprised that Brett Giroir had been part of this change in guidance, and I wondered what kind of pressure he'd been under. From comments Brett made to me later on, I gathered he thought the posting of the new testing guidance was inevitable. He believed he was trying to salvage as much of the public health elements of the document as possible. But his reversal on testing was a disturbing sign of how quickly testing had been lost to politics. In the past, the administration had boasted that our ever-increasing testing capacity was a signal that we were on top of things—when we weren't. Now this boast had been turned on its head: With this new Atlas-driven guidance, Trump had put in place a formalized message through both the White House and the CDC that would reduce testing at a crucial moment for both the election and the pandemic. This had to be countered.

The case numbers pointed to another dire period. "I am at a loss [as to] what we should do," I wrote to Bob, Seema, and Tony on August 13. "We need to stop these infections, or there will be 300K [dead] by Dec[ember]." I was wrong. By November 30, there were nearly 260,000 deaths. By December 15, we would surpass my 300,000 estimate. By Christmas, we were on the verge of 350,000 deaths.

I knew what to do to stem the incoming tide of infections and deaths. I felt handcuffed, but not helpless. I'd figure out a way to manage the crisis and keep the numbers from rising even higher. The UPenn/CHOP model had worked well in Arizona and elsewhere. That was the message I would carry to the governors and public health officials.

IRUM AND I SET out on another round of state visits—this time to the heartland. Iowa, Ohio, Kentucky, Tennessee, and Virginia were on the schedule. In each place, cases were either rising or had briefly plateaued at a high level. Also, many states did not yet have a plan in place for returning college students. We had to push nationwide for weekly mandatory testing of those students, regardless of whether they lived on campus or off.

The testing issue followed me to the states. The new CDC guidance equivocated on who should be tested after being exposed: "You do not necessarily need a test unless you are a vulnerable individual or your health care provider or state or local public health officials recommend you take one." In other words, the CDC wasn't directly saying "Get tested." The media cut through the flimsy language and called it what it was: most outlets reported that the CDC had come out against asymptomatic testing. After all, it was right there on the CDC website—what the Trump administration had wanted for months: fewer tests and fewer cases. But these weren't the only words in the new guidance that could potentially cause confusion. A few lines down, the guidance had retained the phrase "you can be infected and spread the virus but feel well and have no symptoms." Obviously, the CDC's contradictory remarks created confusion.

Meanwhile, the scientific community was in an uproar. The CDC's most recent word on testing was roundly blasted as wrong. Reducing testing was not a solution; it was a problem. But the guidance worked as the Trump administration had wanted it to work: testing across the United States plummeted in most states by 5 to 10 percent and in some by 20 percent, and at the worst possible time. Of course, for some in the administration, it was the best possible time: they were out campaigning. And unless Americans were really paying attention to everything in the news, they wouldn't have fully understood *why* the case numbers were dropping. Those plummeting numbers falsely signaled a trend that the administration, busy out on the campaign trail, heralded as a triumph.

I'd been in the field enough to know that there were governors who would embrace this new guidance, so I had to do everything I could to refute it, to make clear—whether it was internally, within the White House, or externally, in the governor's reports, or while out on the road—that they should ignore it. I would continue to challenge the administration's stated and unstated faulty positions. The fall was coming. We needed to continue to warn people about what the future held. Testing needed to be expanding, access to testing needed to be expanded.

Given everything that had been going on with Scott Atlas, and the chaotic nature of the White House, I thought it best to put those sentiments in writing: "If one is threatened: we are all threatened. I trust we have each other's backs as[,] united[,] we can weather this to ensure we continue to make progress to save American lives." We needed this unified approach. I'd been out there in the states, proselytizing for the UPenn/CHOP model; advocating for mask wearing, expanding testing, and ending in-person, maskless get-togethers. We had to counteract the administration's anti-mask, anti-testing messaging. President Trump, his family, his aides, Vice President Pence, and many others were campaigning in mask-free venues; fortunately, many were held outside. But campaigning with packed rallies drove the perception that all was fine and that if you truly wanted to "Make America Great Again," you wouldn't wear a mask, either. I tried to convince the vice president innumerable times to wear a mask. He said he would. I was very concerned that he would get infected. I spoke with Marc Short, reminding him that it was his responsibility to protect the vice president from getting infected, including at the Atlas event.

On August 25, I wrote a sharply worded email to Marc Short, informing him that I would not be a part of Atlas's roundtable, especially not with a fringe group who championed herd immunity and who believed that the United States could be like Sweden, which had followed the same path of minimal mitigation—but with population far, far healthier than our own. I offered to be out of DC, on another visit to the states, to give the vice president cover for my absence. And again ask him to stop the roundtable. I again cited my three hundred thousand death figure, adding that half a million people would likely die before a vaccine was in full use.

In the end, with the help of many others, I was able to stop the big circus Scott Atlas had planned for his open roundtable with major press coverage. He and the other doctors came to the White House to meet in private with the president, some advisors, and with Secretary Azar, but the press wasn't present. In requesting that no press be allowed, I had been able to diminish some of the public perception that Scott Atlas and his theory had taken control of the response. As much as I would have liked for him to have no voice in the public discussions, and to engage in constructive dialogue with the docs on the task force, it was more important that he stay a talking head.

I had to do more to achieve that aim. I would be out of the White House for a period, on state and university visits. Somehow, I had to cut Atlas off from any degree of influence he might try to exert on anyone short of the president, whether during my absence or while I was still there. Individually, I contacted Marc Short, Mark Meadows, and Jared Kushner. I communicated clearly to everyone: "I won't be in any meet-ings any longer if Scott Atlas is present at them. If that means a meeting in the Oval Office[,] it doesn't matter. If it's at the task force, it doesn't matter. If it's at the Covid Huddle, I don't want him there."

Not only did they indicate that they understood, but they agreed. Scott Atlas no longer attended task force meetings. I would be out on the road for much of the coming months, but Tyler Ann was present for all the Covid Huddles. I would phone in from the road to participate in these meetings, and she'd confirm for me that Atlas wasn't present.

After sending my email to Marc Short, I was finally able to meet with Vice President Pence alone. For days, I'd been asking for ten minutes of his time, but he had been out campaigning. I knew that what I had to say would put him in a difficult position. If we met privately, I'd be able to give him plausible deniability.

I immediately reiterated the points I'd been making to the docs and to Marc Short, Mark Meadows, and Jared: Scott Atlas was a danger. The position he and the president were advocating wasn't working and would worsen the fall surge that was rapidly approaching. When I told the vice president that Scott Atlas was effectively persona non grata in the task force and in the Covid Huddles, he nodded in agreement.

Emboldened, I took things a step farther: "You know that I've been out in the field, meeting with governors, doing press, and meeting with the community. I want you to know that what I have been saying and what I will continue to say in every state contradicts what the president is saying publicly. I want to be clear: I can't support what Scott Atlas is saying, and I will need to say exactly the opposite of what the president is saying."

The vice president's eyes narrowed for a moment. He considered his response for another few seconds. Finally, he looked me in the eye, his voice steady, and said, "You need to do what you need to do."

I read this as permission for me to do what was right. It meant he understood and tacitly supported my position over Atlas's. The vice president was already getting constant pressure internally for what I was saying and doing, but in that moment, I saw he believed that taking the fight to the governors was the right thing to do. From that point, I would continue to infer that Vice President Pence had my back. The administration could always say I'd gone rogue, that I was part of the "deep state," bent on undermining the president and disrupting the election by distracting voters from the "real" issues. I wanted to believe—and this is my optimistic nature rising to the fore—that the vice president knew I was doing what was right and needed. I couldn't give up hope that our arguments had been persuasive, that data and facts and science had prevailed at least to some degree.

When I returned to my office, I received a draft of an op-ed Scott Atlas had written. It looked like I had more contradicting to do. Now that he was part of the White House staff, any of Atlas's communications would be subject to review by the Office of the Staff Secretary. Jared Kushner and a long list of others in the Executive Office of the President were on the distribution list. I cc'd them all in my response, which, once again, was a cogent refutation of Atlas's arguments. I paid special attention to his false claims about lockdowns, reminding everyone that our mitigation efforts—wearing a mask indoors, testing to find silent spread, decreasing social interactions—could not be characterized as part of a "lockdown." Keeping retail spaces and schools open were the opposite of a lockdown. I spelled out again the simple measures that had been enacted to produce very positive results. I reminded them that we'd been balancing the economy with public health. I also referenced the role silent spread was playing.

Hoping to prevent the damage that another Atlas op-ed might do, I wrote a separate email to Jared: Why were they letting Atlas be a distraction with this op-ed during the Republican National Convention? In another email to him, I again refuted the notion of herd immunity, pointing out why it was impossible to say it was responsible for some of the improvements we were seeing. At the end, I wrote that if Dr. Atlas now spoke for the president and his policies, Jared should let me know, and I would "take that under advisement."

Jared didn't respond. As was the case when his father-in-law was similarly noncommittal about the extent of Atlas's influence, I presumed that Jared's silence on the matter meant that Atlas didn't speak for the president. Even if Jared had said that Atlas did, I would have stayed on and fought harder against him. Fundamentally, I was constituted to always do the right thing, even if others didn't see it that way. I couldn't change my stripes; nor did I see a need to.

Being on high alert has always been a part of my nature and my profession. With Covid-19, I was always eager for anyone in higher-risk age groups or those with underlying medical conditions to be as vigilant as possible. In the White House, I spoke with all forms of staff, from housekeeping to senior advisors to the vice president. I was preaching the gospel of vigilance. My concern wasn't bound by any strictures, political or otherwise. Consequently, my concern extended to the members of the Biden campaign and the candidate himself.

On August 29, I saw an opportunity to put that worry into action. I received a text from my friend Chip Lyons. He was contacting me on behalf of David Kessler, the former FDA commissioner, who later served as the dean of the Yale School of Medicine and then in the same capacity at the University of California, San Francisco Medical School. Chip stated up front that Kessler was then informally serving as an advisor to former vice president Biden and was in communication with him regularly. Kessler wanted to be in touch with me, but he understood that I might be reluctant. I let Chip know that I was willing to speak with Kessler.

At about this time, I'd been reading the Biden pandemic plan, which had been made widely available. In short, I was impressed by it, particularly because its emphasis on the importance of testing aligned with my own beliefs on that crucial mitigation tool. I wasn't sure what role, if any, Kessler had played in developing the Biden plan, but now knowing that he was an insider on the Biden team, I was eager to learn why he wanted to be in touch with me. I suspected that it had something to do with vaccine development. I'd learned that the Biden camp was in contact with the various manufacturers directly, so, in my mind, I believed they knew as much as I did. Perhaps they hoped I'd have some additional insight.

I wanted them to know what data we were collecting—so if they had additional data requests or additional thoughts we could incorporate them. I was always open to learning from others and doing things better. Kessler came to my house shortly after the twenty-ninth and I provided him with samples of the daily reports—both my daily pandemic analysis and the four pager we created for the Covid Huddle and governor's reports we'd been producing. I went through the data streams we had created to be able to get eyes on the pandemic across the country—from test positivity, to cases, to hospitalizations, to deaths, down to the most granular level possible—counties and metro areas. Pictures and graphics told the story, some of which are still updated in the HHS

Community Profiles we quietly posted in the late fall. He thanked me and said that he'd review them and get back to me with any questions or requests for additional data he thought would be useful to have integrated into the data stream we had created. I was glad to know that he and others on the Biden team were thinking ahead, and that my usual desire to have a Plan B and Plan C was being met. November's election was still months away, but preparing for either result would ensure continuity of care so essential to proactively managing the next surge.

I felt comfortable agreeing to speak with Kessler for several reasons. The first was that I had always acted apolitically throughout my career. If I could discuss pandemic matters with another medical professional in any small way, I was open to the idea, regardless of party affiliation. Second, because I didn't know anything more about vaccines than the Biden people, I wouldn't risk leaking any information. Speculation about the Trump administration's hope to spring a preelection vaccine surprise was rampant in the media, but I had no secrets to share. And my third reason for agreeing to speak with Kessler was my concern for the health and safety of the former vice president. Biden wasn't actively doing personal campaign appearances, but many of his staffers were out on the campaign trail (along with, critically, the uniformed Secret Service protection), who could be exposed and asymptomatically infected. I wanted to make sure they knew how critical silent spread was, how important testing and quality masks were across the board. I also wanted them to know what data we were collecting so if they had additional data or additional thoughts we could incorporate them. I was always open to learning from others and doing things better.

CHAPTER 14

Where Community Persists

Lessons from Our Tribal Nations

The next day aboard Air Force Two, Irum and I flew into Duluth, Minnesota. We were back out on the road, into the Northern Plains states and the Upper Midwest. We went where weather patterns and Covid-19 outbreaks told us we needed to be to get ahead of the pending surge. Fall and winter would soon be arriving in the Northern Tier states. We wanted to ensure that governors, mayors, health personnel, and tribal nations were prepared for the surge we saw coming; we had to share with them what we had learned from our previous trips, to the Sun Belt and the Midwest: Reopening safely was possible. Staying open was possible with active testing, masking, and avoiding unmasked indoor gatherings. Scott Atlas's anti-testing message put them in peril. Proactive testing would lead us away from another surge.

During this trip, I reflected on our prior time away from the White House and how this upcoming round of state visits, and, specifically, of tribal nations, had been impacted by the previous visits we'd made throughout July and August. Along with the fifty states in the union, the United States is also made up of 574 federally recognized tribal nations, and just as the states were not a monolith in their response, this was even truer for the tribes. The pandemic affected people and communities differently; the damage Covid-19 was doing to people's physical, emotional, and economic health was disproportionate, especially so for the tribal nations. In many places, the nations were ground zero for severe Covid-19 disease. After nursing home residents, the tribal nations were experiencing the highest case fatality rates of any race or ethnic group in the United States.

On this visit to Minnesota, we'd scheduled a meeting with members of the Fond du Lac Band of Lake Superior Chippewa. This wouldn't be our first time on tribal lands. Our tribal visits had become a critical part of all our state visits, starting with our first trip back in June, when we met with President Martin Harvier of the Salt River Pima-Maricopa Indian Community in Arizona. To varying degrees, the 574 tribal nations have historically been underserved by public health agencies in this country. Those early meetings on the ground gave us a crucial window into some of the unique complications these nations were facing in their response to the pandemic.

Several factors were contributing to a high rate of significant disease and population-based fatalities among the tribes: multigenerational households, comorbidities, and economic issues being the most prominent. Poverty was deeply entrenched in many of these communities, contributing to the pandemic's toll in myriad ways. At the start of May, the Navajo reservation across Arizona, New Mexico, and Utah had the third-highest per capita rate of Covid-19 infections behind New Jersey and New York, but the highest fatality rate. This trend continued, in surge after surge. We'd taken steps to intervene, as had the tribal nations themselves—with varying degrees of success. But the work was ongoing, and we needed to better understand the nations' needs and efforts, so we could be more supportive and they could be more effective at battling the pandemic.

On the economic side, the news for all tribal nations was equally concerning. Many of them had improved their economic prospects over the last thirty to forty years by operating more businesses, including casinos. But because of the pandemic, many of these critical enterprises had to be shut down, cutting off employment opportunities. Also, much-needed revenues to fund tribe-operated services (law enforcement, public safety, and social services) had been greatly reduced. Many tribal-led businesses also shut down as the tribes imposed strict mitigation measures to prevent as much community spread as possible. The two-trillion-dollar CARES Act provided eight billion dollars to offset the economic impact of the virus to the tribal nations, but some of that was through the Indian Health Service, at HHS. But money sent directly to the nations, with its many strings attached, was often late and difficult to access. As with so many aspects of this crisis, we knew that what we were learning through looking and listening—what would be written up in a general summary—might not accurately reflect what was truly happening in a given locale.

President Harvier served a population of some 11,000 people who represented three pre-Columbian sovereign Indian tribes, the Pima, the Yaqui, and the Maricopa. Unlike other groups who had been displaced and moved, these peoples had a long history in the area. On our visit back in June, we had met with the president and several council members who helped administer services to their community, half of whom lived on reservation lands.

These officials saw their primary mission as implementing the CDC's Covid-19 guidance, but one of the challenges they faced was the presence of significant underlying comorbidities among their members, including diabetes. Many tribal nations had the highest incidence of that potentially deadly condition of anywhere in the world. Patients needing dialysis were placed in triple jeopardy, with this significant comorbidity increasing their vulnerability to serious Covid-19 disease. Patients had to test negative for Covid-19 before being allowed into the dialysis clinic, but there often weren't enough tests available for them to get negative results in time—a catch-22 that boggled our minds.

The Pima and Maricopa required very specific tests—the Abbott ID NOW or the desktop nucleic acid test, with results in fifteen minutes or under—to match their needs. But much of these critical supplies were going to large metropolitan-area hospitals with many testing options when they should have been going to the tribal nations and other disadvantaged communities across America. This problem was a subject of my running dialogue with the testing czar Brett Giroir. I needed to know why nearly 25 percent of all the test cartridges were going to California's large, well-served hospitals and another 15 percent were going to CVS and Walgreens. In the case of those two retail outlets, none was located on the reservations. Test type had to be aligned with the unique needs of each community or population. Covid-19 underscored the public health crisis these tribal groups had long endured, and was another reminder that the United States needed large-scale reform to its public health system to be able to mitigate multiple future health crises.

The Pima-Maricopa health services people were doing a good job of tracking the infected, but they had encountered issues on multiple fronts: With insufficient laboratory personnel, notification of positive tests was often delayed, sometimes by more than two weeks, allowing the virus to spread. Similarly, there weren't enough qualified, experienced public health officials integrated with the state's health systems, which meant those systems had been operating from a deficit from the outset of the pandemic. This had been true for years before the pandemic. As a country, we had chosen to run two systems, with the public health system as a parallel stream to overall clinical care, creating inefficiencies and duplication of personnel, data, and other information. These systems needed to be integrated, with one common data set and analysis and with the data transparent and available to the public. We had learned from the hospitals that they were willing to provide real-time data to a national system, something that could be done without violating HIPAA rules. There were synergies that could improve the nation's health. We just needed the will to enact them.

The antiquated Indian Health Service was another problem. A federal program created in 1955 as part of the Department of Health and Human Services, the IHS had been designed specifically to address the health needs of native peoples. Unfortunately, it wasn't always as effective as it could have been. At times, it even appeared paternalistic, believing it knew better than the tribal health officials on the ground.

With the Gila River tribe and the Yaqui in Arizona, we had heard variations on the same theme. Tribal officials were being as proactive as possible in the face of severely limited testing kits and supplies, but when tribe members needed treatment, they had few options. Many IHS programs and local hospitals were often felt to be discriminatory and stigmatizing. Native peoples we spoke with talked about overhearing doctors and nurses talking about them outside their curtained hospital beds or at nurses' stations, making assumptions about alcohol and drug use.

As a nation, as those involved in public health, we needed to apply what we had learned in Arizona. There were clear benefits to having culturally sensitive health clinics and hospitals funded by the IHS but staffed and run by tribe members themselves. While I'm certain there are many well-intentioned Anglo members of the IHS, Irum and I heard enough offhand remarks and entrenched perceptions from those providing these services to know there needed to be a change. Expectation and execution match up far better when programs are run by the same people whose needs are being served.

Even where available supplies allowed tribal nations to better track and test, safe spaces for the infected to isolate were severely limited in number. Not only was housing scarce, but the housing that did exist was often substandard. Poorly ventilated structures containing multiple generations were highly problematic in the face of a virus that lingered in the air—just as secondhand smoke had done in bars and restaurants before indoor smoking in such places was outlawed.

Those we spoke with also pointed out an inequity inherent in the tribal system. Their lands were held in trust by the federal government. This meant they didn't actually own the land, and therefore lost out on the other important benefits of landownership. For this reason, building or updating a home became a futile endeavor. After all, you can't accrue equity on a property you don't own.

Irum and I had both spent a lifetime addressing this cascade of overlapping structural issues in Africa and elsewhere. And in the two weeks just before our flight to Minnesota, we'd seen how the same constellation of issues played out in Oklahoma. The Muscogee (aka Creek) Nation, the fourth-largest tribe in the United States, resided in this state after having been forced west from parts of various southeastern states in the eighteen hundreds. In time, they would develop their own college, build a long-term-care nursing facility on their reservation, provide social services to their tribe, and in partnership with Oklahoma State University, establish the first tribe-affiliated medical school to instruct and train doctors who would go on to serve their community. This was a real success story. The Muscogee (Creek) were finding local solutions to the issues they faced. In the near future, with a deep understanding of and respect for the cultural needs of their patients, they'd be able to fully care for their own.

But in mid-August 2020, the Muscogee (Creek) were experiencing a test positivity rate of 21 percent. This would rise at the end of the month to nearly 67 percent—nearly 95 percent higher than the state-level rate—illustrating not just the depth and breadth of community spread but also the fundamental lack of tests. Irum and I saw a photograph of a local, indoor funeral where unmasked people stood shoulder to shoulder, mourning the loss of one life while potentially instigating the loss of others. That image ran counter to what the Muscogee (Creek) had intended with their Protect Our People program, which called for masking, physical distancing, and outdoor gatherings only. Whether this funeral was typical or atypical didn't really matter. It was a possible source of spread, and tribal leadership was aware of it. The leadership knew their issues and were working every day to find their own solutions. Still, our message of heightened vigilance needed to reinforce the risks of this kind of lapse.

We met with the leadership of several Oklahoma tribes in a casino ballroom in Norman, just outside Oklahoma City, sitting physically distanced around an enormous oblong table that filled the ballroom. The Oklahoma tribal leaders expressed some of their concerns and let us know they had just canceled one of their most important cultural events: a triennial dance ceremony. For months, in fact, they had been canceling similar such gatherings. Stomp dancing played a critical role in the cultural and religious life of the larger community. Their not holding this dance ceremony was on a par with the rest of the United States stopping all in-person religious services *and* canceling a major national holiday like Thanksgiving. The dance ceremony was a time for reflection and sharing, when the tribes' sense of community spirit was cemented, a time of renewal. The tribal leaders in Oklahoma knew what needed to be done to limit the spread of Covid-19, and despite the pain and disappointment of missing out on this ceremony, they'd made the hard choice.

Amid all the discussions, Irum and I were struck by a comment one participant made: "Men see challenges, and women want to find solutions." This lingered for us both. One of the solutions tribal leaders were hoping to find was to a problem facing the Iowa Tribe of Oklahoma. The Small Business Administration was enforcing arcane regulations (predating the pandemic) that prevented funds from being disbursed to businesses when employees worked from home. It was yet another area where Covid-19 guidance and other administrative regulations got tied up in knots. This was information we would take back to the White House. The issues we had confronted during our state visits spanned public health, the economy, and agency regulations—basically, all issues that come together at the White House and can be resolved at that level or at a specific agency. Each time I wrote to Intergovernmental Affairs or to the heads of other agencies, these issues were often addressed. The small changes enacted at that level made a big difference.

The Oklahoma tribes were also deeply concerned about where to relocate infected and exposed members of their community. With the casinos the tribes operated mostly shut down, they were using the casino hotels as housing for isolated quarantined individuals. They were finding a way, but overcoming the inequities inherent in the U.S. government's long-standing treatment of native peoples and people of color wasn't going to be accomplished easily. My eyes had been opened to their reality, and I found memories of my experiences in Africa rekindled. It was sad that it had taken this new global pandemic for me to become aware of conditions that existed right here in the United States.

In Africa, we had had to work at the local community and county level to ensure that resources and services were aligned with need. Sometimes, we found that several counties were dramatically underfunded by host governments, almost exclusively in locations where opposition to the government was strongest. As soon as we discovered it, we aggressively addressed this inequity in funding and services, using U.S. funding and resources to realign the balance, overcoming whatever political beliefs or ties were preventing equitable treatment. This is what is possible when you use the most granular data in real time.

What was happening among the tribes in Oklahoma and elsewhere wasn't directly tied to the members' political agency or to their voices on things like elections, but to the long, tragic history of mistreatment native peoples have experienced in this country. While it would be difficult for the United States to move out of this entrenched path, and though the structural barriers history had put in place seemed insurmountable in the moment, I was certain that with constant attention, the right policies and resources, shared partnership, and mutual accountability, the full force of the IHS and its affiliated subagencies at the federal level and similar groups at the state level could surmount them.

As much as most Americans were wishing for a return to normal at this moment in the pandemic, I couldn't wish the same for any of the tribal nations we visited. We should neither accept nor return to the business-as-usual of 2019. Instead, we should expect a better new normal, one that addresses structural barriers to care and uses data to chart improvements and identify new gaps. In a better new normal, we would work toward decreasing comorbidities and ensuring that all Americans, no matter where they lived, could thrive. Yes, this sounds overly optimistic, impossible perhaps, but as I have learned from battling the HIV/AIDS pandemic, often what many believe impossible is actually possible if we listen, hear, understand, and support local communities.

IRUM AND I WERE reflecting on all of these lessons from our prior visits to tribal lands. Riding in a truck in Duluth, I watched as many small lakes flashed past the passenger window. I squinted into the glare of a late-August midday sun. We were about to meet with the leaders of the Fond du Lac Band of Lake Superior Chippewa, our first stop after deplaning. Jostled by the ruts in the dirt road our tribal driver was traversing, I saw a native family in a canoe in one of the lakes, in a patch of tall reeds near the shore.

"What are the people in the canoe doing?" I asked our driver.

"Harvesting rice."

"By hand?"

"Yes. We've been doing it this way for centuries. It's part of our tradition. It's a way to maintain the long roots we have to our land and to our past."

I had had no idea this kind of farming was still taking place in America.

"Perfect," I said, and smiled.

Irum and I looked at each other. I could tell she was having the same thought: This was the first time she and I felt really at home working for the domestic Covid-19 response. For two decades, our work had been focused in Africa and Asia. We'd worked in similar communities, places where the continuity with the past was readily apparent, where people moved forward in unity, sharing a strong cultural identity.

I'd done my homework, but as the truck slowed and pulled into a clearing, I was both eager and anxious. I wanted to see how those two elements, history and culture, were being lived out on the reservations, but I was wary of how a representative of this White House, a member of the federal government, a body that hadn't always dealt fairly with the native peoples, would be received.

In the clearing, several pickup trucks sat parked side by side. Masked and socially distanced tribal leaders stood waiting for us. After a brief

(carefully distanced) round of introductions, I made a few opening remarks. I thanked them for agreeing to meet with us, and then gave a
brief overview of what I'd seen and read. I told them I hoped to learn
much more from them. Initially, my words were met with silence and
dispassionate regard. A few people expressed their appreciation for our
being there and said they were willing to listen. That's when their apparent dispassion turned to passionate engagement.

The conversation began with a summary of what they'd been doing. Prior to the pandemic, the Lake Superior Fond du Lac Chippewa
had created their own culturally appropriate and sensitive health services on the reservation, free from stigma or discrimination. They had
elder care, community centers, early childhood education support,
broad social services, and addiction support services. This tribe had
been solution-oriented before Covid-19, and was still so in the midst of
Covid-19. Over time, the Fond du Lac Chippewa and others we'd visited
had recognized the need for greater agency over their social support
services, public health, community health, and other aspects of their
lives. They didn't need another distanced layer of federal bureaucracy
to manage their affairs. They needed sustained funding for the programs they had but also for programs they had established that were responsive to their unique population. Their accomplishments up to then
could serve as a road map for other tribes to establish their own internal services. I was very glad to hear they had mandated mask wearing
to enhance their physical distancing efforts. But they understood that
these two things alone wouldn't be enough.

They had heard the message on silent spread, and because so many
of them lived in multigenerational households, they were keen to be
as vigilant as possible with testing. As we had heard across the South-
west and Midwest, the tribal nations in the Upper Midwest did not
have enough tests to combat community spread. They needed testing
today, to mitigate their ongoing daily risk, and they needed results immediately, so they could isolate infected family members as quickly as
possible. Tribal members also wanted to test more frequently and more
widely—many of their younger members worked and interacted with
people from the towns surrounding their tribal lands.

Despite the need, in many cases the IHS limited the tribes' access to,
or rationed supplies of, the rapid nucleic acid tests and rapid antigen
tests Abbott had produced. As a result, the lag time between sample col-

lection and test results was far too long. Without knowing who among them was asymptomatic or even presymptomatic, they weren't able to isolate the early infected cases, and the virus had spread unrelenting through entire households, getting to those with underlying conditions that made them susceptible to severe Covid-19 disease. Complicating matters were the limited spaces available in which to quarantine. They had to get the infected out of the household, but had few places to relocate them.

It was heartbreaking to know that the sense of community and family the tribes cherished was a significant part of what was putting them at additional risk. There are many benefits to having several generations living under one roof—but with Covid-19 those benefits became a glaring weakness, with negative, sometimes fatal consequences. We had the tools to change this: testing and indoor masking for ten days if isolation areas could not be found. There were other commonsense solutions, but it all came back to testing and identifying who was positive.

These testing issues were particularly frustrating. We were failing tribal nations (and people elsewhere) who were making the behavioral changes we had hoped they would. The Fond du Lac Chippewa had imposed mask mandates. They were social distancing. They wanted to get tested. Having closed their casinos, they'd sacrificed one of their major sources of income, one that funded many of their programs. As with other tribes, their CARES Act funding, in comparison with state-level help, had taken two extra months to reach them. Instead of fully supporting this vulnerable group of people, the federal government was failing them. Again, we took this issue to Secretary Mnuchin, and he addressed the problem. Ever since the early March travel ban debate, he and Larry Kudlow always saw all sides of the issue and supported public health mitigation.

As the case of the Fond du Lac Band of Lake Superior Chippewa illustrates, producing more tests doesn't mean they'll get where they're needed or that the tests will be appropriate for the particular population. As was the case in the United States overall, the tribes were, effectively, kept from using the rapid antigen tests because both the CDC and the FDA remained united in their preference for the nucleic acid tests, which had a longer turnaround time. The CDC didn't think comprehensive, aggressively proactive testing was necessary, and the FDA didn't think the antigen tests were as effective on asymptomatic individuals. Because of these objections, this critical frontline point-of-care test remained woefully underused. The antigen tests would have addressed many of the tribes' concerns about supply and would have been the most effective at quickly determining the early onset of an outbreak. Rapid point-of-care tests weren't perfect but they were better than nothing, and just then many areas of the country had nothing.

Throughout our travels, Irum and I were able to convince some local leaders to make full use of the rapid antigen tests to identify and mitigate early community spread, but it was a constant battle—one that lasted for months, until the summer of 2021, when the FDA finally recognized the undeniable fact that these tests could detect antigens for SARS-CoV-2 in the noses of the asymptomatic as well as the symptomatic. The tests (the very same ones we had in 2020) were finally made available for at-home use, but we lost twelve months. These tests aren't perfect, but they would have constituted a good, commonsense approach. Once again, the good was lost to the need for the perfect.

Brett Giroir continued to answer my calls and emails from the field and worked to send more tests to the tribal nations and to the historical black colleges and universities throughout the late summer and into the fall of 2020. Aligning the specific type of tests with the unique needs of communities was a problem throughout 2020 and 2021.

During our discussions, the Fond du Lac Chippewa's chairman, Kevin Dupuis, made an important point. As much as they had been able to enact measures to serve the needs of their own people, and despite living on a reservation, the tribe's members weren't completely isolated from the rest of the state. The mitigating measures Minnesota had put in place did have an impact on them. Members of the tribe who worked outside the reservation interacted with members of the larger community. Far too often, Americans seemed to forget that their actions, the personal choices they made, had consequences that extended beyond their own lives and the health and well-being of their families. The ripple effect of their actions extended out into their neighborhood, into the larger community, and far beyond.

Minnesota's governor, Tim Walz, had issued shelter-in-place orders in late March, and these remained until mid-May, despite President Trump's tweeting in mid-April, "LIBERATE MINNESOTA" (just as he'd done with Michigan). I had made several trips to Minnesota to understand what was driving the relentless infection rates in certain areas of the state. Jan Malcolm, the state's health commissioner, and Ruth Lynfield, its epidemiologist, were both extremely dedicated. As was true elsewhere, Minnesota had great public health leadership, smart technical personnel, and brilliant physicians and nurses. The state government had gotten out of the gate aggressively. Still, health officials were never able to find the precise communities of ongoing spread and mitigate against them.

In Minnesota, looking at average case numbers and statewide seven-day averages, it was clear that the state was controlling the virus. But as was often true for this pandemic, those figures hid the silent spread in specific communities, rural areas, and in vulnerable pockets of Minneapolis, and continued to result in ongoing fatalities. We had seen the same thing with HIV/AIDS. When you use high-level data and statewide trends, it can hide pockets of infection, areas where you weren't successful. To identify such places, you need deep data disaggregation—that is, the separation of the data into smaller units to shed light on underlying patterns and trends. The right data can make the invisible visible.

The Fond Du Lac tribal leadership was also concerned about neighboring Wisconsin. It wasn't until July 2020, with case rates rising rapidly, that Governor Tony Evers issued a statewide mask mandate. This was, unfortunately, met with protests and condemnation, a painful reminder that not everyone shared the community-minded spirit necessary to ensure full mitigation against this insidious virus. Masking in indoor spaces made those spaces safer for those with underlying conditions—essentially, making those spaces accessible to everyone, not just the invulnerable and the young. Masks weren't about limiting freedoms, but expanding freedom of access for everyone. If individuals chose to unmask at their friends' or family gatherings, that was their decision, and it impacted only those who voluntarily chose to be there. Taking a personal risk in private is one thing. Making public spaces unsafe for others is quite another.

By placing the welfare and needs of the many ahead of the desires of the few, the Fond Du Lac tribal leadership and people stood in stark contrast to those who chose not to. It hasn't always been this way. I have seen a more unified spirit in the United States many times. When natural disasters hit a state, or even another country, Americans are generous with their time and money. During 2020, I frequently wondered why this crisis hadn't produced that same sense of unity and support—the kind of support we'd seen rallied in the New York City area at the beginning of the pandemic, in March 2020, with volunteer health workers and others coming from across the country to lend a hand. Yet, as the virus moved across the country, into regional surges, this spirit of unity and togetherness seemed to have vanished.

AFTER A WEEK ON the road, Irum and I returned to Washington with insights that would shape the next phase of the response and solutions we'd disseminate to the states through our weekly governor's reports. At the time—but more so in hindsight—we recognized that our visits to the tribal nations had provided a road map for a more comprehensive, culturally sensitive approach, one that didn't cost more but that required a willingness to listen and to do things differently. Whether programs and services were aimed at minority or majority populations, they needed to be decentralized and recalibrated to the diversity and vastness of the United States. One size and one policy did not, would not, fit all.

The road trips were grueling, and during them, I still had to attend to many other issues that cropped up daily, but my exhaustion abated with every meeting with state officials and others. We were learning what was working from each and every meeting. We could make a difference. A few times people in the states recognized me and offered their thanks for the work we were doing. It was a small gesture, but it made a big difference to my morale. Whether or not my efforts produced perfect results, I needed to feel I was making a difference, even if only a small one every day, that what I was doing mattered. And you just couldn't know that from the isolation of the DC bubble.

Before our fact-finding mission to the states, several people from agencies in DC that helped manage tribal nations' affairs told us that the tribal nations "can't" or "won't" do such-and-such, that they were resistant to or incapable of managing their own resources and public health. Our experience was the opposite. Sure, the tribes were sometimes resistant to measures being imposed from above. They'd experienced years, decades, of this management strategy failing them. For that reason, they were finding their own ways, their own solutions, addressing their own needs. They didn't need more federal agencies telling them what to do. They wanted more local control. They wanted to engage in dialogue, a partnership, as Irum and I had done with them.

My experience on tribal nation lands in Arizona, Oklahoma, and Minnesota reinforced what I'd experienced in the military working on the HIV/AIDS crisis at home and then globally: If we talk through the issues together, we can find solutions that will work on the ground. More federal agencies were needed on the ground, not just in Emergency Operations Centers. Back in the 1980s, when Tony Fauci invited AIDS activists to his home, and they sat down at meals together, they'd discuss the issues and reconcile their conflicts. This gesture by Tony totally shifted the perspective that the government and the medical establishment didn't care about finding a solution to that crisis. And these weren't one-off discussions, but discussions that took place over months and years, that built trust. We had used this same approach at PEPFAR across the globe. This is what our federal agencies needed to do— not just for the Covid-19 crisis, but to address long-term health disparities and frailties across the United States. As became clear to me throughout the 1980s and my time confronting global pandemics, governing of any kind works best when we work together at the community level.

CDC personnel needed to be not in Atlanta, but embedded at the state level, working alongside their state and local counterparts, learning from them and ensuring that CDC guidance is clear, culturally appropriate, steeped in common sense, and implementable. The agency needed to see the solutions on the ground and bring those back to the federal level, sharing best practices in public health across the states.

For all the nuances among the various tribal nations we met with, one thing we heard consistently was fear of the approaching fall and winter, a fear that mirrored my own. Prior to the pandemic, the tribal nations had frequently felt isolated. Covid-19 had only heightened that sense. The pandemic was approaching the point at which it would become even more debilitating, especially as winter closed in and we saw the effects of the federally induced lack of preparedness during the spring and summer surge.

Later, in October, Irum and I would expand the scope of our learning by visiting tribal nations in the western United States. The threatened fall surge was emerging, and once again, need dictated where we traveled.

IN THE APTLY NAMED Wind River Hotel and Casino in Wyoming—you haven't experienced wind until you've experienced Wyoming's wind—I saw how, out west, geographical distances are great, but bonds are close. The reservation on which the Northern Arapaho and Eastern Shoshone tribes lived comprised 2.25 million acres. Tribal members dated their tribes' history back centuries, when they were spread over large areas of the Rockies. As with other tribal nations we'd met, their land was held in trust by the U.S. government, so the tribes couldn't use it as an investment or for most entrepreneurial enterprises, which exacerbated the cycle of poverty. The tribe needed a viable economic engine that would help drive better public health care and quality of life. As one administrator put it, referring to the aid the United States provided Western Europe to rebuild after World War II, "We need a Marshall Plan."

A 50 percent unemployment rate across the tribes, seven hundred diabetes patients being treated at one local clinic, and a decline in life expectancy from fifty-three to forty-eight years—all these explained the decrease in the median age for residents to twenty-one. As several local tribal chairmen pointed out, if the comorbidities or Covid-19 didn't take residents, diseases of despair would. Even before the coronavirus pandemic struck, substance abuse (alcohol, methamphetamine, heroin), murder, and suicide—real public health issues that hadn't been addressed over the decades by the federal agencies—were already decimating the population.

As is often the case in the United States, it was a tale of two cities—in this case, two tribes—experiencing the best and the worst. The local tribe-led clinic's special designation by the federal government's Health Resources and Services Administration (HRSA) as a "Tribal and Urban Indian Health Center" got the Northern Arapaho access to federal dollars and support. The HRSA directly funds about 25–50 tribal-led clinics across the country, putting tribal communities in charge of executing their own culturally sensitive health care. Because the tribe operated its own clinic, Tribal Health, they were able to move quickly to get testing supplies and conduct their own mass testing of their tribal nation citizens. The IHS was not able to move as quickly, and therefore was delayed in supporting the Shoshone in their testing program.

At the time of our visit in late October, both tribes were tired, but the Northern Arapaho felt they could do even more proactively to prevent uncontrolled spread of Covid-19 among them. It all came down to more testing, and the Northern Arapaho were moving and could do more. Their very proactive IHS team understood partnership and were supporting the tribe from behind, instead of dictating to them. In contrast, the Eastern Shoshone were weeks and hundreds of tests behind, with greater viral spread and greater fatalities. Managing quarantining and isolation was taking a toll on the Shoshone's limited human resources, including contact tracers, who had to advise how the infected should isolate in multigenerational homes, as they, too, were running out of space.

One thing was clear for both tribes: they believed all members of the tribe were their family. For this reason, they thought holistically about the services they needed and worked with suppliers to make sure there was enough food and shelter. With the Wind River Reservation officially "closed" due to the pandemic, regular communication with the people living on it was crucial for conveying the appropriate actions they needed to take to protect themselves. With the BinaxNOW rapid test recently made available in the state, we recommended that local and state health officials make distribution to the tribes an immediate priority.

As always, the great lessons of our meeting were followed by the reality of the daily risks the tribal members were facing by gathering together. A day after the meeting, we were notified that one of the leaders we met with had tested positive for SARS-CoV-2. Because our masks had never come off during our several-hour sit-down, we were pretty certain we were negative; still, the anxiety was there. Irum and I didn't have easy access to testing on the road, and we didn't want to use the limited state testing supplies. So, we checked constantly to see if our sense of smell or taste had changed, early symptoms of Covid-19.

We moved farther west, meeting with the Wyoming Shoshone's sister Shoshone tribe in Idaho, who reside with the Bannock Tribe on the Fort Hall Reservation. The clinical director and I accompanied Devon Boyer, chairman of the Business Council, the tribe's governing body, to our meeting. Three different agencies managed the reservation's health clinics: the IHS, the Shoshone-Bannock Tribes Tribal Health and Human Services Department, and Federally Qualified Health Services (FQHS). Chairman Boyer was keen on testing, but again, supply issues had hampered those efforts. Testing was especially critical here. Since July, 46 percent of the deaths on the reservation were related to Covid-19—many times the percentage of New York City during March and May 2020.

To help the tribes reverse this trend, we asked the CDC to send staff to show them how to conduct testing on the reservation using the BinaxNOW test, asking for specific staff who had worked overseas in PEPFAR programs, staff whose practical experience matched the needs the tribal nations had expressed. Practical and solution-based personnel willing to listen and innovate with the community. These teams worked alongside each tribe to increase testing and quarantine options and provide isolation support. I had recognized the need for this kind of deployment of CDC personnel early on, even before I set foot in my first White House task force meeting.

We learned in our meetings in Idaho that, as with federal government on the whole, the Indian Health Service is only as good as its frontline representatives, its physicians and nurses. In the case of the Shoshone-Bannock Tribes, the local IHS team comprised progressive, innovative problem solvers who were united in deep partnership with the tribal chairman and other leadership. This team of IHS personnel knew the importance of listening in order to better understand, and of not bringing a standard stovepipe approach to the tribes' unique needs. The IHS system may have been antiquated, but those working within it were, for the most part, passionate people working hard despite the limitations imposed on them by the bureaucracy. This IHS group was willing to learn, and so, the outcomes and impacts were greater.

We saw this across the western states we visited. In places where the IHS representatives were culturally sensitive and willing to listen and adapt, they were able to create a partnership and plans that effectively addressed the specific needs of a specific group of a specific tribe. They weren't stuck on past perceptions or prior instincts.

I began to realize that the cultural trend of celebrating the benefits of failure had some merit. So much had gone so wrong for so long in the indigenous communities that they were willing to be open to new ideas. Within the CDC, the NIH, and many other public health agencies, so much had seemed right for so long that rigid use of past approaches had set in. The CDC excelled at outbreak investigation—tracking the source of a salmonella outbreak to a specific type of lettuce or spinach by tracking and tracing back to patient zero and then to the field. But with the pandemic, the agency was mired in the pandemic preparedness they themselves had set up. It wasn't working perfectly, but no one wanted to admit this and find new solutions. But this inflexibility was keeping us from fully responding to the unique needs and specific barriers to wellness in state after state, county by county, tribe by tribe.

With a raging pandemic, there is no such thing as "good" news. Progress in one area is always met with deterioration in another; or it uncovers a new gap. And any progress you place on the scale can't counterbalance the reality of lives lost needlessly. In examining the response of the tribal nations, I found areas of their approach that the states could learn from. Irum and I both felt an overwhelming sadness at how native peoples had been treated historically and how they were being treated during this pandemic. But after a long history of being underserved, at best, and largely ignored, at worst, the tribes didn't seem to dwell on the fact that the government agencies didn't serve them well. They had accepted this reality and crafted a new one for themselves, adapting and working over, around, and through obstacles that would have thwarted other groups.

We had seen the same can-do resilience in West Virginia and in the Covid-19 response team led by Governor Jim Justice, with critical insights from the state's coronavirus czar, Dr. Clay Marsh. West Virginians believed they couldn't rely on the federal government to support them, and like the tribal nations, they had come up with solutions on the ground that worked for them.

While many other communities felt they were too infrequently seen or heard by the federal, state, or local government, tribal chairmen and leaders—rather than resorting to denial, anger, or resignation—had taken a proactive, comprehensive approach to defending themselves against the ravages of the coronavirus. Certainly, they felt those emotions, but they moved forward as best they could nonetheless. This was especially important with rates of comorbidities in their communities so high.

The tribes drew on cultural memory as well, and had developed the mentality of "We can't let this happen again." According to a 2017 CDC report, tribal nations in the United States experienced to a far greater degree the effects of the 1918 Spanish influenza pandemic. They experienced an astounding 24 percent infection rate and had the highest death rate of any racial or ethnic group, losing 2 percent of their total population. More recently, in the 1990s, the hantavirus killed 75 percent of those infected. Nearly half of those who died were Native American. Yet this data, knowledge of these health disparities, had not transformed the federal response, and by the time the novel coronavirus arrived, the tribal nations were at the same starting point.

To combat the risk from within and without their reservation, the Fond du Lac Band of Lake Superior Chippewa took a culturally appropriate approach to its messaging. Taking the notion of "one blanket"—the historically suspect but metaphorically accurate belief that all it had taken was one blanket weaponized with the smallpox virus to wipe out a Native American population—they re-crafted it to remind their people not to let the coronavirus response be that one blanket. The historical resonance of this idea was clear and effective.

This sense of shared history was reflected in the tribal nations' more integral approach to the pandemic. The tribal nations had maintained their sense of community throughout the pandemic, from one tribe to another tribe, across the country. This was true whether they resided in red states or blue. They valued the wisdom of their elderly and believed that ties extended beyond the family, to the entire community. This desire to protect everyone in the tribe through shared sacrifice was something the rest of America needed to learn.

CHAPTER 15

On the Ground

Governors Innovate

In parallel to our meeting with the tribal nations, we had continued our state-by-state visits learning from mayors, governors, hospital leadership, and communities. After that first trip to Texas, Arizona, and New Mexico, Irum and I made three other trips, visiting seventeen other states, some for the second time. In the Southeast—Louisiana, Mississippi, Alabama, South Carolina, and Georgia—I got an eye-opening glimpse of the level of distrust people felt with regard to the federal government and the vaccines now in development.

Arriving in Baton Rouge, we thought we'd wandered onto the set for a postapocalyptic movie; there were so few people around. At a Starbucks—its furniture was pushed into a corner and walled off, and its bathrooms were locked—I asked a barista about Governor John Bel Edwards and the mitigation efforts in place in the state. New Orleans had been especially hard hit in the spring of 2020, but the summer was different: this time, the virus had spread rapidly across the state, leaving no county untouched. In rural communities and rural community hospitals across the South, the most complex cases like Covid-19 patients are transferred to regional medical centers in urban areas. Because those hospitals were already full with patients from urban communities, they often couldn't accept patients from those rural areas.

Consequently, without more advanced medical care available, those rural patients often died at a greater rate than their urban counterparts.

Now, with classes resuming at Louisiana State University in the next month, I was eager to understand how state officials were going to approach the return of students and the resumption of social gatherings. "People have been doing what they should," the Starbucks barista told me, "but I'm starting to hear more and more about parties at people's places." I thanked her for the caffeine injection and the intel. Later, I had reason for greater hope. LSU's administration and athletic department had instituted a surveillance system so that its highly successful football program and other sports could be a "geaux" for the fall. A few of those parties the barista had mentioned had the football program stumbling out of the gate, but its voluble coach assured me that once he reminded his players that their personal choices had an effect on their teammates and could jeopardize the entire season, they fell back in line, reducing gatherings, parties, and bar activities to ensure that a treasured community activity was kept in place. After a brief blip in rising cases at the beginning of the fall semester, active cases throughout the state declined until October. A month after our visit to New Orleans, in mid-August, while on trip number four, we would see the same situation with high school football in West Virginia. Shared community values could bring people together.

Across the board, southern states either had mask mandates in place statewide or were specifically encouraging local officials to implement them. After our continued engagement, even Georgia had moved to this approach. After our conversation with Governor Tate Reeves, who'd taken office in January 2020, we were pleased to see Mississippi, a state with significant comorbidities, put in place mask mandates on August 4. That was the best way forward: targeting mitigation to the specific areas where the data showed the burden was highest. Letting people know that the need for masks was urgent right then and right there, rather than everywhere and at all times, eased resistance to these measures.

Throughout the South, we saw new solutions, too. In Mississippi, state health officials were reaching younger residents through Instagram. In other cases, public health leaders used texts rather than calls to alert residents to guidelines or the need to test. Tailoring the message to specific subsets of the population and addressing their specific concerns and needs was the way forward. In Louisiana, the use of wastewater analysis to understand community spread allowed the state to alert communities of the need to practice enhanced mitigation. Alabama, under the leadership of Governor Kay Ivey, was aggressively mitigating against community spread using data. In Arkansas, Governor Asa Hutchinson and his team wrote a children's book, *The Kids Guide to the Coronavirus*, to distribute through the schools. It included valuable information about preventing SARS-CoV-2 infection and about the coming vaccines—educating children and the adults reading the book to them at the same time.

Farther north, in the earliest days of the pandemic, Governor Mike DeWine of Ohio and his team had crafted a short video, "Back Up, Mask Up, Wash Up," to educate kids on the three main mitigation strategies. Later on, with the Dine Safe Ohio program, 95 percent of the state's restaurants complied with limited dining capacity, enforced masking, and expanded outdoor dining. You can craft the perfect plan, but if people don't comply with it, it may as well be worthless. Ultimately, it took our being on the ground in all these states to recognize each one's unique demographic, but also, happily, how fully each was using strong mitigation to ensure both health and economic progress.

The approach to masking varied from state to state and within states. What was important was masking indoors, not outdoors, but this message wasn't always clear in a given state's mandate. In Indianapolis in the last week of July, when Irum and I were on our third trip out to the states, a girls' basketball tournament was being held, and as we walked the streets of the capital, we saw more ice-cream cones than masks covering faces. This latter didn't bother us as much as what was happening indoors. There were excellent outdoor dining options, but with so many people in the city, many were pushed indoors just to get dinner. In many cities across the country, mayors had altered traffic patterns and regulations to dramatically expand outdoor dining. It was key for residents to understand that engaging in outdoor activities while unmasked was safe, and that outdoor dining was very safe compared to indoor dining. Still, when mandates were issued, rather than merely limiting indoor dining, state leaders often closed all restaurants. This approach unfortunately drove residents from safer outdoor dining to unsafe gatherings in homes.

Elsewhere—in Chicago for example—outdoor dining was prevalent, as were masks on diners and others, including waitstaff. We appreciated all attempts to protect workers at higher risk of exposure, with diners required to put on their masks when waitstaff approached their tables. Chicago's mayor, Lori Lightfoot, was an exemplary leader in many aspects of the mitigation effort. Also, the entire metropolitan area's private-sector hospital teams had transcended the sense of competition among hospitals that had been business as usual before the pandemic to create a unique, unified dashboard to ensure that every resident got to the right hospital, with the right equipment from ventilators to extracorporeal membrane oxygenation (ECMO) where their care could be optimized.

I wanted every American to see what I was seeing, the parts of the pandemic response that *weren't* being covered in the media, the underlying practical solutions that were saving lives. Private-sector companies, looking past profit to support communities and states in their response, stepped up in unique ways across the country. In state after state, I saw competitors become colleagues, to save more lives. I saw this on the federal level, too: Hospital suppliers such as McKesson, Cardinal Health, AmerisourceBergen, Henry Schein, and Medline Industries used our epidemiologic data to align critical supplies with hospitals, to ensure that all hospitals, independent of their ability to pay, had the supplies and treatments they needed to respond to the health needs of their communities.

To gauge how effective we were, and how we might alter our messaging, everywhere we went, Irum and I queried hotel staff, counter clerks, and drive-thru attendants—anyone who could give us insight into what was happening on the ground and how people felt about the pandemic response. What we found was always a mixed bag of gains and losses. From these encounters, we learned that people perceived the pandemic as an "urban problem." In rural areas outside Des Moines, Lincoln, Tulsa, Little Rock, and Charleston, West Virginia, people believed that, because of their geographic remoteness from urban centers, they were "naturally social distanced." We heard this many times, and included warnings against this faulty belief in all the state governor's reports. We had learned from the summer surge that rural areas were under as much threat from full Covid-19 disease as urban areas. Critically, rural areas often relied on urban hospitals for the care of the severely ill. With the summer surge, both urban and rural hospitals had become overwhelmed, and community hospitals were often unable to move their patients to the at-capacity regional medical centers.

On our fourth trip out, this time to the heartland, post–Scott Atlas, I engaged in a fruitful conversation with a Nebraska state health official who was genuinely interested in hearing our take on herd immunity. He opened the conversation by noting that this was how they handled infectious outbreaks in livestock in the Midwest.

Irum and I went through what we knew and didn't know about Covid-19, including the duration of natural immunity after someone was infected and the unknown long-term impact of Covid-19, which had come to be referred to as "long-haul Covid." We knew that, in the short term, infected young people did quite well, but we didn't yet know if later complications from infection would appear months or even years in the future.

I don't know if we convinced the official that "culling the herd" wouldn't work, but at least he had an open mind. I always appreciated it when people asked questions and shared what was on their mind. Being able to talk through these issues was critical, but with the official's mention of herd immunity, my eyes were once again opened to the damage that Scott Atlas and his group of like-minded doctors were doing.

NO MATTER WHERE WE went, we heard two things without fail. Every leader and nearly every American wanted to protect the old and vulnerable while charting a path forward for the young and less vulnerable. The former meant preventing Covid-19 from sickening and killing the elderly and those with severe underlying comorbidities; the latter meant not jeopardizing the education or future prospects of those in schools, small businesses, and working in the hospitality industry. We put out the message that testing and masking brought both those aims together.

So, with the cold weather approaching and driving up cases, Irum and I had set out on the fifth of our trips in the last week of August with these two groups very much in mind.

If we had been teaching a course, the main bullet points would have been:

• Winter is coming. People are going indoors. Viral exposure will increase.

- Public spaces are becoming safer, but significant transmission is occurring in private gatherings and spreading outward from there to the community.
- Test to find the asymptomatic cases.
- Prevent viral spread from escaping the confines of campus and into the surrounding community, especially into multigenerational homes and care facilities.

In September, conveying this message had become more urgent than ever. Some states seemed frozen in time, still using tests mainly to confirm the presence of the virus in people with symptoms or exposure to someone with symptoms—that is, testing less than half of those infected and spreading the virus. The president's calling for less testing and Scott Atlas's influence with CDC testing guidance were both culprits. Their shift in messaging to fewer tests had produced a significant reduction in testing. That had to be countered.

Scott Atlas's message that the task force believed all schools should be closed also had to be countered. We *wanted* schools open. We also wanted those who attended them and worked in them to be tested regularly to keep them from carrying the virus into the classrooms, dormitories, off school grounds into businesses, and to the multigenerational homes they lived in or visited. If we applied what our previous visits to the states had revealed, what had worked at some universities, to more of the population, we'd be able to offset one major failing: the persistent nearly 10 percent fatality rate in those over age seventy. To one degree or another, nearly everything we'd done had driven that rate down from nearly 30 percent in March and April, but even with all our advances in treatment, the rate of death from Covid-19 in this age group stubbornly remained at nearly 10 percent. Yes, it was a 66 percent decline from the earlier, higher rate, but the death of one in ten people over seventy with Covid-19 was still horrific to contemplate, especially with the mostly deadly surge on the horizon.

In our travels, we'd seen how governors had prioritized aggressive mitigation to protect nursing homes and long-term-care facilities. Irum and I had met with plenty of nursing home directors and heads of associations who oversaw and supported those operations. Following guidance from the Centers for Medicare and Medicaid Services (CMS), they had collected the necessary data. The National Guard had supported

PPE distribution and testing, and CMS, in partnerships with states, had set up site-specific strike teams at the first evidence of an outbreak.

In mid-August, Irum and I had driven to a nursing home in suburban Oklahoma City that was part of Leading Age Oklahoma, a state association of not-for-profit organizations that provided services to aging populations. Pulling into the parking lot outside the main housing area, we saw folding chairs clustered outside windows—a stark reminder of what isolation had meant and continued to mean for those inside such facilities and their concerned family members kept outside.

Once inside the community room, for a meeting with staff and residents, we were all masked, but the staff's eyes told me a lot. Their fatigue, their distress, and the collective trauma they had suffered were all evident there. These people needed to talk. Over the next couple of hours, care workers shared their heartbreak, their fears, and their desire to carry on working 24/7 for the home's residents.

At one point, one of the residents, a woman I estimated to be in her early eighties, spoke. Her voice was steady and sure as she began thanking and praising the staff who had cared for her and others. Her voice rose and fell, halting and then rushing on. "My room became my prison. I ate there. Slept there. Bathed there. At the beginning, I thought, *Well, okay. This is what it is, and it will just be for a little while. We'll get through this.* I didn't see anyone for weeks. Weeks became months. I was behind closed doors and only saw one of the staff a few minutes a day. When they brought in my meal, my medication. I wanted them to stay, but I knew I shouldn't take up too much of their time. They had so many of us to look after.

"Then, as the weeks went on, I was told about my friends and others who'd died. We didn't get to say goodbye. Before the pandemic, others had died, but we'd have a memorial service of some kind to remember and recognize their time with us, but we couldn't [do that] anymore. We couldn't share memories. We couldn't comfort one another. Everything just went flat, gray. It was all kind of matter-of-fact. 'So-and-so died. They're gone.' I noticed, after a while, that the aides and the nurses spent more time with me. They lingered a bit. We talked, and I could hear it in their voices: They had more time because—well, because there weren't as many of us to tend to. I started to wonder if all we were doing was worth it. Was this living when you were nothing but lonely and sad?"

Her words made everyone in the room emotional. Irum and I were there, in part, not just to listen, but to offer everyone another way through this; to deliver a warning, but also hope, in the form of a reliable, testing-based solution. Still, the fatigue was palpable.

As summer turned to fall, and I spoke with more governors and members of communities, I heard expressions of exhaustion like this one. The tone of the conversation had turned. People were beaten down. The mixed messaging, the sense that nothing we'd done before had worked, doubts that any other approach might work, the sense that we were drifting like a leaf on the wind—all these had settled in. This had to be countered. We had to show that strategic testing to identify asymptomatic spread was working elsewhere.

I had learned from my time in the military that you can't stay at "up tempo," on high alert, forever. You need to adjust your level of alert up or down based on the threat at hand. Regularly testing younger people to protect the older, most vulnerable could offer peace of mind, helping older residents get through the hardship they'd had to endure.

Nowhere was this more apparent than at America's colleges and universities.

ALL ALONG DURING OUR visits to the states, under the leadership of Vice President Pence, we'd been meeting with higher education officials, and throughout the spring and summer, we'd been holding conference calls with educational leaders. They would share with us their plans for reopening in the fall and their contingency plans in case of a wide outbreak. We'd provide general insights, most specifically about the importance of regular testing not just testing those with symptoms. By the time we returned to Washington at the end of the first week of September, schools had reopened for the fall. For the most part, we had been impressed by the solutions-based approaches many higher education institutions had put in place to open safely. Now we'd get a chance to study the effects of their level of preparedness and see if theory matched reality.

Despite the success we'd seen earlier in the summer—with southern universities handling sports teams by regularly testing and immediately isolating those infected, thus preventing spread—initially, the full reopening of colleges and universities raised alarms.

What had happened with the North Carolina university system was particularly alarming. North Carolina's university system had paused in-person schooling in late August, going fully remote by September 4 and remaining remote for the entire fall semester. The University of North Carolina at Chapel Hill had opened for in-person learning on August 10. The sheer number of students who tested positive that first week, and who had to isolate and quarantine, overwhelmed the system. The administration moved all classes online. Similarly, North Carolina State University, which had also opened with in-person classes, moved to online only a few days after UNC Chapel Hill; as did East Carolina University. This involved nearly ninety thousand students across these three universities. We wanted to meet with them to find out what had and hadn't worked, and then immediately get to other large land-grant universities elsewhere. We couldn't have thousands of students away from home, in leased housing in the community, without access to testing.

In a meeting with us, the UNC system representatives made their case that they'd done all the right things: they'd modified buildings and classrooms to enable social distancing; moved communal activities like their dining services outdoors or offered meals to-go, limiting the size of student congregations; and had imposed a strict mask mandate—just what other universities had also done since the start of the outbreak.

To their credit, students at UNC Chapel Hill recognized the role they played in viral spread, citing the reckless behavior of some that had put the health and education of others in jeopardy. The student newspaper, the *Daily Tar Heel*, and its editorial board joined many in denouncing the UNC administration for failing to anticipate the kind of behavior some students would engage in and for doing too little to "dis-incentivize" it. I was concerned about how much the UNC system had emphasized symptomatic over wider routine regular testing of all students independent of symptoms as an essential component of mitigation, and I wondered what these students could now do on their own. The data we'd been gathering from the universities that were testing all student athletes revealed a noteworthy trend. Among those in the eighteen-to-twenty-five age range who tested positive, between 90 and 95 percent were asymptomatic and accounted for the majority of student-to-student spread.

After our meetings ended, I called Brett Giroir to discuss what we'd found. I was really concerned with UNC's stance on remaining fully remote, as most students lived off campus—nearly 85 percent of them in apartments with fully paid leases. For this reason, I believed that many students would remain in the community, in the apartments they'd already paid for—but without regular testing. Brett understood, and he immediately sent federal teams to expand community testing to the areas around the now-closed UNC system campuses. He didn't wait until it was a crisis and hospitals were filling; he was willing to be proactive and get federal support there quickly.

The University of South Carolina, Columbia, demonstrated how good planning, linked to continual data collection, produced a flexible program that could be adjusted based on changing data and other findings. The university's agile, responsive leadership tested 1,300 students per day. Based on the data and the university's ability to scale up its own labs' testing capacity, that number rose to 4,000 per day to nearly 50 percent of the student body weekly. With an enrollment of more than 35,000 students, that figure wasn't 100 percent weekly but they were making progress and significantly expanding testing, testing every student, every week, those without symptoms as well as the symptomatic.

Within its own university laboratories, USC Columbia developed a means to test wastewater to detect the presence of SARS-CoV-2 in dormitory complexes. Louisiana State University, Clemson University, and the University of Connecticut also used this method successfully. Critically, they not only used it on campus but in the community to understand on- and off-campus viral spread to provide early alerts to trigger early response. We were encouraged to see universities' proactive approach to asymptomatic spread and testing and their use of their own researchers and vast stores of equipment to find solutions. We actively encouraged other universities to do the same. Relying solely on the federal government or the state to act was not ideal. As with tribal nations, these large universities were nearly closed systems, bound by a sense of purpose and community. Taking agency over their own needs and priorities fit into a pattern of communities breaking free from a one-size-fits-all approach.

Universities are the backbone of much of the country's basic research activities—with many primarily funded by the National Institutes of Health—and would be critical in a crisis like this one. Yet, when universities closed in March 2020, their research laboratories across the country were also shuttered, and research technicians and postdoctoral students went home. As a result, research and innovation across the country declined. We felt this loss throughout the spring and summer of 2020. Our scientists in the public and private sector are the envy of so many other nations, but the core of our basic research is done at our universities.

The contribution of many of these institutions to the pandemic response—the data people at Johns Hopkins being one of the earliest and most helpful—was undeniable, but without a pandemic preparedness plan that made use of the depth and breadth of their research scientists, their influence became limited. In this, I saw early on the evidence of yet another missed opportunity, one that had contributed to our being caught flat-footed at the outset of the pandemic and beyond. Although each university did a small amount of voluntary SARS-CoV-2 research, this effort wasn't organized, it wasn't done in a comprehensive manner, and it didn't ask or answer all the questions we had. We needed all our university researchers available in March through June, but many were at home. For the next national health crisis, university researchers who receive federal funding need to be available to contribute their expertise to finding solutions. This should be part of the pandemic preparedness plan and pandemic response moving forward. All universities receiving federal research dollars should be required to attend a weekly meeting at which leaders assign essential, timely research questions to experts in the behavioral and medical sciences so we can build the new evidence needed to combat new pandemics in real time.

Among the universities that opened for the fall 2020 semester and stayed open, it wasn't just medical scientists, researchers, and students who contributed to the pandemic response. Various universities and colleges enforced isolation and quarantine regulations and practiced contact tracing—which was effective but also time consuming; as was mandatory testing. One benefit of institutions of higher learning is that these schools have enormous information technology departments, with students and faculty who have studied computer science. IT departments at several schools—USC and "Ole Miss" chief among those we visited—developed phone apps to help with contact tracing, and Virginia Tech overcame the testing supply deficit by using 3-D printers to produce their own swabs. There and elsewhere, universities used their own research facilities to overcome the testing and processing shortfall. Across the country, it was all hands on deck, with people rising to meet the pandemic's challenges.

Our trip to the Carolinas taught us something else: If what had happened with the UNC system were repeated elsewhere, a flood of younger people driving community spread would cascade across the country. A good plan required vigilance. It was clear what students were doing while on campus and in classrooms, but what they were doing outside these places likely contributed greatly to the transition back to remote learning. Irum and I were so concerned about the effects of these school closures that we canceled plans we had to go to California. (Steve Hahn and Jerome Adams would go in our place to better understand the continuing viral spread among agricultural workers there and in Oregon and Washington.) Irum and I would again focus on schools in the Southeastern and Atlantic Coast Conferences. Eventually, into the fall and winter, I'd visit more than thirty campuses.

Other trends and insights coalesced during these school visits. Irum and I were enormously impressed by the student body leaders, who were actively engaged in supporting mask use to protect the community. Rather than viewing mask mandates as a regulation being imposed from above, they transformed this perception into a grassroots campaign that felt organic to the mission of a university: to enlighten and expand one's view beyond parochial interests. Simply put, they wanted to take care of one another and of their university's faculty and staff and, in turn, the larger community and family and friends they would return to back home. These student leaders embodied the kind of selfless spirit we all needed to get through this crisis. Talking with these students, I felt that the sense of community was alive and well at our universities.

A roundtable with students at the University of Alabama was especially eye-opening. There, we learned that many of their classmates needed to remain on campus because it was the only place where they had a bed, a desk, and reliable access to food. With the university kept open, local businesses were able to stay open, too, providing these students with work, so they could afford books and tuition. A "virtual," online university experience would have left them homeless, with neither job nor income, making them unable to stay enrolled. In-person schooling during a viral pandemic—with its masks and social distancing and canceled parties—may not have been pleasant, but for many students, it was a must. The students at the roundtable understood this, and many had started food banks for classmates who were less well-off and had become attentive to their mental health needs. This was community. This was what we'd seen with the tribal nations.

Ultimately, though, the role of testing emerged as the key takeaway. In the end, of the thirty-plus schools we visited, only 50 percent employed mandatory surveillance testing of every student. Often, schools tested only 3 to 5 percent of the student body, frequently the "worried well," who got tested voluntarily, and not those who were at greatest risk. A version of what we saw in North and South Carolina was reproduced around the country.

Schools that required at least weekly testing of all students fared much better in the end than those that did not. I saw innovation in testing across the universities in New Hampshire: Plymouth State and the University of New Hampshire required weekly or twice-weekly testing, and they didn't let dwindling resources stop them—spending their own money to keep their students and faculty safe. The University of New Hampshire was one of the most innovative, developing its own testing system on campus out of its genetics lab. UNH believed in its students, and trusted them to self-test twice a week, providing mailbox-like structures on campus where students could drop their nasal swabs. Students understood that this was the way for them to stay on campus and in class. This clear partnership between the students and the university administration showed that open dialogue and shared goals could result in successful Covid-19 mitigation.

And the proof was in the pudding! The data showed that those universities doing regular testing saw much less community spread among their student body. Here was the clear evidence, provided by the universities themselves. They had recognized the problem—that asymptomatic infection among Americans under thirty-five, which on average was close to 40–50 percent, rose as high as 75–85 percent (and in schools with mandatory schoolwide testing, to 95 percent)—and found a solution. Testing often and finding the asymptomatic cases in less than twenty-four hours from swab to notification prevented significant spread across the student body and allowed these schools to stay open.

The truth was simple: without regular testing of the asymptomatic, the effects of all the other good measures universities were instating would be diminished. We took our findings back to the task force and put them in the governor's reports. We needed more regular testing and more rapid, youth-friendly turnaround on the results. We worked constantly to expand not just the sheer number of tests, but also this type of strategic testing. Unfortunately, the ongoing resistance of the FDA and the CDC to the rapid antigen tests limited our ability to move forward on testing at the pace needed. Throughout November and December, I worked with Brett to expand PCR testing and shorten the turnaround times, to make the testing better able to stop community spread. We moved nearly $400 million to support this effort in twenty regional surge sites—monies that could have covered three hundred thousand additional PCR tests per day across the country. This additional testing capacity, if consistently applied, would have brought us to well over two and a half million PCR tests per day. Unfortunately, the new administration didn't prioritize testing, and though the money was there, they didn't spend it, and this surge support testing didn't happen until 2022.

Also, manufacturers planned to double production of the rapid antigen tests to one hundred million per month. With antigen and PCR tests combined, this would have resulted in a total of five million tests available per day. This plan would eventually be discussed with the incoming administration, but it took a backseat once President Biden was in office. Testing rates fell day after day, month over month, decreasing to three hundred thousand per day in June 2021. With warehouses filled to overflowing with unused tests in late spring 2021, rather than continuing to ramp up regular, inexpensive testing, as was done in the United Kingdom, manufacturers had to shut down their lines, and once again we entered the next surge, in summer 2021, blind to the early silent invasion.

WHAT WAS SUCCESSFUL IN the university environment could be used effectively at K–12 schools, offices, and other places. In late fall 2020, we again worked with David Rubin, director of PolicyLab at CHOP, to bring the same concept to a portion of the Philadelphia-area school system, and it was working; there was the evidence. Wherever regular testing was made routine and available, early infections were found and community spread prevented. What had worked in theory also worked in practice. This approach should have been universally adopted at all schools and workplaces in 2021. Unfortunately, it wasn't. Instead, other, deeply entrenched issues hampered the testing effort.

Perhaps unsurprisingly, the failure in testing came down to money and a lack of faith among some at the FDA and the CDC that aggressive widespread testing could change the outcome of community spread. It may seem self-evident that if you test all students, staff, and faculty in a school, you will be able to head off silent spread. And while I got the sense that there was no real resistance among officials in higher education to the concept, the combined authorizations and priorities of the FDA and the CDA made this difficult.

Even the way the CDC reimbursed schools for surveillance testing stemmed from the agency's belief that silent spread wasn't a significant contributor to symptomatic Covid-19 disease and hospitalizations. The CDC operational plan allowed for colleges and universities to be fully reimbursed for *symptomatic* testing, but did not cover the cost of proactive testing to find the sources of new infections. As you can imagine, given the size of a standard university population, weekly asymptomatic testing is quite expensive. (My brother is the president of Plymouth State University, which has an enrollment of nearly five thousand students. His school spent nearly four million dollars on testing in the fall of 2020 and the spring of 2021.) Some schools and states were willing to foot the bill for broad, asymptomatic testing, using a portion of the CARES funds they'd received or whatever other monies they chose to use to offset the cost of their lifesaving vigilance. But not every school had that option. When Irum and I were in Missouri at a governor-sponsored roundtable, the president of Lincoln University (a Historically Black University and a land-grant institution) noted that the school didn't have access to PCR equipment that it desperately needed to access routine testing. I called Brett and organized testing there and for all the Historically Black Colleges and Universities across the country. This was another illustration of how critical it was to learn the unique issues of various constituencies.

We had the testing tools that institutions of higher education needed to keep their doors open, avoid a negative financial impact, and prevent educational deficits in their student body. The CDC wouldn't recommend proactive asymptomatic testing beyond 5 to 10 percent "surveillance" testing of a university's population. This had forced these schools to make a difficult choice at a time when revenue loss was high. Everyone needed to be tested regularly. The CDC should have recommended that states conduct routine weekly testing. It had access to billions of dollars for testing to send to the states for K–12, community colleges, and universities, but it failed to ask states to prioritize the money for this effort—again, because it didn't believe the role of silent spread was very great. This contributed to hidden community spread early on in the pandemic, leading to dreadful consequences down the line for more vulnerable people, as the strains that infected students made their way into nursing homes.

I applauded the schools that disregarded this federal disincentive and absorbed the cost of proactive testing. I also understood why some schools had opted out of this costly approach. They should never have been put in that position. All Irum and I could do in the face of federal agency intransigence was make the recommendation to the university presidents and public health officials, and in our weekly governor's reports, that they should fund a comprehensive routine proactive testing plan. We also advised that, individually and collectively, university administrations exert additional pressure on the chief executive of their state, the person at the top of the hierarchy overseeing public institutions of higher education, the governor, to fund this testing. Vice President Pence, whose care and attention to the needs of such institutions of learning were particularly acute, made clear, in no uncertain terms, that colleges and universities needed to institute mitigation efforts that allowed them to stay open for everyone. Some schools listened; some did not. How much of this was due to funding is impossible to say.

Irum and I had seen examples of appropriate funding decisions producing great results. These, too, were added to the response plan for the remainder of the year and beyond. After the 2009 H1N1 outbreak, many universities across the country had received Global Health Security (GHS) funding. The same was true in 2015, when the potential for an Ebola outbreak beyond Africa's borders was seen. At the University of South Carolina and in Tucson, Arizona, we visited two of the finest clinics and laboratories either of us had ever seen—both facilities were built, in part, using GHS monies provided by the HHS.

As with how CARES funds were distributed and monitored (or not), various public universities spent the money they received from the federal government at their own discretion. Some used the money to prepare for the next pandemic; others used it for other purposes. As was the case throughout, the federal government handed out vast amounts of money but didn't attach any meaningful requirements to it regarding how it should be spent or for reporting those expenditures.

This wasn't a recent issue. Year after year, the states submitted plans and funding requests, and the CDC and other agencies sent monies. There was no required reporting on the outcomes or impact of the dollars spent, as we had with our global PEPFAR funds. Holding the states accountable for the federal dollars they spend seems, at a minimum, a commonsense approach. At the schools we visited, Irum and I saw a range of funding uses—some that were brilliantly innovative and others that supported business as usual and didn't contribute greatly to the pandemic response. As with so much of the pandemic, it came down to finding the right balance between control by the federal government and independence for institutions and states. The only way to ensure progress against public health issues is to hold each side accountable. An annual report on whether agreed-upon outcomes and impacts have been achieved with federal funds isn't too much to ask.

At some schools—Texas A&M and the University of Kentucky, in particular—administrators were keenly aware that a top-down model with regard to the imposition of mandates wasn't going to work with their student body. If you wanted students to comply with mitigation measures, open and transparent exchanges with them would be necessary. Although true in so many cases with this pandemic, ineffective communication got in the way. At these universities and many others we visited, students and administrators worked hard in partnership to craft effective and reasonable guidelines. We learned a lot from the students and the university officials. At many universities, like Texas A&M, we saw amazing leadership among the students—self-policing and ensuring that words were translated into action. In some instances, their wording emphasized what students could no longer do, instead of highlighting what they still could. Similarly, the University of Pennsylvania used pulse surveys (that is, regular, frequent surveys designed to take snapshots of public opinion) to assess how student perceptions about Covid-19 were evolving and devised a marketing strategy to respond to student input. At a time when so much of life during the pandemic felt out of our hands, people, not just university students, wanted to take ownership of the response.

Stony Brook University, in New York State, devised a simple method to get over the "You can't do this" message. Rather than placing a red X or a D somewhere, to indicate where you weren't allowed to sit, stand, or move, they used green arrows and dots to indicate where you *could* go, what you *could* do. Creating a sense of what was still possible helped offset the feeling that nearly everything had been taken away. As time went on, and Covid fatigue set in among us all, these subtle psychological reminders grew in importance.

If there was one universal, it was this: Nearly every campus that opened or that we visited had set up a comprehensive response team. Whether they had been prompted by an incidence of Covid-19 disease or some other catalyst, these teams created websites and other forums for sharing essential information that were unique to that particular campus or community. At the University of Kentucky—which tested all students upon their arrival to campus and required a daily symptoms check—when students tested positive or contact tracing revealed they'd been exposed to someone who was infected, teams leapt into action, delivering food, bringing the students needed school work and materials. When students had to isolate as a result of infection or quarantine after exposure, they weren't completely cut off; they were supported. Examining every aspect of student life, the university did its best to meet the needs of those in isolation and quarantine.

Student mental health was very much a concern for university administrators, as it was for me. That included not just college and junior college students but those in K–12 schools. (More generally, we were well aware of and very determined to limit the mental health strains on everyone in America.) At one point during our preparations for issuing school reopening guidance, Elinore McCance-Katz, the assistant secretary of health and human services for mental health and substance (SAMSA), addressed us. Dr. McCance-Katz and the First Lady spoke eloquently on the pandemic's severe effects on mental health. Frighteningly, there had been a decrease in reporting of child abuse. With schools closed, teachers and administrators, both of whom are bound by law to report any suspicions of abuse in the home, were no longer seeing children in person. I had been concerned about this since March. One of my daughters is a social worker and she knows this world well and how critical a partnership with teachers is in ensuring that children have safe homes to return to in the evening.

McCance-Katz presented deep and compelling data on the deteriorating mental health of our children, documenting increasing calls to help lines and emergency room visits rising for adolescents and young adults with suicidal thoughts and actions. We needed to address these issues and work across the agencies to provide additional guidance and work to get schools fully open safely. No one could argue with that. I understood this and was an ally in getting this information to parents, teachers, and school boards to support a road map to opening K–12 and universities safely. I believed it was possible to keep schools open safely with regular testing programs in place; getting enough tests to educational institutions was important.

Elinore McCance-Katz was absolutely right to raise these concerns. I wanted the CDC, as part of a whole-of-the-child approach, to include in their guidance to schools the overall mental health of our children and SAMSA's concerns about suicide and safety. I wanted parents and school officials to be alerted to this growing mental health issue. McCance-Katz wrote and called and said she wasn't getting any traction with the CDC to include these mental health issues in the school reopening guidance they were developing. I was so concerned that I wrote Bob asking the CDC to reconsider the McCance-Katz document that included data analysis and considerations developed by the NIH and SAMSA addressing the mental health alerts about depression, anxiety, and suicidal thoughts. With that included, parents and school boards could weigh this important aspect of in-person schooling while making decisions regarding reopening schools safely.

More than a year later, I was providing information about the Trump administration Covid-19 response to the special house subcommittee on Covid-19. A staffer pulled out the very email I had written to Bob implying this was yet another smoking gun demonstrating how the White House interfered with the CDC and their guidance. I was dumbfounded. These were smart concerned staffers but they had fallen into the same partisan divide. I responded that they should have been asking the CDC why they would not include the importance of the mental health of America's children in their school guidance to provide a whole child approach. At every level we have turned this into a black and white, red/blue partisan issue, and we have lost any ability to see that even people who don't agree with us can have critical insights and are worth listening to. McCance-Katz was worth listening to. She and her people were aware of the reality of decisions that impacted children and families. What they said was more important than any narrative other groups were trying to spin. It was also important, as we saw on these university visits, that the students themselves be heard at all levels of on-campus pandemic issues. They were key to helping change behaviors.

Another solution to mitigation resistance was direct involvement with student leaders and other influencers in the campus community. Reducing the traditional, top-down instructional hierarchy and replacing it with horizontal, collaborative teams provided a comprehensive approach to pandemic management. By conveying the notion that they didn't know everything, and by getting input from all quarters on mitigation measures, school officials were able to create solutions and get buy-in from students. This further fostered a sense of community and responsibility, sending the message *We're all in this together, and if we take the time to listen to one another, we'll find ways to resolve conflicts and build trust.*

As with the tribal nations, the schools we visited evinced no interest in adapting a one-size-fits-all approach to student needs. We saw this clearly with student employment. It is easy to forget that many students and members of the community rely on schools for jobs, either to pay for tuition or housing or to otherwise support themselves. At Ole Miss, LSU, the University of New Hampshire, and the University of Alabama, the school administration, local county commissioners, and mayors cooperated to ensure that the surrounding towns' commercial ventures remained open, to preserve as many of these jobs and work-study opportunities as possible. Cooperation between an institution and local authorities was critical to how the outbreak was managed. The University of Alabama administration went a step further to address student needs by allocating some of the CARES funds it had received to help students pay for their tuition or rent.

Housing circumstances affect college and university students as much as they do the rest of the American population. In Columbia, South Carolina, the mayor and city council were able to enact a measure banning house parties in rental apartments and houses. With landlords being held accountable to enforce this, you can be sure the message was passed along to student renters. When we visited the University of South Carolina, students there told us they had come onto campus fully expecting their university to go into lockdown—so they might as well party. The university's proactive testing programs prevented this, and perspectives and momentum shifted. As the semester progressed, the students saw that their school was staying open and was committed to a Covid-safe campus environment. They now had a stake in preserving what most of them wanted—to have as full a college experience as possible in very difficult circumstances. These students were seeing that how they behaved had both positive and negative consequences for them and their friends.

Communal housing during a pandemic—in dormitories, fraternities and sororities, and shared apartments—presented unique challenges, and universities devised solutions to keep students safe. In Lexington, Kentucky, and elsewhere, the living quarters themselves played a role in how the outbreak was managed. Over the previous nine years, all dormitories had been overhauled—with two people living in one room and sharing a bathroom. This reduced the level of social contact compared with places like Ole Miss, where the dorms' communal bathrooms made social distancing difficult and most likely contributed to viral transmission, despite all the good work their resident directors did. Similarly, when the University of Tennessee in Knoxville first opened, 680 students tested positive and 2,000-plus were put in isolation/quarantine. Many of the dormitories there had communal bathrooms.

At many schools where Greek life is a major part of the social scene, fraternities and sororities demonstrated their leadership by modeling mask wearing and chapter presidents were required by their national organization to enact contact tracing measures. Greek organizations even adapted their traditional membership rushes, turning them into virtual, online events. Contact tracing, both by university housing and the Greek organizations, illuminated the role of housing in the pandemic: When students wore masks in class and were properly spaced, extremely few instances of viral spread resulted. The vast majority of cases of Covid-19 resulted from social settings on and off campus. Transmission rates were high in Greek houses, not necessarily because their residents were partying, but because they shared living and eating spaces in a way many dorm residents did not. House mothers worked very hard to transform common spaces into safer areas.

From this we learned that adults living in communal spaces similar to a fraternity or sorority house needed regular testing to prevent spread. This applied to those working in shared spaces in offices and retail and service spaces, too. When preemptive testing programs ensured that infection was spotted quickly and the appropriate measures taken soon after, these places could remain open.

In the end, what was taking place at the universities represented a microcosm of the nationwide pandemic response and experience. I was heartened to see the innovation and flexibility at these schools, the willingness to listen and to respect others' beliefs and needs. When everyone—students, faculty, and the administration—had a seat at the table in planning, implementing, and using data to revise their approach, it resulted in joint learning and evolution.

On one campus visit, in assessing their efforts and what they faced moving forward, the university president said, "Human behavior is hard to manage." Yet, instead of giving up, the school had buckled down and brought students and faculty together, as a community, to listen, adjust, and overcome issues—all this in racially, politically, and ethnically diverse settings. They united around a single goal: to attend classes in person and not remotely, as they had had to do in March 2020.

Irum and I were inspired by their adaptability and their commitment to a shared goal and to one another. Whether the schools were in red states or blue states, they put aside their differences. This community-first spirit worked in the heat of the summer in the South and in the cold winter in the North—and provided us with a road map. Their planning was horizontal and not top down. Everyone's voice was heard. They adapted. The administrations running these institutions believed in and trusted their students' willingness and ability to understand both what they needed to do and the consequences of failing to do it. I couldn't help but wonder how different things across the country might have been if the rest of us had acted with the same intention to protect one another.

These students and their community, and the tribal nation leaders and members we'd met, showed us that, now more than ever, we needed to focus on what united us, not what divided us, a point worth sharing—even as we entered the middle of the most divisive election season in modern history.

CHAPTER 16

You Can't Quit

As was too often the case during my time in the White House, my attempts to share information, recommendations, and insights continued to meet with resistance. Communication in large organizations, in any human endeavor, is always difficult. In this case, with this administration, among certain elements, there seemed to be an inverse correlation between the rise in the urgency of the pandemic and the administration's failure to—or desire not to—communicate effectively.

Despite having gathered effective solutions on our road trips, I found it was difficult to convey them, much less rouse a sense of urgency. Some in the administration still considered the deceptive lull in cases in late August a permanent condition. I tried to make it clear that for every trough, there was going to be another rise, another crest—just in a different part of the country. We were gaining greater clarity about how this virus moved through communities and regions. We were seeing the earliest signs that indicated a cascade was beginning. We understood that the greatest drivers of community spread were friend and family indoor gatherings. How precipitous that rise would be depended on many factors, but we had to remain vigilant and stay on message. We had solutions, but they needed to be implemented appropriately and consistently.

Since July, I'd been placing in the governor's reports key common mitigation recommendations for what to do when counties were in the red or yellow zone. As a result of what we'd learned on our trips, I'd refined these reports with the practical, ground-tested solutions that were working. I had also developed a list of mitigation bullet points for the entire state. Those went to every governor and state health official based on the level of community spread. In short: *These are the things you need to do in red and yellow counties.*

While I was still on the road, Marc Short had written to me stating, essentially, that these common mitigation points in the governor's reports had to stop. He wanted the county-specific red and yellow zone language removed. We'd reviewed these with him just weeks earlier, in July, and now he was reconsidering.

For best mitigation in the most at-risk (red) counties—those with the highest spread and a test positivity rate greater than 10 percent—I had recommended that they reduce hours of operation for bars, as South Carolina had done, closing them entirely or at 10:30 p.m., to prevent packed indoor gatherings. Restaurants needed to reduce indoor dining capacity and create more outdoor spaces. Friend and family gatherings in homes needed to be reduced to ten people. Based on the level of community spread in those counties, they also needed to: institute weekly testing of all workers in assisted-living and long-term-care facilities; mandate masks at all indoor retail and personal services businesses; provide targeted, tailored messaging on the risk of serious disease for those with comorbidities and others in high-risk categories; recruit more contact tracers to ensure that all exposed or infected were contacted and that positive households tested within twenty-four hours; and provide isolation facilities outside households if Covid-positive individuals couldn't quarantine successfully. It also included steps to increase testing, specifically in areas with the highest case rates.

These highly specific recommendations and suggested steps had been approved, posted, and used for more than two months now, in state after state. Many governors had told us that federal recommendations like these were critical for them to justify the actions they were taking to limit community spread. These recommendations had, in many cases, been very effective at slowing the spread and reducing fatalities. Why change them now?

I wrote back to Marc Short, telling him that the optics of any change would be horrible. The guidelines were out there, had been out there since July. Any abrupt change to them would look suspicious. I asked if we could hold off for a week, until I returned to Washington. He agreed. But when we did finally meet, his resolve had only hardened. He told me the reports had to be revised and the bulleted action steps deleted.

I pushed back: "There are critical aspects of this, college and antigen testing information that can make a real difference. If you stop the governor's reports, then that information won't get to those who need to know right now."

Marc was adamant: "You're overstepping. You cannot include the common recommendations to red and yellow counties. If you do, the reports won't go out. It's that simple."

My data contradicted the president's message that the situation was better and would continue to improve. But we couldn't relax our vigilance now. The states needed to put in place the effective solutions we'd seen on campus communities and elsewhere.

Feeling thwarted, I agreed to rewrite the reports. I removed the common red and yellow county bullet points in question. I wasn't happy, but I had to balance the data and specifics with the interests of the rest of the state. If the reports didn't go out at all, the governors wouldn't get the county-by-county data they needed to know the precise location and level of spread in their state.

I devised a work-around for the governor's reports I was then writing. Instead of including those recommendations in the common bulleted list, I'd include them in the pandemic summary and state-specific recommendations in the governor's reports, where they wouldn't be so obvious. These weekly reports couldn't go out on Monday without administration approval. Week by week Marc's office began providing line-by-line edits. After the heavily edited documents were returned to me, I'd reinsert what they had objected to, but place it in those different locations. I'd also reorder and restructure the bullet points so the most salient—the points the administration objected to most—no longer fell at the start of the bullet points. I shared these strategies with the three members of the data team also writing these reports. Our Saturday and Sunday report-writing routine soon became: write, submit, revise, hide, resubmit. Fortunately, this strategic sleight-of-hand worked. That they never seemed to catch this subterfuge left me to conclude that, either they read the finished reports too quickly or they neglected to do the word search that would have revealed the language to which they objected.

In slipping these changes past the gatekeepers and continuing to inform the governors of the need for the big-three mitigations—masks, sentinel testing, and limits on indoor social gatherings—I felt confident I was giving the states permission to escalate public health mitigation with the fall and winter coming.

This wasn't the only bit of subterfuge I had to engage in. Immediately after the Atlas-influenced revised CDC testing guidance went up in late August, I contacted Bob Redfield. He confirmed my suspicions: he had disagreed with the guidance, but had felt pressured by HHS and the White House to post it. Also, many on his staff in Atlanta were still comfortable prioritizing symptomatic individuals. Even at this late point, eight months into the pandemic, many at both the White House and the CDC still refused to see that silent spread played a prominent role in viral spread and that it started with social gatherings, especially among the younger adults. We had to find a way around them. Recognizing the damage to public health the Scott Atlas–driven testing guidance could do and was doing with testing rates dropping across the country, Bob and I agreed to quietly rewrite the guidance and post it to the CDC website. We would not seek approval. Because we were both quite busy, it might take a week or two, but we were committed to subverting the dangerous message that limiting testing was the right thing to do.

On the first day of September, while working on this rewrite, I received an email from Alyssa Farah, the White House director of strategic communications and assistant to President Trump since April. She asked—as if the governor's reports were a new development—why two sets of data were being used. Governors were complaining to the White House about the source of our data.

I understood the subtext: No governor wanted their state classified as a red zone. Apart from indicating that the statewide positivity rate was more than 10 percent, a sign of broad community spread, the "red zone" appellation, some governors no doubt feared, would convey the perception that the state wasn't handling the crisis well. Again, politics and public health were entwined. During a pandemic (or other crisis), political concerns, like an upcoming election, can do one of two things: either they can motivate a leader to do the hard things to mitigate the crisis or they can prompt that leader to question public health efforts at the expense of their constituents' well-being.

It was easy for me to answer Alyssa Farah's actual question, and I did, strongly emphasizing that I used only one set of data across the board—that data in the weekly governor's reports and the daily reports to the administration. Throughout my career, I have never played one set of numbers against another. The numbers are the numbers. On their own, they neither lie nor distort. How they get interpreted, however, can be variable.

Alyssa Farah's question foretold a deeper concern about why the governors were complaining about the source of the data. Did that mean they were receiving data that ran counter to ours? Or was this complaint a shorthand way of challenging my results. And who was behind this questioning?

This was the first time I'd heard of the governors' data concerns. Prior to this, my experience had been that if a state official had an issue with the numbers, they contacted us, and our data team worked with them to resolve it. Specifically, Alyssa seemed interested in the figures for California, Arizona, and Florida. For example, the governor of California believed his state's test positivity rate was 9.9, putting it in the orange zone. But the state was actually at 10.1, which placed it in the red zone. Both the timing of this inquiry and who specifically was mentioned were concerning. Governor Newsom had complained about this before, so it was easy to understand why he was among the three governors registering dissatisfaction with a difference between 9.9 and 10.1.

More puzzling were the complaints from the other two states. I'd worked closely with Governor Ducey of Arizona and Governor DeSantis of Florida to manage the serious outbreaks in their states. Both had experienced success. I'd even been using Arizona's and Florida's (particularly Miami's) success as a talking point.

Governor DeSantis's complaints to the White House brought this political-versus-public-health divide into stark relief. It also vividly illustrated Scott Atlas's damaging influence. From the outset of my tenure, I'd spent an enormous amount of time and energy supporting Florida's management of the crisis, whether in securing PPE or providing staffing support. We had learned a lot from Florida in our weekly calls with the Miami mayor, on the governors' calls, in private phone calls, and during in-person visits. We learned about the role of family gatherings in continuing community spread from the two mayors of Miami and Miami/Dade county. We saw health care innovation creating "hospitals without walls" from our Tampa visit. DeSantis had agreed to allow local officials to issue the needed mitigation including mask mandates, due to the diversity of case rates across the state, and he supported community-level enhanced testing in neighborhoods with elevated rates.

Those approaches had worked. From a summer peak of more than 15,000 new cases per day on July 12 (two weeks after my meetings with the governor) to the first of September, the rate of daily new cases fell below 2,500. Unfortunately, on September 13, Scott Atlas was in Florida, speaking with Governor DeSantis. Irum and I were in Texas when Tyler Ann informed us of this. As we drove from Texas into the Southeast, through torrential rains from Hurricane Beta—a constant cloudy, looming presence—our mood soured. We knew that elected leaders and communities had been working there for months trying to control their community spread. We wondered what effect this latest contact with Atlas would now have on Governor DeSantis and his vulnerable state.

Less than a week later, Bob and I had finished our rewrite of the guidance and surreptitiously posted it. We had restored the emphasis on testing to detect areas where silent spread was occurring. It was a risky move, and we hoped everyone in the White House would be too busy campaigning to realize what Bob and I had done. We weren't being transparent with the powers that be in the White House, but we were being transparent with the American people.

On September 18, I was still on the road—in Arizona again, for a meeting with those conducting proactive testing at the University of Arizona—when Mark Meadows's name and number flashed across my White House–issued smartphone.

"What the hell do you think you're doing? You rewrote and posted the CDC testing stuff."

"Yes, I did, but—"

"There's no 'buts' here. You went over my head."

I explained why I had done it. We'd already seen the drop in testing numbers resulting from Scott Atlas's dangerous guidelines. Those few pages we'd rewritten would change how states could test, and we'd prevent even more community spread going into the dangerous winter ahead.

Mark Meadows took this in and then, biting off each of his words, said, "You went over everyone else on the task force's heads. You went around the whole approval process. You do not make unilateral decisions. It's that simple. Period. End of sentence. Understood? Don't ever do this again."

"Understood. I did what I needed to do."

"Don't do that again without talking with me first."

Some of the edge in his tone had softened. I suspected he believed Bob and I were right. He never said this, but in this case, his actions spoke louder than words. He allowed the new testing guidance to stand. He didn't have to; he could have removed it and replaced it with the Atlas–influenced version.

This was the dichotomy that defined work in this White House and had helped justify my decision to stay. Yes, the chief of staff had every right to be angry at me for what I had done. But he had listened to why I had changed the guidance and he had let it stay up.

People are not one-dimensional; everything isn't in black or white— even if the media often paints a picture in these tones, with catchy sound bites and headlines. In the end, at certain moments, just as Mark Meadows did the right thing here, so did others. Still, this near miss didn't completely offset what I saw every day—how Covid-19 communication efforts we so desperately needed and that should have been priorities were diminished by this White House.

I had to continue to ensure that science was at the decision-making table inside the White House. If I hadn't been there, along with Bob, to subvert the process that had empowered Scott Atlas, this occasion when the Trump administration did the right thing almost certainly would not have happened. Spread this principle across the entirety of the response—and all the critical victories like this one might not have happened. I couldn't quit. I couldn't leave Bob, Tony, and Steve alone on that fulcrum. (We were joined by so many others who worked in supporting roles, and I couldn't abandon those people, either.) Each and every day the weight of all the physicians together on the task force could help offset the enormity of the disparity that existed between doing what the science and data dictated and what political considerations required.

I clearly recall standing in the Arizona heat while on that call with Mark Meadows. I was sweating—from the heat, from anxiety over possibly being late to my meeting, and from the knowledge that the guidance gambit was only the tip of the iceberg of my transgressions in my effort to subvert Scott Atlas's dangerous positions.

Ever since Vice President Pence told me to do what I needed to do, I'd engaged in very blunt conversations with the governors. I spoke the truth that some White House senior advisors weren't willing to acknowledge. Censoring my reports and putting up guidance that negated the known solutions was only going to perpetuate Covid-19's vicious circle. What I couldn't sneak past the gatekeepers in my reports, I said in person.

I'm grateful that our CDC test guidance efforts paid off. On August 31, the day the CDC released the new Atlas–inspired "anti-testing" guidelines, 702,320 new tests were administered. By the end of September, after Bob and I had revised the guidance, we were back to more than 1 million tests a day, and we would achieve 1.7 million per day in January 2021. Over the four months after that, testing would hit a free fall, decreasing to 300,000 per day in May 2021. From the time we changed the guidance until the election, we doubled the number of people being tested, restoring some order to the universe. Yes, with more testing, we'd have a better sense of just how dire things were. This was never good news, but at least we'd have more accurate information to help direct attention and resources to the most critical areas.

On September 19, the day after Mark Meadows's angry call, I received a call from the surgeon general of Florida. The news was not good: Scott Atlas had told the governor that the state had achieved the level of infection that met the criteria for herd immunity, that it wouldn't experience a significant surge again, and that it didn't need to keep mitigating as it had been. Scott Atlas was wrong. Florida had not achieved herd immunity. It would see another significant surge, the holidays would come, gatherings would increase, and the virus would spread again.

After ending the call with Florida's surgeon general, I immediately wrote to Jared Kushner. I informed him that we couldn't afford to have another state reverse course like Florida had. If Atlas was going to travel to states like I was doing, then we needed to send him someplace like South Dakota, where the outbreak was low and where he could do less real harm. Governor Kristi Noem wasn't mitigating in any serious way, and Dr. Atlas would find a kindred spirit there. We couldn't have Atlas telling governors just before winter that testing, masking, and other mitigation were no longer needed.

Jared wrote back to say he got it; he'd see what he could do.

When it came to Florida, though, whatever Jared could do didn't really matter: Governor DeSantis appeared to believe Scott Atlas and had begun to act on those beliefs. On September 24, the governor proposed a reckless college "bill of rights" for students to party. This was in the aftermath of Tallahassee police breaking up dozens of large gatherings at Florida State University. The next day, DeSantis issued an order that effectively reversed what he'd told me he supported when I visited with him—local control and local mitigation as needed based on the local data. His new edict prohibited local governments from enacting comprehensive mitigation measures that were working (these included fines for violating mask mandates), allowed restaurants to open at full capacity, and prevented local governments from ordering restaurants to operate at less than 50 percent capacity.

Each of these moves countermanded the three basic mitigation measures that had proven so effective and that Scott Atlas had challenged as unnecessary—he had obviously convinced the governor they were no longer needed. At the time, daily cases in Florida had fallen from an end-of-July peak, but as we'd seen time and again, a relaxation of vigilance if not reinstated at the time of a surge would result in cases skyrocketing. Just shy of a month after the governor's new edict was instituted in Florida, daily case figures there doubled. Hospitalizations and fatalities increased. The same pattern we'd been seeing was at work.

By the end of September, Governor DeSantis had lifted all restrictions and prevented local counties and cities from adopting their own restrictions based on ground-level assessments. His about-face wasn't a result of a significant decrease in cases; instead it seemed to be the result of embracing Scott Atlas's beliefs and positions. Apparently in a single visit to Florida, Atlas had undone months of work.

At the end of September, Florida was seeing just shy of 3,000 new cases per day and had lost 16,490 of its residents to Covid-19 disease. Three months later by the end of the year, with the holidays just passed and without the needed mitigations in place, these figures rose to 21,000 new cases in one day and 23,349 deaths.

THE SEARCH FOR GOOD news in the pandemic data continued in the White House. Throughout August, the president had made the claim that we were doing better than Europe. He did so by misusing an excess mortality figure someone in the White House had given him, likely the same person who had given him the graphs for the *Axios* interview. Put simply, "excess mortality" aggregates all causes of death during a period into a single number that can be compared to a prior period. For example, March 2019–July 2019 compared with March 2020–July 2020 (read: prepandemic to pandemic). Someone in the White House who knew that public health agencies around the world tracked excess mortality figures was analyzing them. Believing that this figure showed that the number of deaths was higher in Europe than in the United States, they had used it to prove that we were "doing better" in response to Covid-19. I didn't believe the numbers supported this.

For a long time I'd been urged to provide the White House communications people with positive news the administration could use to bolster its message of "We're doing great!" I guess they'd tired of my not fabricating anything usable. They knew the United States was coming out of a surge, but we were heading into fall and winter, and I knew the worst was yet to come.

I enlisted the support of HHS secretary Alex Azar, and together we asked the CDC to review the administration's calculations. They did so, and came back with an excess mortality figure that showed we were only 8 percent better than all of Europe. As a scientist, I knew this was statistically insignificant. The administration had not quantified their "better than." In citing their figure, the White House had obscured the fact that the most relevant data in a pandemic is data that confirms whether you are preventing or limiting community spread. Excess mortality rates, and final fatality rates—all these had their place, but these measures were more useful for analyzing a situation *after* a surge or after the full pandemic was over. They were more useful for telling the story of *what happened* rather than guiding you in the present to do what was necessary to alter that story's ending. Sometimes they told a tale of the very recent past, but they always gave a glimpse of what was in the rearview mirror, and not what the view was through the windshield.

I wrote to the Staff Sec, the gatekeeper of all White House communications to the president, to inform the staff there that this fruitless use of data distorted the truth, shifted priorities and perspectives, and fundamentally altered the course of the response. The data misuse and misunderstanding had to stop.

I didn't get a response. The distortion of data and miscommunications in all forms, intentional or otherwise, continued.

There were times when I received clear signals from the outside world that I was not alone in my thinking. They often coincided with a low point I'd experienced and the world had witnessed—after the "disinfectant" briefing, for example, or when the press or the president had written or said something unfavorable about me. These messages from outside the White House didn't come to my direct line.

Once, Condoleezza Rice used the White House operator to be put in touch with me. We had met a number of times during her tenure as secretary of state in the George W. Bush administration. Though she seemed to have a sixth sense about when to call, she never cited these low points as her reason for reaching out. Instead, she'd ask me about an issue related to PEPFAR—among her many ongoing commitments was her passionate belief in the importance of global HIV/AIDS programs. Sometimes she would call to ask about global vaccination programs or other public health issues, like malaria. She was always upbeat and expressed her gratitude for the work I'd been doing in my other capacity. It was as if she were sending me an encoded message. She wasn't working directly as a cheerleader, encouraging me from the sidelines as I took on my role as task force coordinator. She was more like a coach, reminding me that I'd done good work in the past, that I had navigated difficult times, and that, in the end, it would all be worth it because of the lives we had saved.

I also believed that every time she called, she wasn't speaking just for herself but also for the former president. PEPFAR had been launched in 2003, during George W. Bush's administration. The president and Mrs. Bush had a deep commitment to improving the health of others, including in the face of HIV/AIDS and malaria, particularly in Africa. This was in keeping with their compassionate conservatism. I also knew that in contemplating a run for the White House, President Bush had consulted with Condoleezza, who advised him that Africa should play a large role in his foreign policy. I sensed that he couldn't contact me directly—he had to avoid the perception of meddling in the Trump administration's handling of the present crisis. Still, each time I spoke with Condoleezza, I felt I was hearing President Bush saying, *I'm glad you're there. Thank you for staying there.*

I was struck by the difference in tone between President Bush and President Trump. Through my PEPFAR interactions with President Bush, I'd seen him as a very positive force for action. He believed that people could accomplish great things if they were supported—even when they doubted their own abilities. I had worked in Africa before PEPFAR, and I had stood witness to the death and despair permeating every village and household as AIDS claimed friends and family members. President Bush had a vision: He believed we would be able to put two million people on HIV treatment. *Oh my gosh,* I thought, when I heard this. *That is a really aggressive goal.* At that time, fewer than fifty thousand Africans were being treated for HIV infection. I knew the reality on the ground: limited physical infrastructure and personnel. But President Bush had a way of making all of us confident in taking on the really difficult tasks and completing them. He created a space where people could succeed, supported us to make the impossible possible.

Trump's White House was the opposite in many ways.

As September rolled on and many in the White House moved into campaign mode, I saw the distortions and deceptions only get worse, not better. Back in March, I was led to believe that Google had developed an app that would allow users to find out the nearest testing location and that the app was operational. When I found this wasn't true, I knew I had to question everything I'd been told. Now, the number of untruths, partial truths, distortions, and blind spots were coming at us so fast and furious that dealing with them all in real time was a near impossibility.

By mid-September, I was worn out from dealing with the politicization of the entire enterprise. Fighting both the pandemic and Atlas's influence had used up nearly all my resolve. For the first time, I began to question whether it was worth it for me to be there. Prior to this, those thoughts had been fleeting, but as it became clearer to me that I was being pushed to the side, they became more persistent. For the first time, I considered resigning. I am not a quitter, but I had to weigh the costs and benefits of my remaining. I'd come into the White House fully aware that there had been a revolving door of officials who said they'd stayed at the White House as long as they had out of fear of what would happen if they left. It was a common trope: they had to remain there to be the "adult in the room." This was less an issue, in my case, than my concern that whoever took my place would be a huge setback for the larger pandemic response. I felt certain that Scott Atlas would be tapped to become the new task force coordinator. All signs pointed to it. He'd been the only person added to the roster. He'd played an active role in the writing of the CDC testing guidance. He seemed to have allies among others in the administration and at the agencies. The extent of his influence was difficult to fully discern. If I left, I believed that I'd be removing one element of resistance in his way toward shaping the response.

The consequences of their bringing in someone more sympathetic to the administration's positions on testing, masks, and the data would be disastrous. The data team I had constructed with Irum was the comprehensive daily truth that transcended the White House and the State Houses. Maintaining that and pushing for more and better data would be not only key for this pandemic response but also for creating the new pandemic preparedness strategies. They would need comprehensive data to review and I wanted to make sure that existed. It wasn't just that this new person wouldn't push as hard on the things I prioritized, or would give President Trump his way more, it was that this person would almost certainly actively work to undo much of what I'd helped accomplish. It was one thing for progress to have slowed as it had, but entirely different to think about someone leading the response *backward*: misrepresenting federal data to skew the message toward whatever story President Trump wanted to tell; dismantling the national reporting systems that were created; telling the public that Covid-19 wasn't a big deal, that we should just go about our lives; accepting hundreds of thousands of lives lost as a natural and unavoidable cost of a successful economy, an acceptable form of collateral damage.

When I thought about how much a new response coordinator, one aligned with President Trump on the most important issues, would risk American lives, the stakes for my quitting suddenly felt incredibly high. From where I sat, the picture was plain: if I left, more people would die—not because I was irreplaceable, but because the last two months had shown the willingness of the Trump administration to bend data to fit their desired ends, to misrepresent reality. I wouldn't bend data and I could provide that daily analysis based on all the integrated data.

Still, in spite of all that logic, I couldn't escape the feeling that quitting would help me breathe a much desired and needed sigh of relief. I'd been at it nearly round the clock for months. It felt as though something was about to give. I needed a break, a fresh perspective, a reminder of what was possible, and of the successes of the past. Proactively, I had called a member of the Bush team and said I would be coming through

Texas on our next road trip and that I'd like to meet with the president. And on September 19, I was in the Bush family home in Dallas, getting a tour of the president's painting studio. I opted to follow Condoleezza's strategy, and when we sat down to speak, I briefed the president on PEPFAR, letting him know that Covid-19 had not gotten in the way of people's treatment. The work we had done to build the capacity of indigenous groups for service delivery had been critical, as they were carrying the ball. Most Americans who had been working there were back in the United States, and most international partners were unable to fly into Africa. Treatment access was being sustained, but we were losing ground on our prevention activities. The former president and I reviewed the impact surveys in Zimbabwe and Lesotho, seeing the evidence of our controlling the HIV pandemic. We talked about how the investments made in infrastructure and human capacity for HIV were serving the broader cause and now supporting the overall Covid-19 response there.

I paused, and I then laid it all out on the line, letting him know why I had actually come to speak with him. "I am not having the impact on the White House with the current pandemic that I would like. I should focus on the other pandemic. I think I should return full time to my PEPFAR position at the State Department."

The president regarded me for a few moments, his eyes glinting. I knew he understood what I was really saying. Finally he said, "So, Deb. What you're saying is you *want* to *quit*."

Hearing those two words paired that way made me drop my gaze from his eyes to my own hands. We were heading into the worst part of the Covid-19 pandemic. He knew those words would stop me in my tracks.

I looked up and saw his trademark grin. Nearly chuckling, he said, "Well, you know you can't do that. I know you know that's right. You've got to do this. You need to finish. You can never quit."

I felt an infusion of hope and energy. Okay. Good enough. I'd gotten another set of orders.

"Never again" versus "never quit." What a difference a single word makes.

President Bush has remained an anchor for me. He is a good and decent person with unlimited empathy for others. His and Laura Bush's compassionate conservatism had driven them to found PEPFAR, the

Millennium Challenge Corporation, and the President's Malaria Initiative, each of which positively impacts the lives of millions around the globe. He continued his work to improve global health through Go Further, a campaign to prevent HIV-positive women from succumbing to cervical cancer.

I liked that, in dealing with me, President Bush didn't pull any punches. He didn't throw any, either. I respected him and his insights. On this occasion, he wasn't telling me what I wanted to hear. (If he had said, "I get it. It's an impossible task. You should leave," I would have done that.) He also didn't paint a rosy picture of the days ahead. Without being directly critical of the present administration, he let me know he understood what we were all up against. All the more reason to carry on.

That was it. We chatted a bit more about his art and his present life and PEPFAR. It was September 19, and in about four months, as it turned out, a new administration would be sworn in. I knew we were in for a rough patch with the upcoming election, but I had no idea just how awful political events in the country would become.

CHAPTER 17

Outsider

With those inside the White House thinking of me and treating me as an outsider, I was always puzzled when people outside that world believed I was a White House insider. The effects of being an Oval Office outsider extended in many directions, most notably in my not being privy to advance information about the president's health when he tested positive for Covid-19 the first week of October. Like the vast majority of Americans, I learned of his diagnosis by watching the news.

I was immediately concerned. He was among the many Americans with contributing comorbidities that might complicate his treatment and recovery. Out of personal respect for his privacy, and because of HIPAA regulations, I never asked anyone anything about the president's illness beyond the information the White House released. But, I did learn, shortly before it became widely known, that he was being transferred by helicopter to Walter Reed. Tyler Ann heard this through the White House grapevine and shared it with me.

She was alarmed as was I to hear this. The president struck me as someone who would want to avoid being treated in a hospital at all costs. To me, it seemed, he always wanted to be in his own space and his own bed. When I learned he was going to Walter Reed, I presumed he was in need of treatment and monitoring beyond what he could receive at the White House. Most likely, he was in need of more than monoclonal antibodies and remdesivir. Later, Mark Meadows, after initially telling the press on Friday that the president had mild symptoms, said that the president's vital signs were concerning. In the book Meadows wrote later, he states that the president's oxygen levels had dipped into the mid-eighties, a serious harbinger to respiratory failure, and that he needed proactive access to significant respiratory support. In the same memoir, Meadows reported that the president had tested positive on September 26, four days before the first presidential debate with Joe Biden. He then tested negative after taking a more "accurate" test (Trump has denied Meadows's account). Prior to this 2021 revelation, the administration had claimed that the president had entered the debate, as required, having tested negative for SARS-CoV-2. At the time, and until Mark Meadows revealed it more than a year after the fact, I had no knowledge of this potential deception.

At the time, in 2020, after the media reported that Hope Hicks was sick with COVID on October 1, a rumor had been circulating in the White House that she could have potentially infected the president. All of us in the West Wing worried about being the person who potentially infects the president. But if Mark Meadows is right about the earlier positive test of the president he could have actually been the one who infected Hope.

That the White House and the president might have engaged in deception around the timing of his infection doesn't surprise me. The White House couldn't keep its own Covid outbreaks under control, and contact tracing and transparency were limited. The White House never wanted to acknowledge the depth and breadth of each outbreak. (No White House would.) I found out about most White House outbreaks from the press or by seeing empty desks. Between the mixing that occurred at campaign events and indoor White House gatherings and the incredible numbers of people the Trump campaign was exposed to on the road and back at the White House, it was all but guaranteed that someone acutely infected would penetrate the sole mitigation of testing. Each of these outbreaks was a threat to the president, the vice president, and anyone with underlying conditions who worked at the White House.

I wasn't surprised then that the president was infected. It had been one of my greatest fears for him and the vice president. This was part of the reason I routinely masked, and often distanced myself substantially from the president, whenever I was in the Oval Office. No one in the

White House wanted to be linked to infecting a senior leader and go down in history as the Typhoid Mary of the 2020 White House.

I believed then, and still do, that outdoor gatherings were far safer than indoor ones. The media had categorized Amy Coney Barrett's Rose Garden announcement gathering as a superspreader event. It's true that eight or more attendees did later test positive. The real story was the frequency of unmasked indoor activities, both at the White House and at hotels and other locations near the White House, at that time and at other times throughout Trump's last year in office. I am sure there were multiple intimate unmasked indoor gatherings before and after the Rose Garden event that were truly responsible for the superspreading.

Days before we learned of the president's infection, I'd written to Marc Short imploring him to return to using Abbott's ID NOW, a point-of-care nucleic acid test with a lag time of fifteen minutes from swab to results, to test all White House personnel, both in the West Wing and the EEOB across from the White House. I'd become so concerned about the number of infected within the White House that I never took off my mask while in the building except when alone in my office. I had very little company in this. Even with the president himself infected and hospitalized, I noticed no increase in the number of people masking up.

I tried to set an example, to make clear just how important I thought masking was to protect oneself and others. The president's counter message and the top-down consequences of it were obvious—following his example, few people in the White House wore masks, and many of those who didn't got infected.

I'm walking proof of the efficacy of masks and other precautions. For all the positive cases in the White House, and for as much as I traveled then and have done even more since leaving the White House, I've never tested positive for asymptomatic SARS-CoV-2 infection or developed Covid-19; no one in my family has, either. Though we all still worked, and were in offices and other workplaces throughout 2020 and 2021, we all followed careful guidelines.

The two people I spent the most time with in the White House, Tyler Ann McGuffee and Irum Zaidi, haven't been infected. Masking indoors when in public, restricting any unmasked social gathering, and getting tested routinely protected my close circle of family and friends and me throughout surge after surge. These simple precautions worked for us, and they had worked elsewhere. In the example of the White House, we all saw what not adhering to them could bring.

I wrote to the task force docs, advising them to stay away from the White House or always wear a mask while there. (Even in this relatively small community, contact tracing was inadequate.) I did the same for the vice president, urging Marc Short multiple times to limit his and his boss's exposure to that environment. I did the same in person with the vice president. Both men appreciated the warnings I was issuing and the precautions I was recommending, and the vice president did begin to wear a mask publicly.

I brought the same message to the White House security and maintenance people, to uniformed Secret Service personnel, and to the CIA employees who frequented the building and who were already masking. To anyone I came in contact with at the White House, I said, "Wear a mask. There is virus all over this complex. You need to be really careful. I can't tell you how dangerous this place is." I told one support staff member who I knew had inherited the trait for sickle-cell anemia that they should never take off their mask inside and that the only way to eat or drink safely was to do it outdoors. This staffer was among several others with vulnerabilities to whom I issued these strict directives. I wanted to make sure they knew how to protect themselves. All those with whom I was able to communicate directly were grateful that I had intervened and spoken honestly and that I was reinforcing the behaviors they'd already adopted to remain safe. The irony of this was not lost on me: my direct, face-to-face messaging was having a positive impact in my own workplace while the same message was either being ignored or altered for the rest of the country.

The White House was a microcosm for the rest of the nation. Those of us who worked in the West Wing were tested every morning. Despite this, people were still getting infected and passing along the virus. Why?

A test for SARS-CoV-2 is only a snapshot of your viral state at the exact moment your cells are swabbed. You could arrive at work at 9 a.m. and test negative. However, the virus could already be in your cells, using their machinery to quietly replicate itself. By 2 p.m., when the cells have burst open, shedding contagious virus, you would then test positive. This was a known limitation of testing for this virus. Testing without masking, without social distancing, and without reducing indoor gatherings wasn't as effective at containing spread, resulting in the repeated Covid-19 surges in the White House.

I loved the way Alabama governor Kay Ivey described it: The mitigation measures were like Swiss cheese. One layer of cheese (testing) had holes, but when you added another piece of cheese (masks), some of those holes were covered up. When you added yet another slice of cheese (physical distancing and reduced gatherings), all the holes were covered. And with the arrival of vaccines, we'd be able to add another layer. (Still, vaccines alone would always have their own holes and could never be a stand-alone substitute for the more layered approach when virus was active in the community.) We had worked so hard from the outset to establish this concept as a baseline, but with a White House that, at different times, either actively defied or implicitly undercut the importance of each of these "layers," the message was nearly impossible to get across.

The president's quick return to the White House from Walter Reed only heightened this apparent contradiction. While his recovery was good news, his performance on the South Portico steps, the dramatic removal of his mask as he gasped for breath, provided more mixed messages. His speedy turnaround was a testament both to what medical professionals had learned about how to more effectively treat Covid-19 disease and to the therapeutics that had been developed to save lives. And yet, President Trump's public appearance less than ten days after diagnosis spoke volumes at a time when we were asking infected Americans to isolate for ten days after symptoms and always to wear a mask around others.

Back at work, the president reverted to form, tweeting upon his release that the American people should not be afraid of the coronavirus—this when more than two hundred thousand of them had died of Covid-19. It was a blow to everything we'd been working to do and to everyone who had lost a loved one or their own life—losses that millions would continue to experience for decades. Despite benefiting personally from the best care in the country, the president was contributing to the belief that the more people who got infected, the better; that this disease wasn't particularly deadly. Time and again, the president and the White House acted without adequately accounting for the consequences of their actions and beliefs. We were fast approaching the staggering figure of 240,000 dead, even before the fall and winter, and the White House was suggesting by their words and actions that Americans had little to fear. Indeed, openly defying commonsense public health guidance appeared to have become a point of pride.

I had hoped for more from President Trump in this moment. With the fall and winter surge coming, Bob, Tony, and I had all hoped that his experience with the virus and developing significant Covid-19 disease would serve as a wake-up call for the president, the White House, and the senior advisors to take this virus seriously and do the right thing. I knew it was far too much to expect the president to come out and say he'd never again go without a mask while in close contact with others indoors. Or, that he'd never again participate in superspreader events, like his campaign rallies. Or, that he'd never again refer to SARS-CoV-2 as "the China virus."

Bob and I were both shocked and dismayed when, after his release from Walter Reed, Trump strode up to the South Portico, turned to the waiting press, and defiantly removed his mask. This symbolic gesture was akin to his thumbing his nose, not only at the task force and all the work we'd done for the last seven months, but also at all those Americans who'd spent months following our guidelines. I was saddened to think that *this* was the lesson President Trump had learned from his illness: *Double down*. If he was hoping to lead by example, his example terrified me. He'd dodged a bullet, and now many Americans would believe they could dodge one, too.

Trump's defiant attitude belied the reality that his survival had been dependent on care that was not yet immediately available to most of America. The good news was that we had developed effective treatments. This meant that, from here on out in the life of the pandemic, most patients could recover from the infection. The monoclonal antibodies the president had received were made available to him only through an FDA compassionate use provision. This very effective treatment would shortly thereafter be approved for wider use for other Covid-19 patients. Trump's "nothing to fear" remark was supported by one fact: we had come far in terms of treatment. However, while this was true for many of the infected, it was still not true for the most vulnerable. We were losing more Americans to Covid-19 disease *every month* than we lost during an entire annual flu season.

From the outset, we had talked about building a bridge to vaccinations. This was a very worthwhile goal, and we had used testing and early therapeutic interventions to build that span. We needed to do everything we could to help people survive until those vaccines—a preventative measure, not a treatment—came along. Also from the outset, I had warned against putting too much emphasis on the message that once vaccines arrived, all would be well. Vaccine development and deployment was only one prong of a multiprong approach to managing this pandemic, or any pandemic. At the time the president was infected, the White House was still signaling that vaccines would be available soon. Dangling the vaccine carrot while simultaneously taking a stick to the three important preventative measures was misleading and reckless then, and still is now.

As early as July, in my role as a board member for Operation Warp Speed, I had adamantly stated that it needed to be made clear to the American people that vaccines would not end the threat the SARS-CoV-2 virus posed globally. It would take at least two years to get the world immunized. Also, reports from around the world were beginning to show that prior natural infection was not preventing reinfection. The evidence was yet to be fully realized. We didn't know the durability of protection from either infection or severe disease with either natural infection or from vaccines. The vaccines being studied were to protect against severe disease, hospitalization, and death.

From the coronavirus vaccine trials and later, we would be able to gather data on a vaccine's effect on the path Covid-19 disease took in a person, but not on any protection it offered from asymptomatic infection. This was the case because even in the clinical trials, only symptomatic individuals were tested for Covid-19, and because the length of follow-up in the original trial was too short to evaluate the durability of immunity. As a result, from the outset of distribution, we still didn't know if the vaccines protected from silent infection—the very thing that created the first invisible community spread.

Vaccines that protect from all infection produce what is called "sterilizing immunity," a very high bar in the world of vaccine development. In some cases, like measles and mumps, natural infection in childhood leads to long-lived immunity and protection from any reinfection in most children. Some vaccines are designed to mimic the immune response that accompanies natural infection of measles and mumps, and these vaccines also result in long-lived immunity, known as sterilizing immunity to both infection and disease. Sterilizing immunity is almost impossible to achieve if it isn't possible with natural infection. You are asking a vaccine to do better than a natural infection. Many vaccines work by protecting against disease; any infection is often dealt with silently, is cleared up rapidly, preventing the vaccinated person from getting seriously ill.

The vaccines in development for this pandemic were designed to provide "protective immunity," limiting the virus's effects (Covid-19 disease) on us. Everything else about these vaccines was an open question that would need to be answered. We also didn't know how long that protection would last and this needed to be studied over time in real-world situations.

I then emphatically summarized: All individuals who were immunized could still potentially get infected and, even if asymptomatic, could still infect others. The vaccines if effective would provide protection from severe disease and death, but they did not create an impermeable bubble around the immunized. In time, after the vaccines were in use, we could use population-level data to determine the full extent of their protection against infection versus disease. In the meantime, if the virus was actively circulating in the community, people still had to test, wear masks indoors, socially distance, and limit the size of indoor gatherings. We couldn't give up on public health mitigations that we knew worked while we accumulated the evidence of the new vaccines' efficacy at the individual and population level. We needed to know the durability of protection, especially the suggestion that reinfection was happening after initial infection. I didn't know these answers then, but I knew they needed to be addressed.

With the vaccines still months away, and with so many other matters to attend to, and believing that I'd made my points abundantly clear—I *still* didn't let the matter drop. Over time, in the governor's reports, I reinforced the message that vaccines may offer only limited protection from infection; sterilizing immunity would be a high bar. There was no genie in the vial.

It was also important for people to understand that the pharmaceutical companies and the FDA had to look at not just how effective the vaccines were, but how safe. That October was when politics entered the vaccine discussion to a greater degree than before. Prior to this, Steve Hahn had frequently shared with the other doctors on the task force the enormous pressure he felt from the administration and the enormity of the decisions his agency, the FDA, had to make during this crisis. Mark Meadows, Jared Kushner, and the president had all called Steve, urging him to speed up the emergency use authorizations for vaccines and treatments. The numbers of deaths and cases were rising, and vaccines would be a critical intervention. The FDA needed to act, but it was caught between a rock and a hard place: okay the use of vaccines that had not been fully tested, with all its potential consequences, or stick to the established approval protocols and endure the president's wrath.

When Steve discussed the pressure he was under, it became clear he was distressed by being placed in this position by the president and his advisors. He felt Secretary Azar was conferring all the "blame" on him and not on the fact that even if things were sped up, essential procedures still needed to be followed. Once the pharmaceutical companies submitted their application to the FDA, he said, the FDA staff would work around the clock. But he went on record stating unequivocally that when it came to vaccine approval, the FDA would not sacrifice safety to speed. Whatever blowback there would be, the FDA wouldn't budge.

The blowback was immediate. Even as the president was recovering from Covid-19 in October, he was angling for voter support by tweeting, "New FDA rules make it more difficult for them to speed up vaccines for approval before Election Day. Just another political hit job!" He wanted his base to believe that if it weren't for the FDA, he would be able to make good on his promise to deliver vaccines at warp speed—in time for voters to go to the polls knowing that a literal shot in the arm was not months but only days or weeks away. Vaccine makers Pfizer and Moderna had also both been under personal pressure from the president, to speed vaccine production. All had remained steadfast in the face of this, stating they were putting safety first. Trump was angry that the FDA later recommended the gathering of at least two months of safety data after full immunization as part of the last phase of the vaccine trials. This would delay the submission of the data from the pharmaceutical companies by approximately fifteen to thirty days.

This data, from all the volunteers who were a part of the vaccine clinical trials, would be derived from follow-ups after they had received their second dose. It would take those two months to get a better indication of the vaccines' benefit-risk profile. Among other considerations, we needed to see if these volunteers experienced adverse effects from the shots or if any of them developed severe Covid-19 disease. The safety window could have been made even longer, to allow for even greater caution about the vaccines' safety, but at this point—with roughly a thousand people dying of Covid-19 disease each day—the FDA was walking a tightrope between the need to ensure safety and the need to be expedient. This was a delicate balance to strike.

Yet, in pressuring Steve Hahn and the manufacturers, the president was ignoring a very large elephant in the room—vaccine hesitancy, especially among adults. Sure, you could produce and approve an effective vaccine, but its efficacy was also dependent on our getting it into the arms of the American public. Every year, public health officials could see this substantial problem in the rates of adult vaccine uptake, including the flu vaccine. Whether it's because people are afraid of needles, have concerns about vaccine safety, or view getting vaccinated as inconvenient, as a nation, large numbers of Americans decline to get vaccinated against the annual flu variant. The CDC, which expends a great deal of time studying the quantitative data, has noted an emerging trend. Young children and the elderly get the flu vaccine at a much higher rate than adults. When all age groups are combined, and with variations from state to state, it found that only between 33 percent and 56 percent of adults got the flu vaccine in the 2018/19 flu season (the latest data available). But the CDC didn't study the qualitative reasons behind the low vaccine uptake and didn't do substantial or effective work to alter the rates. We didn't change adults' perceptions of vaccines by age, race, ethnicity, and geography. We didn't study the different reasons for lack of uptake of current vaccine utilizations and didn't understand hesitancy. Without doing these things, we'd never make an impact on vaccine uptake then or now.

We couldn't just rely on therapeutics, either. The supply of remdesivir, monoclonal antibodies, and other drugs for which we hoped to obtain expanded access authorization was finite. To be effective, many treatments required use within the earliest of days after infection— and this meant testing regularly. Also, the delivery systems and other aspects of treatment and vaccine distribution could be slow. Many of them had not been tested for use at this scale. Wearing masks, employing good hygiene, appropriately social distancing, and limiting indoor gatherings during surges were the main buttresses for the cathedral of care and common sense we'd put in place. The rest might collapse, but these structural components could not.

In my mind, by relying on therapeutics and the promise of vaccines, what the administration and the president himself were doing was dangerously akin to signaling to drivers of automobiles, *You have antilock brakes. You have airbags. You have safety crumple zones. You have other state-of-the-art driver safety features. Go ahead and drive as fast as you like and take as many chances as you like. We're betting you'll survive the inevitable wreck.* All the while, they were ignoring the fact that, to extend the metaphor, not everyone's cars were equipped that way. But: *Too bad for them. Too bad for the ones we might hit.* That attitude was also a reflection of our public health care system's bias toward treatment over prevention, something that drives up costs—in this case, the cost of human lives.

Between October 5, when the president was released from Walter Reed, and December 15, when the first newly approved vaccine dose was administered, another 105,000 Covid-19 infected Americans would not survive long enough to choose to be inoculated—and another 10 million new cases were reported. This didn't have to happen. We knew what early community spread looked like and the sequence of events that led to inevitable hospitalization and death. When test positivity began to rise locally, instead of doing what was needed (mask up, socially distance, limit indoor gatherings), the country waited until the hospitals filled up again, failing to believe the early warning of rising test positivity in those under thirty-five and saying to ourselves, wait and watch. *This time will be different.*

As a result, days to weeks later, more Americans succumbed to this virus. This kind of magical thinking was what had gotten us in trouble in 2020, and it would do so again in 2021. We were very good at making excuses, at highlighting the fact that we had vaccines, at claiming that this variant was not as deadly here in comparison to other countries. We made claims that this wave would look like the wave in such-and-such country, that this time the surge would not cause our hospital beds to fill, would not cause thousands of Americans to die every day. We seemed to prefer to believe that this time would be better than before, instead of using the effective tools we had to ensure that it was.

While I was touring the states, I took every opportunity to make Trump's Covid-19 scare a teachable moment: If the White House was experiencing outbreaks, the same thing could happen, *was happening*, in that state or locality. Nothing about being an American, working in the White House, conferred any special immunity. If the virus were allowed to freely circulate in your community without any mitigation, it would find its way to your vulnerable great-aunt, your grandmother, your brother with underlying conditions, your child with Down syndrome. The only way to maximize your chances of staying well, and keeping others well, was to consistently employ those three simple measures: tests, masks, social distancing by reducing indoor gatherings.

More to the point, members of the Trump administration and in the Scott Atlas camp had continued to put obstacles in front of our efforts to make clear the dangers that lay ahead as fall turned into winter and Thanksgiving and then Christmas celebrations were held. Fortunately, there were people—from those in the White House communications office, to Tucker Obenshain and others in the Office of Intergovernmental Affairs, to the vice president—who helped me get that message out.

IN THE FIRST WEEK of October, Marc Short handed off the editing of the weekly governor's reports to John Gray, because he was busy with campaign matters. Gray was a former advisor in the Office of Management and Budget and then deputy assistant to the president and director of policy for the vice president. In his first notes to me, Gray actually asked for "a bit more ambiguity" in our next governor's reports. He and I engaged in a line-by-line review of the wording, Gray challenging, changing, and even threatening not to send out these key reports meant to communicate a clear message about a situation that was costing lives.

At one point, Gray communicated to me not only that we needed to be more ambiguous, lest the governors think we were telling them what to do—never mind that, as I'd been reporting since spring, the governors were *asking* for exactly that—but that we weren't allowed to mention the possibility of reducing indoor dining or closing bars when counties were in the red zone. When I wrote to Marc Short, explaining that all these edits were harmful, he responded that Department of Education secretary Betsy DeVos was complaining about any mention in the reports of school closures even when the pandemic was raging in that community and no one was vaccinated.

We were recommending that schools in the green and yellow zones fully reopen and stay open and that we use other mitigation in the yellow zones when the counties became red hot with the highest level of community spread and there was a risk to children's primary care providers: grandmothers, teachers, bus drivers with underlying conditions. (A Black councilwoman had made this critical point to me during a roundtable in Virginia.) Governors and school boards needed to take this issue into account. They needed to prevent movement into the red zone by employing commonsense mitigation.

I never believed children couldn't be infected, as was suggested early on. Indeed, not only could they be infected—we saw this in the summer of 2020, when Americans were moving around the United States on vacation and rates in children rose—but I believed they would be part of the community of primarily silent spreaders. True, children might be more likely to have mild initial disease, but we didn't know the full impact of Covid-19 on them or the potential for long-haul Covid.

I reminded Gray that the task force didn't want schools to close or stay closed: we wanted each school to have the mitigation measures they needed to stay open safely—as so many universities had done through testing. Marc Short had also specifically mentioned to Gray that Governor DeSantis was the most vocal critic of the reports.

No longer surprised by anything that came out of this administration, I wrote back, emphatically asking, "Does [the governor] honestly need to see what unmitigated spread looks like?" We were in the lull before the storm, and we needed to empower the governors and Americans to do what they needed to protect their families. There was a clear way to put protective measures in place and save lives, one that wouldn't destroy the economy.

Those in charge of reviewing the weekly governor's reports seemed not to want to acknowledge the state data or the recommendations triggered by that data. They were far more concerned about public perception of White House overreach. While I was warning of far worse days ahead, they had expanded their censoring of key public health recommendations in the lead-up to the election. Week after week, month after month, September to November, the edits escalated. In September, their edits had focused on only the common mitigation points that appeared in each and every state report. Now I was being told to remove state-specific bullet point recommendations relevant to the level of viral spread in that state, laser-focused recommendations and solutions to use at the county level.

I dug in my heels and protested. This made no sense at all—these bullet points served as a reminder to governors of what they could do either to prevent their state from slipping over into the yellow or red zone or to get their state out of those zones. Again, I was told that we couldn't be perceived as telling the governors what to do. I tried to work around this with references to successful, strategic moves specific governors had made to effectively control local outbreaks—a classic effort to show and not tell. But these references to specific states were shot down as well. And I was back to my laptop to devise another strategy.

It was no coincidence that as Election Day approached, their censoring took on a new urgency. I couldn't come to any other conclusion. As the seriousness of the fall surge became apparent, so did their attempts to decrease our impact. They wanted to protect the governors who didn't want to implement the recommendations, and those governors didn't want a record of the actions the White House thought were critical to controlling community spread in their state.

While I was engaged in this censorship battle, I reviewed the draft of a proposed national strategy document. Generally, my edits consisted of inserting the words *safely* and *fully* in front statements about returning to work and reopening child care facilities, schools, camps, and universities. Justifiably, many people had criticized the federal government's response for a lack of clarity. That I was still having to revise documents, at this late stage, to emphasize safety reflected the White House messaging problem that, like a bad case of poison ivy, just would not go away. With the doctors on the task force having to twist ourselves into knots to release any public communication, messages became garbled and the simplest things—such as an emphasis on *safe* reopening—nearly fell through the cracks. I spent so much time creating work-arounds, that it was often impossible to see all those fissures. Steve Hahn and Bob Redfield were doing the same each and every day.

To combat John Gray's edits, I used the same method as before to disguise specific language. Whereas before I sensed the reviewers were not searching for key words to find and eliminate any objectionable terms, I knew that John Gray was too busy to read the entire report. I wrote to the three other report writers and asked them to move the key recommendations to different bulleted lists and not start those lists with the recommendations. This ran the risk of making it more difficult for the intended audience to find the message quickly, but if the reports didn't go out, then valuable information wouldn't get disseminated.

I engaged in more open confrontation with regard to the worst purveyor of misinformation: Scott Atlas. While the president was still being treated at Walter Reed, I wrote to Mark Meadows to tell him that Atlas should be removed from the task force. He could say what he wanted to the president, but I couldn't have him as part of the task force. Every meeting at which he was present was devolving into a fight over his faulty pronouncements. He never presented any supporting evidence and continued to invoke the president's name to bolster his argument. The doctors were in agreement. Atlas, as he had done from the outset, was still taking my daily reports to the administration, stripping my name off the distribution list, and providing analysis that ran counter to the message I was delivering: Here are the facts. Here is what this means for the foreseeable future. Here's what we can do to head off this surge.

Meadows asked me to give him forty-eight hours.

I found it strange that Atlas wouldn't send his critiques to me directly. Why was he playing this childish game? He might as well have folded his arms, rolled his eyes, and said, *I'm not talking to her*, an attitude more appropriate in a situation comedy than a Situation Room during a national emergency.

If I couldn't get Atlas off the task force immediately, I wanted him off the screens from which the American people got their news. I went to my allies in the White House Comms team responsible for booking media spots. (To protect them, I won't name them here, but I have to acknowledge the work they did to help this country.) Unlike some at higher levels in the administration, and unlike the Staff Secretary and the president, the Comms people got it. Scott Atlas's position on this virus ran counter to everything else they'd seen, read, and heard, and they agreed not to seek media opportunities for him. In fact, they told him he needed to do his own press outreach. They wouldn't silence him, but they weren't going to hand him a microphone to promote his as the voice of the White House Covid-19 response. Downstream, their efforts paid off in a way I couldn't have anticipated.

AS OCTOBER 2020 PROGRESSED, my sense of déjà vu became overpowering. It didn't help that while on that Northeastern college swing, I received communication from Vanderbilt University Hospital that they were out of remdesivir. This couldn't be happening, not at a facility in the Southeast, where for so long the pressure of this crisis on the health care system had been relentless. This kind of crisis management intervention was common, and exhausting for all involved. I asked Dr. Bob Kadlec, of the Office of the Assistant Secretary for Preparedness and Response (ASPR), to investigate. We learned that the supplies were there, but that an administrative snafu in processing orders within the hospital's internal pharmacy had produced the false shortage. Thank goodness. Still, with each and every such alert, the problem needed to be tracked down and resolved.

The supply chain alert system, called "Green Light," was working. The system that Rear Adm. John Polowczyk had envisioned but the CDC wouldn't implement was now operating effectively under HHS supervision. This was extra work for every hospital, but essential for ensuring they were getting the supplies, staffing, and therapeutics they needed. This clear and comprehensive hospital data from every hospital across the country was critical to saving lives but also became essential in understanding the full impact of each surge, the variable impact of the variants, and how to improve our pandemic response.

I also continued to go where I believed I still had some influence: the Covid Huddles. I wrote to Jared and the others in the Huddle that serious warning signs were emanating from the Upper Midwest and the Northern Plains states. The data showed that the spring surge had been driven by spread within workplaces and on public transport to workplaces, and the summer surge from friend and family gatherings. The fall surge was shaping up to be like the summer surge, as workplaces and transport became safer with masking and other precautions and the risk shifted to social gatherings indoors.

We had to make everyone aware of this. Because I couldn't successfully do that through the censored governor's reports, and because I couldn't be in all places at all times throughout the fall, doing local media, I reached out again to Jared and the Covid Huddle. Like nearly everyone else in the Trump inner circle, Jared had found his attention divided between the ongoing pandemic and the upcoming election. Like me, he was often only a voice over the phone in the Huddles. Just as the number of task force meetings had dwindled, so had my level of personal interaction with him. From September on, it wasn't so much that he and Mark Meadows and Marc Short had disappeared as much as slowly faded from view.

Still, in terms of support, Jared remained a constant, along with others in the Covid Huddle: Brad Smith, Adam Boehler, the White House Comms team, and Paul Mango from HHS. The intense coordination continued. The Comms team continued to set up hundreds of local media hits for me, Jerome Adams, and Secretary Azar. Brad and Adam helped support the coordination with FEMA and ASPR to make sure states were getting the supplies they needed in the moment and proactively in advance of the critical moments to come.

Tyler Ann remained vigilant to White House maneuvers and always had my back while I was on the road. She made sure I never missed an in-person meeting, over Zoom or the phone. The virus was out there, in every region of the country, and by the end of October, I'd travel to an additional twenty-three different locations in the Northeast, the Midwest, and the West.

Notable among the places I didn't visit was South Dakota. Governor Kristi Noem refused to meet with us. This came as no surprise. She was the leader of a rural state and, early and often, had bought into the message the Trump administration espoused. In July, the state had hosted an Independence Day celebration at Mount Rushmore, which the president attended. Governor Noem, a rancher's daughter, went on record as saying that, according to the science, it was "very, very difficult to spread the virus when you're asymptomatic." No data supported this core Atlas belief. Data did support the presence of the same viral load in the noses of the asymptomatic and symptomatic alike. Too frequently, those without symptoms were more likely to engage in social activities and spread the virus unknowingly. Noem acknowledged that mitigation was important, but insisted that it wasn't actually possible to stop the virus from spreading.

The outcome of this misinformed message wouldn't have been so bad if Governor Noem hadn't also encouraged people around the country to attend the annual motorcycle rally in Sturgis, South Dakota, in August. While a motorcycle ride was low risk, what happened before, after, and around the ride was not. Many of the 366,000 people estimated to attend the rally crowded into bars from August 7 to 16, participating in a national superspreader event. Two weeks after the rally, cases of

Covid-19 in South Dakota more than doubled, and hospitalizations tripled. To make matters worse, rally participants returned to their home states across the country, many of them bringing SARS-CoV-2 along for the ride. It is difficult to accurately determine the total effect of this event, but researchers at San Diego State University's Center for Health Economics and Policy Studies projected that 260,000 cases of Covid-19 could be linked to the Sturgis rally. Governor Noem labeled the study "fiction." The projections may have been "fiction," but during my travels across the Northern Plains and Rocky Mountain states, hospitals there reported patients whose Covid-19 disease was specifically tied to their attendance at the rally.

Despite rising cases and fatalities, Noem had held the line against any stay-at-home orders or masking, promoted the use of hydroxychloroquine, and—a month before we called to set up a visit—announced that she would spend five million dollars of federal Covid-19 relief money on a campaign to boost tourism in her state. We had never recommended she issue stay-at-home orders after the 15 and 30 Days to Slow the Spread campaigns, but we did recommend masking and testing to see the silent invasion and prevent spread.

On October 7, freshly out of the hospital, President Trump retweeted a clip of a session of the South Dakota Legislature in which Governor Noem described lockdowns as "useless." He captioned his tweet "Great job South Dakota." Out of fifty states, South Dakota currently ranks among the top twenty states in deaths per capita. We continued to send Governor Noem our reports, but she never converted our recommendations into action.

Actionable steps were precisely what was needed, not just for South Dakota, but for the nation. To that end, Tony, Bob, Steve, and I worked to formulate a strategic plan for the remainder of fall and into the winter. Regardless of the outcome of the election, we needed a scheme in place to get the response back on track. The main points were familiar ones: We had to communicate directly to the American people, let them know unequivocally and with no ambiguity, that the success we'd seen at universities and colleges could be translated into a national Covid-19 response. This, of course, meant expanding testing to the asymptomatic—moving testing from convenient to impactful.

We also had to end the battle of words. The governor's reports needed to be restored and preserved as the source for accurate data to be used in the states' decision making. We had to continue the weekly phone calls with the governors that the vice president led while always setting the standard for professionalism and using data for decision making. Internally, I asked for the task force meetings to be held three times a week—especially after the election, when I was certain (and later had confirmed) that cases, hospitalizations, and deaths would be on the rise. As for the four core doctors on the task force, we needed to meet daily to ensure that we all agreed on what we were seeing and what needed to be done and that we continued to maintain our united front.

I also included an anti–Scott Atlas component in my postelection strategy. We needed the administration to see, once and for all, that what Atlas had been advocating was wrong, was dangerous. We had to show Atlas and, by extension, the president and vice president that our balanced approach had been working—and point to Florida, and its sevenfold increase in cases three months after Atlas's visit with Governor DeSantis, as a cautionary tale. If Atlas's influence spread outside Florida, an additional tens of thousands of Americans could lose their lives.

I suggested an "Atlas Summit," a meeting during which Atlas and his herd immunity colleagues could explain what was transpiring in Florida—the rising cases, hospitalizations, and, soon, deaths occurring despite their having assured the governor his state wouldn't experience another deadly surge. Sadly, the Atlas theory was in action. I had said his approach failed in theory; now it was failing in practice. There was a middle ground, a middle ground created by adapting the UPenn/CHOP model—keeping businesses and schools open through masking, outdoor dining, and testing. Florida's total cases would go from about 750,000 in October 2020 to nearly 3.7 million in October 2021. Similarly, deaths in that same period would increase from 16,500 to just shy of 58,000. From June 1, 2020, to the end of October of that year, just over 15,000 Floridians succumbed to Covid-19 compared to nearly 24,000 in 2021 in that same time span. That 50 percent increase in fatalities occurred when vaccines were available.

This wasn't a case of hindsight being twenty-twenty. It was a case of what wasn't done in 2020 extending far into 2021—and, likely, beyond. Atlas had to be stopped. I'd later suggest that he be reassigned to work with Larry Kudlow and others on the economic side. The economy seemed to be what Atlas was actually most interested in preserving, so why not put him where his interests and abilities were best suited? Anywhere but as part of the response to a raging pandemic. We knew what worked. It was about ensuring consistent implementation across the country.

Throughout October, the task force worked toward finalizing the postelection strategic plan document. It wouldn't be easy to convince the senior White House operatives that continued mitigation with decreased social gatherings indoors and increased testing to find the silent invasion was the way forward, something the document would lay out explicitly for them. I pushed Meadows, Short, and Kushner on the main points, probing for reactions to each of the document's elements. It took longer than I hoped to get responses. I don't know if they now saw things the way I did, but I was pleased when I didn't receive any real pushback. We were hoping to institute in the fall and early winter much the same response we had instituted in late March and April. Whatever name you called it, we were once again asking Americans to "slow the spread."

It was Halloween, another holiday on the road away from friends and family and my two amazing granddaughters, Abbie and Addie. Irum and I were in Salt Lake City, Utah. Following our meetings with the health care leaders, we decided to go for a hike. Our loss of fitness and the altitude made it rough going, and seeing other, much older hikers cruising past us up the slope soon overcame our resolve to get some exercise. Bone-weary and in no mood for Halloween pranks or merry-making, we made an early night of it.

The sight of those sturdy Utahans on the trail didn't surprise us. The day before we arrived, neighbors of Utah epidemiologist Dr. Angela Dunn had risen to her defense. After her personal information was leaked online, a dozen anti-mask protestors gathered outside her home. Her neighbors rallied, turning on lawn sprinklers and parking their cars in the street to discourage the protestors and block their access to Dr. Dunn's home. Governor Gary Herbert also acted to defend her, saying that it was acceptable to protest an elected official like him, but not a state employee, and particularly not at her home. Irum and I were encouraged to see community engagement in defense of a public health official and the care and compassion Dr. Dunn's neighbors had for one another. The protestors certainly had their right to register their displeasure, but as the governor (and Dr. Dunn's neighbors) pointed out, a line had been crossed. Unfortunately, this wasn't the only place in the country where that same line had been crossed or where public health officials were targeted for only trying to do everything they could to save lives.

The following morning, we woke to news that Scott Atlas had done our work for us.

In life, timing is nearly everything, and Atlas's timing could not have been worse. Mere days before the presidential election, he had allowed himself to be interviewed for *Going Underground with Afshin Rattansi*, appearing from the actual White House complex without (it was later revealed) White House authorization. Worse, the television program was aired on RT, the disreputable state-controlled Russian propaganda network known for spouting anti-American rhetoric. With this faux-pas, Atlas had let his ego, his desire to be visible on TV, any TV, overrule his reason—but it was what he said that will be remembered.

While he hadn't gotten approval for his media appearance on the Kremlin-backed news outlet, he did repeat the assertions he had been making since March 2020. Citing the media's "gross distortion" of the pandemic, Atlas said that "there's this frenzy of focusing on the number of cases when we see a lot of reasons to, you know, be cautiously optimistic here." Before the deadliest surge this country would experience, he was saying he was optimistic. He also claimed that "the disease is deadly only to the elderly and the high-risk people"—in other words, millions of Americans. Yes, many of the victims of Covid-19 were elderly and at high risk, but they weren't the only ones at high risk. Some of the highest-risk individuals were in their thirties and forties, moms and dads in the prime of their lives. Was he implying that they were expendable? Atlas then claimed, offering no rationale for his statement, that the White House Coronavirus Task Force was responsible for 233,000 excess deaths. He also said that the models predicting as many as 500,000 deaths in the United States were wrong, and generally tried to dismantle and discredit the task force. He was the one who was wrong. It wasn't a model that had given us the figure of half a million Americans dead, but my own projection. And we reached that ghastly total by February 2021, from the deadly winter surge about which Atlas was so "optimistic."

In a parting shot, on Twitter—which he had thought would help promote his TV appearance but that, instead, alerted more people to his dangerous influence—he tweeted, "New interview. Lockdowns, facts, frauds . . . if you can't handle the truth, use a mask to cover your eyes and ears." I suppose he thought this great political theater, but his appearance had the opposite effect from what he intended.

After Atlas had issued an apology for appearing on a television network that, according to the Department of Justice's National Security Division, was a registered foreign agent representing the interests of a foreign power, he didn't go underground, exactly, but he would never again exert the kind of influence in the White House he once had.

I had wanted Atlas gone from the task force, but I hadn't counted on his committing this careless act of self-destruction. It wasn't lost on me that whatever desire for attention had led him to appear on Russian-backed television was probably a function of his having been off-camera for a while, thanks to my White House Comms Office friends. They had trusted that I was right. As a result, they had helped me stanch the worst of the bleeding by keeping Atlas off as many screens as possible.

To ensure that the moment wasn't lost on anyone in the administration, I wrote to Marc Short, Mark Meadows, and Jared Kushner, telling them that Scott Atlas had just given them the perfect reason to send him out of the White House and back to where he came from. I reminded them that they didn't need that kind of liability just then, not with the election around the corner. This may have been gilding the lily, but in a time of crisis, you do whatever is necessary, and then some.

As it turned out, Scott Atlas faded into oblivion after the election. I didn't know then and I don't know now where he went or what he did. The important thing was he was no longer a physical presence in the White House. Still, I couldn't say that sad chapter in this story was over. The impact Atlas had had in his three months in the White House was still being felt around the country in the misinformation circulating across media platforms.

Because of my fear of misinformation specifically and the state of the pandemic generally as the fall surge turned into a winter one, I was heartened when David Kessler from the Biden camp stayed in touch and inquired about my sense of the rise in cases, its trajectory, where we were, and where we were going. He wanted to know what he was looking at.

I was concerned about current White House staff and support personnel, the president, the vice president, vulnerable Americans, and everyone else in the country—and my worry included Joe Biden and his team. As a result, after not hearing from David Kessler for a little more than a week, I reached out to him. I let him know that those on the Biden team who were most at risk were his on-the-ground campaign staff. I imagined that most of them fit the profile for asymptomatic individuals who could become silent spreaders: the under-thirty-five crowd. I made the same recommendation that Irum and I had followed while on the road—Biden campaign staff across the country should stick to drive-thrus and takeout. A week later, I sent Kessler a similar reminder. Both the campaign and the pandemic were heating up.

We all needed to be vigilant and do our part to lessen the damage that had been done over the last nine months.

CHAPTER 18

Why Not Now?

On November 2, the eve of Election Day, the *Washington Post* reported on an "internal memo" I'd written. The article, headlined "Top Trump Advisor Bluntly Contradicts President on Covid-19 Threat, Urging All-Out Response," led with a line about my sounding the alarm: "We are entering the most concerning and most deadly phase of this pandemic."

At first, I wasn't greatly concerned about having my words cited in a story by writers with whom I hadn't spoken—mostly, because what they'd quoted was accurate. I learned of the story while I was in Denver, on the last stop of a multi-week driving tour of the Upper Midwest and Intermountain West. I'd been using similar language when letting governors and health officials know exactly what our future held and how each of them needed to prepare for it.

I had also written those exact words for the last couple of weeks in the daily report I routinely sent out before 6:30 a.m. I had never shared these reports with anyone not on the distribution list, which had remained essentially the same for the prior 220 such reports I'd prepared since coming to the White House. I had never leaked any of the reports to the press, and to the best of my knowledge, no one else had, either.

My words were not unique, of course. I had used similar language many, many times before in the previous month, when I'd seen first-hand the emergent crisis in the Northern Plains and Rocky Mountain states. So, their meaning wasn't news to me, nor would it be to anyone on the task force or who had received these daily reports. I'd made the same point clear, using the same language, in previous daily reports and in personal conversations with senior White House leadership—in particular, with Short, Meadows, and Kushner.

I was dismayed that the writers of the *Post* piece had chosen to point out only the alarm bells I was sounding and had not included the strong solutions and recommendations I'd made. I understood that bad news and sensational headlines gets more eyeballs. If only the reporters had gone a step further and clearly laid out that my words were not a denunciation of the administration, but a call to action. We had to immediately shift attention away from the voting booth and the ballot box and back to what was happening in our own homes. This was what our postelection strategy was all about—getting the White House fully engaged, so that we could be that critical voice prompting Americans to prevent the fall and winter surge from becoming as horrific as it turned out to be.

I had again produced a balanced plan that would save lives while keeping businesses and schools safely open. Having key aspects of it made known to the public could have been the kickoff the new campaign needed. As I'd also written, after reiterating the need for mitigation measures, "This is about empowering Americans with the knowledge and data for decision making to prevent community spread and save lives." Of course, none of this made it into the *Post*.

After a day of meetings in Denver with the governor and his staff and the mayor and his staff, Irum and I needed to head to the airport very early the next morning to fly back to DC on Election Day. I was still packing late Monday night when my White House phone rang. It was Marc Short.

"You've seen what the *Post* put out." His tone was clipped. It sounded as if he were biting off the top of each of those *t*'s.

"I have."

"Why did you do that? You had to know how damaging this is."

"You know I'm the only one that doesn't leak. Ask anyone in the Press Corps."

He wouldn't let it go.

I tried logic, explaining that if the *Post* piece was damaging to the president and the administration, it was also damaging to my ability to move forward with the new strategic plan. "We have a new strategy you've all agreed to. I've been effective in working with the governors. Now we can do the same things nationally. Why would you think I'd blow all that up now?"

"People do surprising things all the time," Short said. "They don't think things through. They let their emotions get to them. It doesn't matter why."

It was late; Marc had vented his frustration. We'd pick up the matter later.

On the flight, Marc's comments about people letting their emotions get to them finally hit me. My reaction to his remark was delayed because I'd heard such things so many times before. Plus, I had needed to focus on refuting his allegation. If a man made a decision to leak information, it would be seen as "calculating," a product of internal logical debate, not emotion. Not so with a woman. And yet, if I *didn't* get bent out of shape, scream and cry and respond emotionally to what I was being falsely accused of, I would be seen as a cold bitch. That I handled the matter professionally and reasonably would be counted against me. I'd faced this no-win situation so many times, as had so many other women, that I had to let it go, just let it stream behind me like a jet's contrail.

Instead, I thought about something Marc Short wasn't aware of. On Sunday, November 1, David Kessler from the Biden team had texted me, asking if I had any information that I could share about a vaccine announcement. He was worried about an Election Day vaccine development surprise, and he did ask if I had any information on that front that I could share. I told him I didn't. As the rest of the country knew by then, no vaccine surprise would be sprung before Election Day. The voters who had yet to cast their ballots or make up their minds had to do so with the vaccines' date of approval still an unknown. While some might have lost sleep over this, I wondered instead whether the vaccines' arrival would be accompanied by an appropriate level of understanding among Americans about what vaccines could and couldn't do for their immune defense.

After I'd gotten Marc's call, I hadn't slept—not because of what his language had implied, but because I was worried what this recent dustup would mean for the postelection Covid-19 communication campaign I had worked on with the task force doctors. While out on the road, I had remained in contact with Tony daily, and with Bob and

Steve multiple times each week. I believed this was the moment we could unify the administration into a single consistent voice to combat the surge I could see building.

The leak to the *Post* had been meant to damage my standing in the White House; I was sure of it. But going to this length to criticize the administration's record on the pandemic response on the eve of the election felt redundant: many people had already voted. I wondered if it was not the president, but *me* the leaker was going after. Maybe the leaker was someone protecting their turf, someone fearful of dismissal. Maybe it was someone hoping to be elevated in status, who wanted to edge me out. Maybe it was someone like Scott Atlas, who'd lost status after his Russian television embarrassment and wanted to see the same happen to me. I didn't know who did it then, and I still don't know. In that environment, nearly everyone was a suspect.

One of my hesitations in coming to the White House in the first place had been President Trump's history of attacks on others, which had produced a culture of retaliation. While still in Africa and elsewhere during the early years of his administration, I was aware that policy disputes in the White House often resulted in those on the losing side going to the press. If you felt wronged, or were on the wrong side of the debate on a given issue, you leaked to the Associated Press. If you were aggrieved, you might leak, for example, an internal memo about the possibility of the administration's deploying a hundred thousand National Guard troops to assist in rounding up undocumented migrants; or a draft of an executive order about a plan to reopen CIA black site prisons for the interrogation of terrorism suspects. If you had a problem with Stephen Miller, you might leak nine hundred of the senior advisor's emails about his views on immigration and minorities.

Because I was an outsider and had an aversion to (and no record of) going to the press with information, I was kept (and liked being) very much out of the loop. Even before my arrival in Washington as response coordinator, I had been so aware of how information might leak that I didn't discuss the possibility of my coming to the White House with my own husband or family members. In task force meetings, I was so concerned about leaks that when we devised our plan for the additional thirty days to Slow the Spread, I did not fully disclose the details among the task force members, for fear that someone—most notably, Marc Short or Mark Meadows—might leak it in advance of my meeting with the president to help ensure that President Trump wouldn't follow my recommendations. This was the level of suspicion and distrust that ran through every action at the White House.

But if the leaker cared about warning Americans, why hadn't they leaked my words weeks ago, before the Northern Plains and Rocky Mountain states were in full community spread? This leak wasn't about raising the alarm.

Why now? Who was the target?

Tyler Ann shared with me a rumor that was circulating suggesting that if Trump won a second term, he'd appoint me to a cabinet position in his administration. I had to laugh about anyone taking that rumor seriously, and I dismissed it just as I'd dismissed so many other rumors, about me and others, circulating through the West Wing. Because others were playing a political power game, they might have assumed that I was, too. Everyone seemed to view my actions and motivations through their own lens, projecting their own ambitions onto me. This was not a White House that understood the nature or importance of public service for its own sake. This administration wasn't used to working with people who had no political ambitions or who lacked that "killer" business instinct. Service without ulterior motives was a foreign concept to them.

I may have been a political appointee to President Obama and President Trump, but for forty years I had served my country. I had already decided to retire from federal service in the spring of 2021; consequently, I had no interest in a cabinet position then or now. I'd kept my cards close to my vest throughout my career, but I'd also made it clear that I didn't have that kind of ambition, or any desire, to advance to that level.

Of course, shifting from public service to the private sector was anxiety inducing. Career change is always stressful for anyone. I was committed to bringing integrated big data analysis for decision making and high-impact solutions to scale to the underserved in America, including the tribal nations. I wanted to engage the private sector to improve the country's pandemic response and translate what I had seen and learned into actions to improve our readiness. Although I didn't know what opportunities there were, after the last months in the White House, I was excited about the unique and essential role the private sector had played (and would continue to play) in our response to Covid-19. I believed then (and still do now) that the private sector needs to play a key role in pandemic preparedness, and I knew I could do something there. That was my personal future; in the meantime, I had the future of this pandemic response to return to.

Back in DC on Election Day, I had my doubts about an outcome in the president's favor. During our visits to forty-four of the fifty U.S. states, Irum and I had heard the people speak. We were certain that the president and his team had underestimated the level of discontent and distrust his rhetoric about the pandemic had produced. I could see it in the eyes of every woman who worried about her elderly parents or her own young children. I knew the same fear and concern. People may have been going to the rallies and acting like they would still vote Republican, but I suspected that many would express their pandemic fears at the ballot box. The twenty- to thirty-somethings, Generations Y and Z, had accepted the scientific basis for the sacrifices we had asked them to make and had made it clear that the pandemic was the major issue of the day. And they weren't happy with the federal response to it. Most weren't voting for President Trump.

AFTER LOBBYING HARD THROUGHOUT the West Wing for most of October for the fall and winter strategy, I felt I had been persuasive enough and I fully expected that, the day after the election, regardless of its outcome, we'd be able to hit the ground running with it. We'd received a green light from the administration. Now, preparing the American people for the dangerous weeks and months ahead could begin. I could not have been more wrong.

I reported for work the day after the election. Despite the plan to improve communications and actions, I could do neither that day. Instead, I walked into a scene right out of a postapocalyptic thriller. The West Wing offices were nearly empty. That day, I got more voice mail greetings than had actual conversations. It was difficult to get responses from senior leaders, and no one was available for meetings that first day. I initially chalked it up to a long night watching the election results.

With the vote tallying not yet complete, I expected a certain level of preoccupation by the administration—but preoccupation with the election continued not just the day after the election, but for the remainder of my term in the White House.

During that first week back, I assumed that concern over and attention to the pandemic would be restored to March and April levels. I continued to send out the daily reports, pointing out that fifteen states were in the red zone and a further eight had verged on that worst category. This was even before Thanksgiving, when people would be taking public transport and gathering together in homes across America, risking further exposure and spread. Unless we made the moves our new strategy called for, the disaster I'd characterized wouldn't remain possible, but would become probable.

Unfortunately, the leak to the *Washington Post*, combined with the distractions of the election, continued to do damage to my plans. Instead of concentrating time and energy on effectively testing to find silent spread, the administration was concerned about how the election truth was being spread. On November 5, I took the narrow steps up from my office area on the first floor of the West Wing to the second floor, where the most senior advisors, the vice president, and the Oval Office were, hoping to find Marc Short. I wanted to ask for a ten-minute press briefing to deliver the message that we needed an aggressive response to the approaching winter and holiday gatherings.

Before I could get to Marc Short, I spotted Mark Meadows.

"Mark," I said, "you don't think that *I* gave the report to the *Post*. I mean, that's just—"

He looked up from the document he was reading. "What I think doesn't matter. It's what you made people think that does. You saw the exit polls. You've seen what's going on in the critical races. You think that your words didn't matter? They did. Big time." He edged to the side to move past me.

"You can't pin all of that on me."

"No?" He turned back to face me. "You can believe what you want, but you need to know this: your reputation is toast. What you did or didn't do isn't going to change that."

What struck me most was how so matter-of-fact he was about it. *You did it. We know it. No one believes you or what you have to say anymore.*

He went on to say that given the report's large distribution list, and the provocative language I'd used, it was inevitable that it would leak.

Of course, it was. That was how this White House operated.

I reminded him that I'd been voicing the same level of concern in the same unambiguous language for the past four weeks, and my words were never leaked before. If he had concerns about what I was saying or how I was stating it, he could have raised the issue with me earlier. What I didn't say, but thought, was: *This reaction to the leak is evidence of the very problem to which I have been giving voice from the beginning.* Prior to this, Meadows and others hadn't really paid attention to my daily reports. Only after one of them was leaked were he and others roused from their campaign fixation. My tone, which they now objected to, had been consistent for quite a while. Indeed, from day one—in my daily reports, in my communications with everyone. I had been direct. I told them when things were better. I told them when things were worse. I told them when things were going to get much worse. From the moment I set foot in the White House, I spoke and wrote the truth.

Yet, instead of focusing on the harsh reality my words revealed, instead of understanding what this reality meant for the lives and safety of Americans, the administration was preoccupied with finding whoever had leaked the truth. It didn't matter that these awful things were happening—that people were dying in the hundreds of thousands. What mattered was that *now the public knew it was happening.* To this White House, solving the problem simply meant finding the leaker. And they believed it was me. More important, they couldn't believe that I would state starkly and clearly that the United States was in an ongoing and worsening crisis. The last week of October, the Friday before the leak, we saw nearly 100,000 new cases nationwide *in a single day.* By the beginning of November, even before we'd entered the full winter surge, 220,000 Americans had died. We still had time to act aggressively.

That they'd censored the governor's reports, that they now objected to my *tone* and *language*, was another example of the collective desire to alter the data to fit a predetermined picture of how the response was going, how it should be handled. All this to support the series of fairy tales they told themselves at night. Engaged in magical thinking, they'd insisted that *This time will be different.* They weren't alone in this misguided belief. Prior to each surge, every governor and mayor had said the same thing: *This time will be different because of X or because of Y.* But this virus didn't care about fairy tales or X or Y. It mindlessly did one thing: replicate. And people's behavior allowed it to do so. We had to stay in a constant state of vigilance about the SARS-CoV-2 virus until we had the vaccine and could immunize those at the greatest risk for severe disease. It could mutate, transforming into an even more infectious variant. We had tools, solutions, and the capacity to change the course of this pandemic, but the virus wouldn't go away miraculously. It wouldn't just "disappear." Hot weather would not, in fact, have a negative effect on it; hot weather drove people indoors. Disinfectants were not a treatment. Hydroxychloroquine had not been shown effective against the virus in clinical trials.

The White House economists were worried about the economy. But death by economic devastation would not exceed death by Covid-19. In June, there were no signs that the pandemic was "dying out" or "going to fade away." In early July, the pandemic wasn't getting under control. Ninety-nine percent of Covid-19 cases weren't "harmless." In fact, 8 to 10 percent of Covid-19 cases resulted in death in those over seventy. Seventy-five percent of the deaths were in those over sixty-five. In late 2021, it was reported that 1 in 100 Americans over sixty-five had succumbed to Covid-19, and life expectancy had dropped by two years in specific racial and ethnic groups. We didn't have "the lowest mortality rate in the world." We weren't "rounding the turn" or in the final stretch. This was the case in late 2020 and was still true in late 2021.

Tony, Bob, Steve, Jerome, and I were in agreement. We'd seen how effective mitigation approaches were. When they weren't in place, cases rose, hospitals became stressed, and people died in greater numbers. The leak had put yet another nail in the pandemic response coffin. The chief White House officials and the president turned their backs on me and on measures they themselves had agreed to implement. I quickly realized that, as time went on, and so many senior White House people became focused on overturning the election results, the critical post-election pandemic-control initiatives we'd hoped to enact were also likely to be discredited. Maybe this had been the leaker's goal: Damage the messenger, damage the message.

I pressed on, letting everyone know that I would limit the number of recipients of the daily report. It still went to all the most essential people—it absolutely had to—but I limited the number of lower-level personnel receiving it. I couldn't afford another leak. I figured that making this gesture would ease some of the pressure on me. I also wrote to all the doctors on the task force and asked them not to have their staff read their emails or forward content widely. Still, I continued stating in the report that the situation had worsened, that the window on having a significant impact on the virus was rapidly closing.

While it hardly mattered to most White House senior advisors, two important pieces of science came in during the first week of November that substantiated the plan we had developed. The first was the Japanese study showing that masks did provide bidirectional protection, protecting both the wearer and those around them. This contradicted what had been the CDC's official position since April of only protecting others, not the wearer. The second was an article published in the *International Journal of Infectious Diseases*, titled "Asymptomatic SARS Coronavirus 2 Infection: Invisible Yet Invincible." I didn't agree with the "invincible" part, but I did agree with the study's conclusion that "Asymptomatic individuals carrying SARS-CoV-2 are hidden drivers of the pandemic" and that "a second wave was expected."

I like to imagine how my cause could have been helped if this research had landed on my desk before the leak to the *Washington Post*. Ultimately, I don't think it would have mattered; minds had been made up, and then closed, for so long. Even though I shared the results of these studies, it didn't make any difference. Our new national communication plan would not be implemented. The president wasn't going to resume the modified press briefings I'd asked for. (I had suggested that President Trump make introductory remarks, so that he'd get his moment, and that the vice president then take over.) Task force meetings weren't going to significantly increase in frequency. The White House would not endorse the new silent spread testing plan. The White House would not return to the level of focus and energy we'd seen in March and April to face a crisis that had become much deeper than it had been in those early months. None of this was going to happen. And it didn't.

My time at the White House was coming to a close. This outsider would soon be lumped in with all the Trump insiders. I had known that going into the White House would be a terminal event for my long federal career, but that didn't make the stark reality of my current situation any easier. Some of my long-term PEPFAR colleagues were sure to conflate their feelings about President Trump with their feeling about me. Still, I hadn't given a thought to quitting since my meeting with President Bush. I'd stuck with it this far, and too much was still at stake. As so many others in this administration had done, I had to direct my considerable concerns and frustrations into action, not recrimination. I wouldn't let myself sink to the level of putting my own interests before those of the rest of the country, as some in the administration had done. Aggrieved over the election results, some in the West Wing appeared willing to engage in a dereliction of duty. I wasn't about to let my personal grievances further weaken our response to the approaching surge. Jared and the vice president remained my go-tos; we continued to make progress behind the scenes.

If I found consolation in anything in this period, it was that with the West Wing's emphasis on overturning the election results, no one seemed to believe it worth the effort to stop me from doing anything I'd done before. I still wasn't able to do national media, but whenever I put in requests to meet with local press in hot spots, I met no resistance. As long as I didn't show up on any of the four presidential TV screens, I was free to talk to whomever I wanted and, essentially, to say what I wanted. The White House Comms team continued to facilitate these local press hits. The Huddle team continued to help me align supplies and therapeutics with need, and the Intergovernmental Affairs team continued to facilitate my calls and visits with governors and mayors across the country. My work was actively supported at one level and completely shut down at another—all within the same West Wing. If, in my dealings with some of the senior White House people, the message I was getting was *Talk to the hand,* at least I was still able to reach out to the American people.

Just as going on the road had gotten me in better touch than from a distance, now, from the studio in the Eisenhower Executive Office Building, I was getting in touch remotely. Along with talking about the guidance and safe approaches to holiday gatherings, I spoke to people about what I'd seen on the ground in communities very much like theirs. I got it. People were tired. They were frustrated. They were nearing the end of a long year and hoping that it could end as it had in the past, with a celebration among family and friends.

On one of these local news hits, I shared a story of meeting an intensivist at one of the two main hospitals in Billings, Montana, a city of more than a hundred thousand people. At first, this critical care physician was responsible for twenty-four Covid-19 patients in the hospital's now-full ICU. Then the administration set up an additional eight ICU beds around the hospital, and she became solely responsible for those patients, too. One doctor and a small team of nurses had to take care of *thirty-two* critically ill people scattered across a hospital. The intensivist spent every day running from one end of the building to the other, just to ensure an optimal level of care for every one of her patients. Day after day, she was finding a way or making one.

This was the image I wanted to implant in people's minds. Just as the image of that overrun hospital in Wuhan had galvanized me and contributed to my returning to the United States, I wanted this image to galvanize the American people. Months after my visit to Billings, our frontline health care workers were still in the trenches, still doing everything they could to save lives.

Because of the high regard so many people had for health care workers, and because I believed in the power of personal stories to get people's attention and spur them to action, I contacted various hospitals around the country, asking administrators to enlist the aid of their health care staff to do media spots. With the holidays approaching, these doctors and nurses were in the best position—certainly in a better position than any statement the CDC or the White House could put out—to explain how early mild cases resulted in hospitalizations a few weeks later as the virus was transmitted to vulnerable and aged family members. These health care workers could share the stories of patients who were hospitalized now because they had gotten infected at a gathering of friends or family and help advance the message—at times and in ways that were more impactful than anything I or their states' governors could do. My hope was that, as members of their local community, they would instill trust in the neighbors who heard their stories on the local news, perhaps offsetting some of the damage done by the administration's silence.

WHEREAS THE TRUMP ADMINISTRATION heard only one clock ticking down to what would be an eventual defeat, I heard multiple clocks, and wanted to be as effective as possible with the time I had left. I wanted the new administration to enter into power with the winter surge in decline, not one still in full exponential growth. One way to communicate this sense of urgency would be to get back out to the states that were in the gravest danger, but continuing to push for our national strategy was the greater priority. I knew that December would be the earliest I could get back on the road. With that settled, I focused on several other priorities—the first of which was to meet with the vice president to warn him that this fall surge was the biggest crisis we had faced and that we were not moving fast enough. Vice President Pence and Jared had the most direct access to the president, and the vice president had a track record of listening to me and trusting me.

I planned to tell him that we needed to resume regular task force meetings. If not every day, then at the very least three to four times a week. Also, the president needed to address the nation at a press briefing, as he had done so frequently in the spring, at the height of that surge. The vice president had to tell the president what was becoming abundantly clear: that behavioral change in the form of wearing masks indoors in public spaces and limiting the size of family gatherings, approaching the upcoming Thanksgiving holiday and then Christmas with the utmost caution, was critical to slowing the spread and keeping your family healthy. The risk was no longer so great in public places; the most dangerous infection zones were in our own homes.

Normally, to get to the vice president, I went through Marc Short—as I'd been trying to do since I returned to DC on November 3. But Marc hadn't responded for a week. I'd expected to have the opportunity to hold press briefings after the election, to convey the seriousness of the crisis we were facing.

Instead, on November 5, the president held a White House press conference claiming voter fraud. A day later, he tweeted that the lawsuits were just beginning. I couldn't get through to anyone about enacting our new communications strategy because they were busy filing lawsuits to halt the vote counting in Michigan and Georgia and requesting a recount in Wisconsin. This pattern of inattention continued throughout the week. Accusations of fraud, a refusal to concede, multiple legal challenges—all took precedence over the response to the pandemic and proactive messaging to lessen the effects of the fall surge.

While the Trump administration waited for recounts, the body count was rising. Between November 3 and 10, nearly nine thousand additional Americans died of Covid-19. On November 10, I abandoned the established protocol of requesting time with the vice president through his chief of staff. Instead, I contacted Zach Bauer, the vice president's "body man," the aide with the most direct contact to him, telling him I was desperate; the situation was dire. I needed just a few minutes of the vice president's time. Pressing for a televised task force briefing, I wrote Short and others: "We can't remain silent on this." I offered an alternative: I would go back on the road. Instead, I was told we'd have a task force meeting two days later. There'd be no press conference, but at least the task force would meet for the first time in weeks.

In the intervening day, I wrote to Jared Kushner, Matt Pottinger, Mark Meadows, Tucker Obenshain—everyone and anyone I could think of who might have access to and influence with the president. Again, I made the case that this was the tipping point. We were within days of a public health crisis that would spin out of control. We needed the president to deliver the message that we had solutions, but that they needed to be enacted. Eager to get my governor's reports out, I wrote to John Gray, Tucker Obenshain, and Marc Short that twenty-six states were then in full resurgence. I asked that we restore the previous bullet points and a summary of what had been done over the summer to control the surge. The governor's reports came back fully approved. I had to wonder if the administration had even read the points I'd made in the reports, or simply rubber-stamped them.

In that first task force meeting in November, I laid it on the line. Every American needed to wear a mask indoors. To bridge the gap until we had a vaccine, social interactions for the next sixty days needed to be limited to immediate household members as much as possible. Retail businesses should remain open. Limit indoor dining and expand outdoor dining. Schools could be kept open. Testing in schools should increase.

Resistance came from all sides, and in familiar guises. Even with what we now knew with great certainty about masks, mandating the use of them just wasn't practical or enforceable, I was told. Sixty days of social "isolation" with only your family was too much like a lockdown, they said.

In response to my direct tone, Marc Short said, "You don't have to lash out like this. You're clearly very distressed."

Without directly addressing the antiwoman language, I responded that I was neither distressed (read: too emotional) nor lashing out. Though I was sending out a clear distress signal, trying to get the attention of key people in the administration who could alert the American people that the time to act was *now*.

Frustrated that my sense of urgency was, again, being viewed as just another woman's failure to keep her emotions in check, and finding no traction in the task force meeting of the twelfth, I contacted Covid Huddle allies Jared and his associates Brad and Adam: "We cannot remain silent." The president needed to address the American people. He had to set the tone. He had to tell the truth.

Jared wrote to let me know he'd try. But to no avail: The president wouldn't agree to lead a press conference. Instead, Vice President Pence got permission to hold one, on November 19, the week before Thanksgiving. This would give us time to alert Americans of the seriousness of the situation and the need to show prudence regarding decisions about holding family gatherings. This was some headway, but I feared it wouldn't be enough.

From November 10, when I first got confirmation that the task force would meet again, until November 19, when Vice President Pence would speak, we saw another 15,000 Covid-19 deaths. The rise in infections was more precipitous—from Election Day to the vice president's press conference, an additional 2.3 million cases of Covid-19 brought the U.S. total to nearly 12.5 million for the year. This raised ramp would launch these totals into an even sharper rise after Thanksgiving. For this reason, Thanksgiving messaging was key. We had seen surges triggered by family and friend gatherings, and Thanksgiving was the biggest one such gathering on the horizon.

The efforts to marginalize me or remove my voice from the conversation were ongoing. On November 13, the president was briefed on vaccine progress. As an Operation Warp Speed board member, I'd attended every presidential briefing before. This time, I was excluded.

Prior to the *Washington Post* leak and the "distressed" message I'd delivered at the first postelection task force meeting, I'd nearly always been the first to speak and to raise the issues that I believed, as response coordinator, needed the most attention. But when we met again on the seventeenth, I found that the vice president's staff, who created the agenda for task force meetings, had placed me eighth out of the eight presenters. Discussion of vaccines now took top priority. This was a harbinger of things to come. It was certainly important to monitor the progress of vaccine production, but for too long now, the administration had communicated the message that the coming vaccines constituted a silver bullet fired at the virus. This posed a real danger.

When it was my turn to speak, I once again shared the bad news that cases were rising and the virus was spreading everywhere. I reminded all task force attendees that vaccines were just one part of the plan. Without these other elements in place, the vaccines' very real limitations, vaccine hesitancy, and inequitable distribution would seriously diminish their impact. Once again, I got looks that said, *You're an alarmist*. In the eyes of Marc Short and others in the West Wing, I was crying wolf and had been since March 2, 2020. Yet, in the last thirty days, cases had increased by 276 percent, new hospital admissions had increased by 81 percent, and daily death rates had increased by 73 percent.

They didn't want to hear what I had to say, and they didn't want to act. They had the solution: vaccines were needed, and nothing else.

Meanwhile, it was no longer just postelection absenteeism that was making it difficult to connect and communicate inside the White House. Ever more staffers were becoming infected and working from home; crucially, Mark Meadows was among them. The White House suffered wave after wave of outbreaks.

So, there we were, on November 19, at the first task force briefing since July. Vice President Pence announced to the press that we had gathered again as directed by President Trump—but Trump himself was not present. The president had made a concession to my request for a briefing, but his absence spoke loudly and forcefully. *I've got other matters more pressing than this. I've got an election to overturn. That's where my commitment lies.*

While the vice president, at my urging, did mention several times that case figures were rising, he began his words with a nod to the imminent approval of the vaccines. The tone and intent of his remarks was encapsulated in one statement: "America has never been more prepared to combat this virus than today." This was true. We knew what worked—but we needed to ensure every American was using what worked. I kept my expression neutral, but inside, I felt dismay. I'd pressed the vice president to speak honestly, to raise the alarm about what we faced in the coming weeks with families across the nation gathering. His tone implied hope for the future and acknowledged our better state of preparedness, but with the worst surge yet to come, I had wanted a call to action!

When it was my turn at the microphone, I tempered his optimism by beginning my remarks with the need for the American people to increase their vigilance. For nearly every reference the vice president had made to the promise of a vaccine, I mentioned the need for mitigation. We had always talked about building a bridge to the vaccines, but I needed to make clear: that bridge could carry only so much weight before it collapsed. I also made clear that the coming surge was likely going to be worse than previous ones, but that we already knew what to do to keep that from happening.

Steve Hahn, Bob Redfield, Jerome Adams, and I followed up this first press briefing in four months by heading out across the country to deliver our message in person. We wanted to spread the good news that even without vaccines available, we could slow this aggressive surge if we met it with mitigation measures of the same intensity. Vaccines were welcome, but they were no substitute for personal, institutional, and governmental will. We would travel to as many places and cover as many miles as necessary to inject the proven components of our plan into the minds of those willing to listen to the truth.

We were three weeks too late. While "Better late than never" applied to some things, like belated birthday greetings, it did not apply here. The rise in incidence we'd seen on this trip, eleven months after the beginning of the pandemic, would soon show that we were starting from too far behind—and quickly losing time. Again and again, across this country we had failed to remember this simple bit of pandemic math: Lost time equals lost lives. And no calling for a recount would ever enable us to make up that deficit.

I wanted to ensure that no mixed messages were sent to the Biden team. After David Kessler's Election Eve query about the status of vaccine approval, we hadn't been in touch since. While a formal process existed for the transition, outside of it, on November 29, I alerted him to the fact that now more than ever, he needed to keep the Biden team on point and safe. Kessler and I would soon meet with the Biden Covid-19 transition and data people, but I felt this particular message couldn't wait. The same was true for how I felt about the rest of the country. They needed to know that vigilance was everything.

Two weeks after Thanksgiving, by December 10, when infections from Thanksgiving gatherings revealed themselves, the United States would confirm a total of 16.2 million cases. By the end of the year, when the virus was passed from person to person along with the Christmas gifts and eggnog, we'd see that rate rise to a total of 21 million cases. In those three weeks, more than 5 million more people would be infected, daily new cases would rise from 145,000 to 249,000, and an additional 58,000 people would lose their lives. There would be no happy New Year.

CHAPTER 19

Winter Is Here

While the Oval Office was singularly focused on the election results in the run-up to the holidays, the Covid-19 response team remained laser-focused on the pandemic. The White House senior leadership's distraction created a lot of internal disruption, most especially in preventing the adoption and execution of a new national communication strategy. We couldn't afford to take our foot off the gas pedal powering our mitigation message. We'd face an additional, ongoing struggle with the vaccines.

Throughout the vaccine development process, I had expressed cautious optimism (read: realism), presenting a hopeful face to the public. Privately, in conversations with the other docs and in the governor's reports, I was more guarded. We could not, should not, rely solely on the vaccines to be the silver bullet many people, within and outside the administration, hoped they would be. Had the pharmaceutical companies and the definitive final trials done their jobs? With the new surge, we would soon have our answer: cases were rising in all the areas of the country where vaccine trial volunteers had been immunized. Despite the pressure being placed on it, we knew the FDA wouldn't alter any of its long-standing approval processes. Its approval would come only when the vaccines were shown to be safe as well as effective.

The real problem lay with how the vaccines, once approved, would be distributed. The decision making on this was complicated by outside guidance groups at the FDA, another outside advisory group at the CDC, and finally by the CDC director. Having these parallel groups involved added hours and days to the approval process; this must be streamlined for future pandemics. Independent of those process issues were implementation considerations. Among the myriad issues on the board over late spring and into the summer were: On what basis should we determine the number of doses delivered to the states? And who within each state should be the first group inoculated?

One aspect of the vaccine rollout plan was clear from the outset: The federal government was responsible for supporting the development and approval of the vaccines. Under General Gustave Perna, chief operating officer of the federal vaccine and therapeutics response, resided the logistics of moving the vaccine supplies and the vaccines themselves to the states. Vaccinating an entire country's population was an enormous undertaking. General Perna would allocate the weekly available doses to the states and ensure that the specific allotments got to them. The CDC was responsible for working directly with states on the on-the-ground plans for vaccine distribution and immunization. The states had to do the very hard part of determining where each vial would go within their state. The CDC had asked each state to submit its provisional distribution plan and overall immunization priorities by mid-October, for CDC review. Within the same time frame, each tribal nation had to decide whether they wanted to receive their doses directly from the federal government or through their state's allocation. The more than 15,000 nursing homes across the country were to receive supplies directly from the federal government through two immunization groups, CVS or Walgreens. In theory, this broad outline of requirements and recommendations sounded relatively straightforward.

I had some notion all along that scaling up an approved vaccine to immunize the entire country was going to be difficult; I had worked on vaccine development for decades, including a phase-three HIV vaccine trial in Thailand with 16,000 volunteers. As a board member of Operation Warp Speed, I had attended the vast majority of the planning meetings in person or by phone while on the road. But it was clear that additional, internal meetings were being held at HHS with Secretary Azar. Until the very end of the Trump administration, Secretary Azar was engrossed in shepherding the vaccines from development to distribution. But because most of what was happening internally was only being summarized at the board meetings, it was difficult for me to follow the decision-making process. Each time I participated in an OWS meeting, it felt more like General Perna and Secretary Azar were announcing what had already been decided rather than conducting an open discussion. To make matters worse, the closer we got to the key moment for FDA approval, the fewer OWS meetings were being held. Even so, I was able to glean this: OWS team members were struggling to get accurate accounts of how many doses were going to be available for immediate use once the trials demonstrated to the FDA that the vaccines were safe and effective.

This was enormously frustrating. Secretary Azar, Dr. Moncef Slaoui, and General Perna all told me they didn't have any real visibility into Pfizer, one of the manufacturers. This was unacceptable. We had contracted with Pfizer for three hundred million doses, at a cost of nearly six billion dollars—and yet we couldn't get specific information about how many doses were being produced and at what rate. Eventually, Moderna would receive nearly five billion dollars to produce the same number of doses, but with that manufacturer, too, we didn't have complete clarity on these same critical questions.

Vaccine development and production involve many moving parts. Production of vaccines is always more like cooking than manufacturing. If the vaccine doses were going to be in significantly limited supply for the first four to six weeks during this critical surge, then—like a pot of soup ladled out sparingly to feed a large family—we had to make optimal use of every dose to save more lives. My not being engaged in key meetings where this was discussed kept me somewhat in the dark about what the manufacturing and production stumbling blocks were and whether they were being overcome. I wanted to make clear that I trusted the OWS team to be on top of all the issues inherent in scaling up a vaccine to the volumes needed in the United States. Ultimately, I just wanted to know how many doses were available now and would be available through the end of January.

Without that estimate, managing the second of the two main issues, distribution, would also be more complicated. The CDC, relying on advice from its outside group, the Advisory Committee on Immunization Practices, had to determine not only which groups would be the first recipients of these initial doses, but how those doses would be allocated in the most equitable manner. These discussions had been taking place since the summer.

Back in the summer, and still in the dark regarding supply, General Perna had decided that "equitable" meant that each state should receive a percentage of the initial doses produced based on *population*—which, in his mind, included people under eighteen, who were not even eligible for a vaccine yet. Fortunately, I became aware of this at a mid-November briefing and was able to point it out. General Perna altered the allocation to properly reflect the number of vaccine-eligible residents—those over sixteen for the Pfizer-BioNTech and over eighteen for Moderna. I also advocated for another age-related position. States with the highest percentage of residents above age seventy should receive more of these initial doses. In this way, distribution would be based on greatest need—not on whether someone just happened to live in a certain location. The general-population approach may work for a disease that affects every age group more or less equally, but we'd known for some time that, with Covid-19, it was older people who were disproportionately impacted. From the outset, General Perna's formula would never precisely target greatest need.

The vaccines' approval depended on many factors, but fundamental to their approval was whether they prevented severe disease. Even before the vaccine discussions of the summer, it was clear that older people were more likely to develop serious illness. So, it wasn't just how *many* adults there were that should have determined how many doses were allocated to each state, but the number of adults in the highest-risk group for severe disease.

The other main question was: Which groups should be prioritized to receive the initial doses? The two options were the over-seventy group or health care workers; on this, there was complete agreement. The disconnect lay in which of those two should take preference. Those involved in the decision-making process—the CDC, Secretary Azar, the other doctors and I, the OWS team, and the rest of the task force—never arrived at consensus on this. I believed the doses should have gone to the nursing homes first, and only then to those over age seventy. I was in the minority, along with Secretary Azar, Bob, Steve, and Tony. It would be hard for our views to be adopted over those of the larger CDC contingent, who felt otherwise.

Collectively, the CDC pushed hard for prioritizing health care workers—in this case, following the recommendations of ACIP. The CDC staff and ACIP also relied on precedent: When the country was not in a state of emergency, ACIP protocol stated that health care workers would be first to be immunized. Clearly, we were in a state of emergency and had been for some time. Neither the CDC nor ACIP saw it this way, but Bob did. Those groups supported an approach that would not address the unique reality of this pandemic. Their choice didn't reflect the present reality: We were in the middle of a growing surge that would target and threaten the lives of elderly Americans. Also, there were nearly the same number of Americans over seventy as there were health care workers. If the doses were available for health care workers we could substitute those over seventy and save more lives in the surges and then immunize the health care industry worker. Prioritizing the elderly would have delayed vaccinating health care workers by only a few weeks, not months, based on production of twenty-five to forty million doses per month.

The ACIP recommendations were based on a common public health concept: exposure risk. Immunize first those people who face the greatest risk of being exposed to a pathogen. Historically, science and statistics have shown that health care workers face the greatest degree of exposure risk by being around so many infected people. This was the case with Covid-19 as well. However, the exposure risk factor is more nuanced than that; it's not just a matter of how great the risk is, but also the *nature* of the risk. Yes, health care workers faced a greater risk of exposure to the SARS-CoV-2 virus, but they had a lower risk of developing severe outcomes (hospitalization and death) from it. Why? Because they were, for the most part, younger and therefore not part of the age group being hospitalized and dying at far greater rates: those in the over-seventy category. In fact, in the case of Covid-19, health care workers had some of the lowest rates of hospitalization and death throughout 2020—whereas those over seventy infected with the virus had the highest death rate, consistently ranging between 9 percent and 10 percent.

A CDC study had revealed that, at the height of the initial surge, from March 1 to May 31, 2020, health care workers made up 6 percent of adults hospitalized with Covid-19, while people over seventy made up 25 percent or more of hospitalizations. (Remember the size of the two populations was nearly the same.) Those over sixty-five accounted for more than 75 percent of fatalities. It made sense to me, then, based on the available data and science, that the CDC should recommend inoculating this highest-risk for death group first. This was especially important in the midst of the viral winter surge. I believed many health care workers would have been willing to wait an additional four weeks for their jab if it meant saving more lives.

This didn't seem to matter. In late August, the CDC released its outline for vaccine administration and distribution: health care workers would get the first doses.

I want to be very clear: I have enormous respect for the medical personnel in this country, particularly for those who have worked on the front lines during this ongoing crisis. In saying that I favored vaccinating those at most risk first, it doesn't mean health care workers weren't at risk and shouldn't be prioritized ahead of other groups—everyone but the over-seventy cohort and, critically, residents of nursing homes, who were dying at the greatest rate within that over-seventy group.

Just because the CDC had issued this guidance, it didn't mean we halted our efforts to have them rethink it or to overcome other vaccine-related issues.

From the beginning, the White House regarded, and spoke publicly about, vaccines as a potential quick-strike intervention to end the pandemic. Consequently, many people looked forward to the day they could be immunized.

I always knew that changing people's behavior would be an extremely important element of virus mitigation, especially as vaccinations rolled out across the country. Vaccine hesitation was no different from resistance to mask mandates or reductions in friend and family gatherings: It would be the job of the task force to institute another PR campaign to get enough people vaccinated so that we could achieve population-based protection and, potentially, herd immunity—but in the best and safest manner possible. But it wasn't going to be easy.

Since early October, I'd been stressing in the governor's reports, in internal communications within the White House, in meetings with the task force docs, and in conversations with governors and public health officials the need to anticipate the response of two groups with distinctly different views on vaccines.

The first group comprised those who wanted vaccines immediately because they believed they were at the greatest risk of serious illness. Given that there was a limited supply, we had to manage their expectations.

The second group comprised the vaccine hesitant and anti-vaxxers. We were going to have to convince both groups that vaccines were safe, make the vaccines readily accessible, and emphasize that the success of the vaccination program relied on a high percentage of compliance. All were problems, but vaccine hesitancy was a very real, very substantial obstacle.

With such presumed variable attitudes toward vaccines (we didn't have data on the reasons behind adult vaccine hesitancy), a one-size-fits-all approach to our messaging wouldn't work—especially given the numerous and vastly different reasons people were either hesitating to get the vaccine or outright refusing it. If, as we suspected (and as eventually played out), a substantial percentage of people proved resistant to getting vaccinated against Covid-19—for example, in numbers similar to those for the seasonal flu—we were in real trouble. Secretary Azar and Bob knew this issue existed.

Historically, flu vaccination rates in America have been nowhere near high enough for the population to develop herd immunity. Before the pandemic, in the 2018/19 flu season, the overall flu vaccination rate was a little above 45 percent. No demographic cohort of adults was above 50 percent, and minority populations such as Blacks, Hispanics, and Native Americans were under 40 percent. Yet, year after year, the CDC merely observes and reports the numbers, doing little in the way of public health interventions to understand or change this dangerous trend.

Long before the pandemic, in the case of the seasonal flu, in addition to counting the numbers and parsing the data into categories, the CDC should have worked to understand vaccine hesitancy and, more important, how that hesitancy could be overcome. If we had increased annual flu vaccinations over time, we could have been very near or have exceeded the number of people needed to achieve vaccine-based immunity. With that past experience, we would have known what worked and would have established trusting relationships with marginalized and vulnerable communities. We had done this elsewhere, with different diseases. For decades, we had been taking this proactive approach to improving HIV/AIDS and TB programs in our own country and around the world. But behavioral science—the kind needed to study vaccine hesitancy—and its critical quantitative and qualitative research require funding. More crucially, just reporting without analyzing the specifics of a problem and without developing solutions for it doesn't adequately advance the ball down the field. Once again, we were flying blind on how best to reach everyone. A report alone might be useful for academic papers, but we needed data to drive solutions, not just identify problems.

The evidence from the trials of the Moderna and Pfizer-BioNTech vaccines were showing positive results. The Janssen/Johnson and Johnson trials had stumbled for a bit, but were now back on track. By the second week of November, Pfizer released data showing that its vaccine was 90 percent effective. By mid-November, Moderna announced an efficacy rate of 94.5 percent for its jab. This was terrific news, especially given how "quickly" (albeit safely) the vaccines had been developed— thanks to earlier work on the *Coronaviridae* family of viruses, such as SARS, researchers developing this vaccine had not had to start at zero. Yet, an efficacy rate of 90–94 percent is not surprising with a highly controlled trial, one whose volunteer participants are motivated to contribute to a historic breakthrough. It would be quite a different story with the rest of the U.S. population, especially if historic flu inoculation rates continued.

Getting as many people inoculated as quickly and equitably as possible remained one of my priorities. In addition to "emergency use authorization," or EUA, the FDA also has the authority to allow the use of therapeutics and vaccines (and the use of experimental drugs to people outside clinical trials) under what's called "compassionate use authorization," or CUA. Lacking the holy grail of emergency use authorization (which was pending), I continued to try to find a way to get the highest-risk group immunized as quickly as possible. In early November, I asked Tony and Steve to approach Moderna and Pfizer and urge them to apply for CUA while their vaccines' efficacy was still being determined but safety was fairly clear. With a CUA in hand, we could inoculate any nursing home residents who wished to be. Whether they volunteered for the jabs or not, at least they'd have the option.

We had a narrow window, and it was closing. Fifteen hundred nursing home residents died in the first week of October. The vaccine manufacturers, I learned, had already stockpiled three million doses. If we could draw from that supply through CUA, thousands of lives could be saved.

This didn't happen. Pfizer and Moderna declined to pursue compassionate use authorization. They believed the process would be a distraction. Their eyes were fixed on the EUA, another complicated process; taking on both simply wasn't possible.

I believed it was—it just wasn't part of the plan these manufacturers had envisioned.

Take a moment to imagine that they *did* apply for compassionate use. And imagine that 1.5 million of the 3 million stockpiled doses went to nursing homes in November, and another 1.5 million at the end of November, for a second dose. If this had happened, the nursing home residents would have been fully protected in December, at the start of the surge, and not, as it turned out, as late as February, after the surge. An additional six-thousand-plus nursing home residents died in mid-December. They all could have been fully immunized and protected before this happened and we could have saved thousands of lives. If this had been done, literally thousands of lives could have been saved. Great good could have been done, and at low risk to these vulnerable people. In a pandemic, you need to innovate on the fly in response to the reality of the moment and not be locked into a rigid plan.

I was never privy to the details in the contracts negotiated between the federal government and the pharmaceutical companies. If there weren't provisions in them forcing these companies to pursue every avenue of use beyond EUA, there should have been. When you are in a crisis, you use every tool at your disposal—not just the ones you've used in the past, when you traveled a familiar path. With the SARS-CoV-2 virus, a *novel* virus—which means "new" to the bodies of *Homo sapiens*, better known as *us*—we were in unexplored territory. We needed to be smarter about what to expect and more creative about what to demand.

If my imaginary scenario had been enacted in real life, it would not have depleted the majority of doses and would not have resulted in lengthy delays for health care worker inoculations. Without the compassionate use option there was still a way to immunize those at risk for the most serious disease. Moderna's vaccine received FDA approval on December 11, and immunizations with it began on December 15. If we had prioritized the elderly over health care workers, the latter would have begun to receive their vaccines only three to four weeks later, in the middle of January—not too long a wait for this younger, less vulnerable group.

Secretary Azar, Bob, Tony, Steve, and I were still in agreement on putting the over-seventy population ahead of health care workers. Others thought immunizing health care workers would build confidence in the safety and efficacy of the vaccines. In reality, most people over seventy had no hesitancy and believed in the efficacy of the vaccine; it was the young health care workers who were hesitant.

Tony was in a difficult position: if he supported a vaccination program that put him at the head of the line, he could be viewed as self-serving. Alex Azar was similarly in a tough spot. As the director of HHS, he was the one who approved the appointment of members of the ACIP group, who were making the recommendation. To his credit, Secretary Azar continued to strongly side with the rest of the docs on this issue. We all held out hope that we could get the CDC to reverse its position and go against ACIP's recommendation. We pushed Bob to overrule the ACIP recommendations, but he told me his staff said he couldn't do that. Still, in just a few months, that is exactly what the new CDC director did—so it was always possible. This is another lesson we need to learn for future pandemic planning and implementation: those at greatest risk for severe disease and death should be the first to receive lifesaving vaccines.

DESPITE BEING SEVERELY CHASTISED and discredited by Scott Atlas at various times in the fall for speaking the truth in the most direct terms possible, I had continued to escalate my language in the governor's reports and in all my communications throughout November and December—my degree of urgency increasing in direct proportion to what the data was saying and what we were seeing on the ground. Then, in North Dakota, my push for a mask mandate during a local press conference got picked up by the national media—and the White House canceled all my remaining press hits. *It's too close to the election to have Debbi out there causing controversy for our governors.*

At the beginning of November, we had crossed a grim mile marker: a staggering one hundred thousand new cases per day. Less than three weeks later, we broke that record with two hundred thousand new cases per day. With rates this high, I knew we could lose hundreds of thousands of Americans over the ensuing weeks, before vaccine rollout even began.

Declaring victory over this virus now would only escalate the number of deaths. That's why any message of hope—and there was hope: of a safe, effective vaccine and more therapeutics being made readily available to the sickest—needed to be leavened with a message of extreme vigilance. We now needed to ensure that vulnerable Americans survived long enough to get vaccinated.

With FDA approval imminent, the White House scheduled a "vaccine summit" for December 8. On December 6, fearing that this summit would become just another opportunity to prematurely declare victory—we were already seeing the dramatic impact of superspreader Thanksgiving gatherings, and Hanukah, Kwanza, and Christmas were just around the corner—I wrote to Bob, Steve, and Tony, warning them to keep the tone of celebration from dominating the summit.

"Look at the daily report. We cannot have ANY happy talk." I also asked for their continued support in the task force meeting with the vice president planned for December 7. We couldn't have another two hundred thousand Americans die while waiting for a promised light at the end of the tunnel. That light, I wrote, was actually just another train, one on a collision course with millions of Americans.

Tony responded by asking me to do two things in the task force. First, if I detected the vice president veering toward any kind of happy talk, I should interrupt him and set him straight. This was not my style, but I agreed. Second, if I felt the vice president wasn't acknowledging the seriousness of the moment, I should counter his points and also ask Steve, Bob, or him to chime in. Tony would have my back, he told me. He knew we'd long since run out of time for any more magical thinking or further distortions of reality.

At the task force meeting on December 7, I planned to let everyone there know that, despite the next day's meeting being called a "summit," we still had a long, hard climb ahead of us. With vaccine approval imminent, we had to counter what was sure to be an unrealistic message from the administration—one that was sure to convince the American public that it was okay to relax their vigilance during Christmas and New Year festivities. I knew I was repeating myself, but I still wasn't sure the right people were hearing me.

I also planned to let the vice president know that he had to be the one to set a more conservative, cautious tone. Vice President Pence is an optimist, but when talking to the American public, I needed him to be the tough guy. He had to warn the American public that the next few weeks would be critical to saving as many people as possible before we were able to get shots into arms. Hearing that from just me wouldn't be sufficient.

In another email that week, to Mark Meadows and Jared Kushner—subject line: "Things Continue to Deteriorate Across the Country"—I reminded them that any positive message about the vaccine needed to be tempered with strong reminders about behavioral changes. The initial vaccine administration, to health care workers, would do very little to slow the increasing hospitalization and death rates.

I was distressed by Meadows's response. He asked me: What kind of behavioral changes?

I wanted to scream.

After all this time, and after all the ways I've made this plain before—

I stopped myself from ruminating. I'd never stop trying to convince them—though I knew I never would. I always felt the need to make one more attempt at educating them. It wasn't in my nature to give up, to admit defeat.

Mark Meadows's response typified what I'd felt all along: the White House was always distracted by other matters. Nearly a year after the onset of the pandemic, I was almost back to square one.

IN HIS OPENING REMARKS at the December 7 task force meeting, held in the Situation Room, the vice president was gracious as always in recognizing everyone's hard work to save as many lives as possible. He thanked everyone, too, for the hard work on vaccine development, outlined the agenda for the meeting, and stated how pleased he was that vaccine rollout would soon begin. I thought of interrupting him at that point, but I didn't. Something about his tone struck me as odd, and my gut told me this wasn't the time. Though the vice president was normally on a very even keel, he was not without affect—as he is sometimes characterized. That afternoon, he seemed to be going through the motions. He looked exhausted, and his matter-of-fact manner troubled me.

Steven Hahn updated us on the status of the EUA request for the

Pfizer-BioNTech vaccine. Secretary Azar spoke next, about the vaccine distribution plan. All the heads nodding in unison worried me—communicating the kind of quiet, obedient consensus the president and vice president seemed always to be looking for. Knowing that their tenure in the White House was soon to be over, people appeared to be rubber-stamping everything, as if they felt there was little else to be done but put a bow on the not-very-attractive package they'd present to the next administration.

I felt I needed to do something. We couldn't sleepwalk our way through these final weeks of our response. Far too much was at stake.

When it was my turn to speak I said, "I'd like to echo Vice President Pence's thanks to many of you who, along with Operation Warp Speed, have worked so diligently to bring us closer to adding another mitigation strategy to complement all the others we have in place." I then reminded the group that it was essential to convey to the American people that the vaccine program and the vaccines now available would not get into vulnerable American arms until after the approaching holidays. They must continue to test, mask, and protect their vulnerable families now and over the next weeks and potentially months until they could be vaccinated. We know what measures work and we need to use them now.

I scanned the room. When I didn't see the same kind of head nodding that had met the vice president's words, I pressed on: "We are still in the midst of a national catastrophe, maybe one of the worst this country has ever faced. As we—"

"How can you say that!" Vice President Pence interrupted me. His dark eyes were wide in anger, his tone clipped. "Are you comparing this to other crises we've faced? You think this is worse than all the horrors this nation has faced? Armed conflicts—Vietnam, the Civil War, World War Two—and other events, like the Great Depression, weren't as bad as this? Do you honestly believe that?" The vice president's face flushed and a look of incredulity became chiseled into a stony expression of disbelief and disdain.

My throat constricted and then loosened. He'd never spoken angrily to me before, and I wondered what I'd said that had triggered him. It was obvious to me that this pandemic was among the worst situations we'd faced. "Yes, Mr. Vice President, I do. It is one. When you consider the depth and breadth of the loss of life in this short of a period of time, this *is* catastrophic. It is only going to get worse. When you think of the Spanish flu epidemic, we were nearly in the Dark Ages of medical knowledge and practices. This monumental loss of life and what we project ongoing is—"

"But those are *projections*!" He said this last word as if he were describing something foul. His face briefly twisted into a disgusted grimace. "To say that the risk of this virus is comparable to what our men and women in uniform face is a real disservice to them. You dishonor the sacrifice of our veterans going all the way back to the Civil War when you talk this way!"

I was so taken aback that I was rendered speechless for a moment. I had struck a nerve in him. It was December 7, Pearl Harbor Remembrance Day, commemorating the loss of 2,300 Americans dead on that very day. At that moment, I thought of his son, Mike Pence Jr., a U.S. Marine pilot serving aboard an aircraft carrier; the vice president was a military father. Still, I was shocked by how upset my words had made him. In all the months of our hard work on the task force, he had never responded to me like this. My intention hadn't been to anger him, and certainly not to do so in front of our colleagues. Still, with the vice president this upset, I asked myself how I'd ever find a way to urge him to offset the president's overselling of the vaccines. I continued emphasizing the perilous position we were in in combating this building surge and how deadly I thought it would be. I said we would lose 500,000 Americans in total by the time the surge receded. We needed to get Americans to do more.

Though there were only about twenty people in the Situation Room, and another ten in the overflow room, I knew that word of this would leach out across the White House. *The vice president finally took down Debbi.* But this wasn't what troubled me: A number of the vice president's staffers were among those in attendance. They naturally felt very protective of their boss and his interests. My upsetting him would now be added to their long list of reasons to shunt me farther aside.

Tony had encouraged me to take on the vice president, but while I was being yelled at, he didn't intervene. Bob tried to provide some support, but not the depth of support I needed in that moment. After the meeting ended, I sent both of them a concise "WTF was that?" email. Why had they let me flap in the wind like that? They both replied that they didn't see an intervention as necessary.

I wondered if Bob and Tony felt I'd verged into hysteria, succumbing to emotion rather than reason. I hadn't been hysterical, but as forceful as necessary to get the vice president to understand we needed to seize this moment to save Americans. We also needed him. He was the only one I could trust to convey that message of vigilance the next day.

After this incident, I had a very hard time reconciling myself with what I felt was a lack of support from Tony and Bob in that moment, not only to stay true to our private compact to have one another's backs, but also to rise to the occasion the moment had called for. I still respected them both, but I continued to puzzle over their lack of engagement in that moment.

Later, I couldn't help but think back to Tony's call for a respectful sit-down with Scott Atlas. Tony doesn't like confrontation, especially not those like the one between the vice president and me. Tony possesses an unwavering belief that cooler heads will prevail and that divisiveness won't solve problems, only exacerbate them. This has helped him succeed throughout a long career in public service. So, in the end, his hesitancy to intercede in my behalf hadn't been entirely out of character.

What hadn't been out of character at all were the words I had used and the sentiment I had tried to express. This wasn't solely about the numbers. I had been trying to convey the deep sense of loss I was experiencing. As with conflicts I had witnessed overseas, the trauma of this pandemic—the fatalities, the isolation, the fear that the virus was spreading rapidly across the country, the fear of more suffering and loss—was very real. As a veteran, I don't say this lightly, but to me, this was a war, and we were on a mission to win the war and save lives.

In his moment of anger, I had only a vague idea of what the vice president had been going through. Like me, he had been away from DC for much of the fall. Whenever I saw him, either in person or in video clips on the news, he appeared to be the same robust man as before. But ever since the election, I could tell something was weighing on him. Only later would I learn that while I was pushing him to be *the one*, others inside the West Wing were also pushing him to be *the one*. While I was asking him to take a stand to save American lives, others were pushing him to take a stand on a very different issue. He couldn't have been sleeping much; he looked gaunt and exhausted. Stress had sunken his eyes and hollowed his cheeks.

Later, after everything that occurred over the next four weeks, I understood what had contributed to his momentarily pushing back at me. His vehement tone of indignation seemed now to have been a response to the mixture of burdens he was then carrying—none greater than his constitutional duty to certify the election results.

By this point, December 7, despite all the other legal avenues the president was pursuing to overturn the election results, it was becoming clear that the administration's strategy would have to rely on Vice President Pence's refusing to certify the Electoral College results. Throughout the pandemic, the vice president had been going into the ring in my behalf, to get the president to listen to sound scientific reasoning and take the pandemic more seriously. Now he had been forced to do battle on behalf of a free and fair U.S. election. As vice president, Mike Pence had a constitutional duty to perform—and a president demanding that he shirk it. In some circles, at that moment, the vice president was no doubt being vilified. This is speculation on my part, of course: throughout my time at the White House, I was never present for any political discussions, and the vice president and I never spoke about internal politics or any specific pressure he was under. Yet, whatever emotions my remarks in the task force meeting had incited in him, I was sure they were a function of so many things happening both inside the White House and out—and this was on top of a deadly, disruptive pandemic. In the end, none of us could really, fully grasp the nature or degree of his burden.

Up to that point, the vice president and I had had a very professional relationship, and following his blow-up, in our more limited contacts, neither of us raised the subject again. Both of us treated it as a discrete moment, nothing that would knock out of balance what had come before or would come after. I said what I said, and he responded as he had. I was grateful that he'd gotten the message about the seriousness of what we faced. Naturally, if I had known more about what he was embroiled in, I would have handled things differently.

The next day and in the weeks remaining of his term, Vice President Pence would continue to deliver a message that balanced vigilance (mitigation) with victory (vaccines). And, in the face of death threats to his family and him, he did his duty and certified the election results.

Despite all my efforts to convey to the president, through emails to Mark Meadows and Jared Kushner, the seriousness of the crisis we still faced, the president's remarks at the next day's vaccine event were a victory lap for the hope and promise the vaccines held—though they were still weeks away from delivery. He used the word *miracle* to describe the vaccines' rapid production. And it was a miracle—a miracle that volunteers had selflessly come forward to test an experimental vaccine for their fellow Americans, a miracle that the cases rose when they did, providing the endpoints needed to determine vaccine efficacy. Many had predicted that vaccine development would take 25–100 percent longer than it did. That additional time would have translated into even more lives lost.

The remarkably short time frame in which they were developed demonstrated what is possible when you work in public/private partnerships, bringing together the best of each. The design of Operation Warp Speed, the use of the Defense Production Act to ensure raw materials needed for manufacturing, the dedication of the scientists whose earlier work led to the discovery of mRNA vaccines, the prior investment by NIH in basic research funded by prior administrations and Congress—all these had brought us to this point.

Being in the White House during this time as an outside observer allowed me to witness the direct involvement of the Oval Office in the push for vaccines and therapies. From March 2020 forward, President Trump removed barriers and sped up processes. The administration should get credit for this innovative approach, which hastened the development of SARS-CoV-2 tests of multiple types including point of care tests that could then be used in the home, novel therapeutics, and vaccines—all of which illustrated the importance of public/private partnerships could aggressively address gaps and obstacles in real time.

That being said, I was continually struck by the disconnect between words and actions with this administration and this president. When the president dodged questions about the large indoor holiday gatherings taking place in the White House, where most attendees were unmasked and didn't socially distance, he revealed this gap. When pressed further on this flouting of CDC guidelines, the president briefly acknowledged their importance—then immediately shifted gears: "But I think this. I think that the vaccine was our goal. That was the way—that was the way it ends."

True, there was great hope for the vaccines and their lifesaving potential—those who had an effective immune response to the vaccine would be protected from severe disease and death—but we were months away from their use. Also, questions remained. How long would the vaccine response last, and whom would they not protect? Not every American would mount an effective immune response. I had two parents at home who were well into their nineties. Americans their age didn't participate in the clinical trials; Americans with chronic cancers weren't well represented in the trials, either. We didn't know yet if the aged or those with diminished immune systems could mount a protective response.

We also had yet to account for all the logistical issues. The CDC had decided that health care workers should get the first doses of the vaccines. How distribution would proceed after they were inoculated was still fraught with potential problems. The final prioritization guidance, with its multiple tiers, was too complex to be easily understood and executed. In reviewing some of the state distribution plans and the CDC's plans, I saw that neither had a real mechanism in place to assist states when they encountered problems executing their programs. There were also no plans to send in CDC teams to work alongside state and local officials. For this reason, CDCers wouldn't be on the ground, learning and making continual course corrections.

I'd seen this before, time and time again. The National Guard had been called up to assist in the initial response to the pandemic, when Covid-19 took us by surprise at the beginning of the year. These troops had worked tirelessly to intervene when the states needed help, and I knew we could count on them again.

On December 11, when the FDA granted the Pfizer-BioNTech vaccine—and later, on December 18, the Moderna vaccine—their long-awaited emergency use authorization, I remained realistic. I was certain the vaccines were safe and effective, and would positively impact the pandemic, but to what degree depended on many factors related both to the vaccines themselves and to aspects of the response. This was why, on each of those dates, I was on the road—in Camden, New Jersey, and just back from Asheville and Raleigh, North Carolina, respectively—to manage overly high expectations.

Yet, as with so many things distorted by politics during this pandemic, it still wasn't being made clear enough that these vaccines might not be the final answer. A vaccine can be effective only if someone gets inoculated with it. A vaccine program can be effective only if large numbers of people roll up their sleeves and stick out their arms. Even if you achieve a high vaccination rate, it doesn't mean you will end a pandemic. I have spent a lifetime studying immune responses to vaccines and other invading organisms. Much remains that we don't understand about why these vaccines might not produce sterilizing immunity across all communities in all age groups. We also don't fully understand the vaccines' impact on community spread. We still didn't know if those with underlying conditions or those over age eighty—two groups that aren't as healthy as the volunteers in the clinical trials—would mount the same protective immune response once vaccinated.

We also didn't know how long it would take us to vaccinate the world. Were the other vaccines developed by other countries equally effective and durable? How many variants would escape from natural immunity and then vaccine-induced immunity? Natural immunity escape was already happening across the globe. How durable was the vaccine-induced protection against infection versus disease? Did the very elderly and those with chronic cancer develop a protective immune response? Would immunity wane, and if so, how fast? All these questions needed answers before we reverted to 2019 behaviors. I kept my family updated. With very vulnerable family members across states and households, we united in our diligence and prevented Covid-19 infections. It was possible with diligence.

In my version of the Vaccine Summit, I would have articulated all these questions and more. Too many people believed that vaccines conferred invincibility. That message started out being conveyed by the Trump administration and was carried over into the Biden administration. Sure, individually, the vaccines would perform well. But when used along with preventive measures, they did far better. They needed to be added to our existing arsenal of solutions, not replace those solutions. For me and so many other Americans, this was our reality. But divisiveness and distortion continued to emanate from others. We had to find a way to develop immunity to that.

CHAPTER 20

Out but Not Done

When former vice president Biden was declared the winner of the 2020 election, I'd set a goal for myself—to hand over responsibility for the pandemic response, with all its many elements, in the best possible place. I therefore felt duty bound, by regulation and by character, to continue to do my best, even in the midst of the tiered exodus that is part of the orderly transition of power culminating in the swearing in of the new president.

If precedent had held, I wouldn't still have been in the White House for the task force meeting scheduled for January 7. As a detailee, I would normally have been among the first tier to move out of the West Wing and would normally have spent that week packing up my office, relinquishing my White House–provided computer and phone, and returning to the State Department. But the winter surge was raging, and even one day lost was not one we could afford. I wasn't about to stop providing my daily update, nor the weekly governor's reports, so I lobbied Matt Pottinger and National Security Advisor Robert O'Brien to be able to ignore the customary protocol for departure. They cleared my request, and I remained in the White House until January 19.

For the first few days of the New Year, I went about my usual work. I wrote to a key person at the CDC and copied Brett Giroir on an email I'd sent to that agency the previous week, about expanding testing and making better use of the university systems' successes in remaining open. I pointed out areas of significant deterioration across the Sun Belt and the cold northern regions. Texas and California were particularly worrisome: 25 percent of the nation's Covid-19 deaths were occurring in those states.

On January 3, 2021, I thought briefly of the BBC piece I'd read exactly one year before: "China Pneumonia Outbreak: Mystery Virus Probed in Wuhan." What had been described as a "mysterious illness" had, over the past twelve months, lost much of its mystery. We knew so much more about the virus now—and crucially, so much more about how to combat it.

And yet, so many were still getting sick, so many dying.

A year ago, we had been caught off-guard. Whether from the FDA, ASPR, BARDA, the CDC, or the White House, there was enough blame, enough missteps, to go around. Pointing fingers earlier or even at this point would only derail my efforts and dishonor the work so many had done so faithfully. By August, we had had a firm grasp on the solutions that worked. By fall, we had demonstrable evidence of the importance of sentinel testing from the universities and various other businesses—film production, restaurants and bars, sports (for example, Tony Fauci's favorite, baseball)—all of which had had to shut down but now remained safely in operation using the tools we had. Governor Doug Burgum of North Dakota—always on top of the data and forever innovating—found that if all students were masked, you wouldn't need to quarantine all students just because one had tested positive. Students were able to stay in school. Tools we had developed in partnership with states were working, and vaccines would protect from severe illness.

For every news report of what went wrong—for every overrun hospital, every superspreader event—I was privileged to be able to see, across the country at the local level, what went "right." I had stood witness to what state and federal workers were doing every day to adapt to the evolving crisis and adopt innovations—something we were able to include in the governor's reports. I wanted to be able to trumpet these efforts to the world, to acknowledge the herculean efforts of the private sector—rolling out tests to the tune of nearly two million per day; developing, delivering, adapting, and adopting therapeutics like remdesivir and monoclonal antibodies—and the courageous, self-sacrificing labors of the country's essential workers. Now, with the addition of th vaccines to the mix of mitigations, we were closer to managing this virus, yet that goal remained just out of reach.

One area where we were still lacking was in our fundamental messaging. Weeks short of the new administration's taking over, the Trump White House stubbornly refused to acknowledge that we had the means and the methods—if not the motivation—to bring down case numbers and deaths, to bring an end to this crisis with routine, clear, consistent communication to the American people, simple, straightforward, commonsense actions that would save lives.

The behavior of the White House continued to confound me. I figured that, by this point—with Congress due to count the electoral votes on January 6 and the results a virtual certainty—they would want to fully address this crisis with best practices and hand the effort off to the Biden administration with the necessary momentum to carry our best practices forward. I thought that, if for no other reason than to provide some baseline to the president's legacy, this would be the aim. Unfortunately, it wasn't.

In late December, I'd lobbied Mark Meadows again to let me hit the road in January to go where I'd seen bumps in the vaccine rollout. He refused to let me go because there wasn't staff to support it. What was there to gain from forbidding me to visit with state officials? An image from my childhood kept intruding—kids with their fingers stuck in their ears, chanting, "La, la, la! I can't *hear* you!"

The Oval Office didn't want to hear me, had failed to listen. I would struggle to understand what motivated this right up until my last day in the White House. As much as I vehemently disagreed with their decision to prevent me from communicating nationally the urgency of our situation in the fall, at least that rationale was rooted in a clear, election-related, politically driven strategy. They wanted to win, and my pointing out the wave that was about to hit the country would have made that harder. It may have been calculating, but it was discernable. Much harder to comprehend was why, once the election was over and former vice president Joe Biden was declared a winner, they continued this behavior. I'd assumed that the senior advisors would stop trying to censor me and would further support the comprehensive response they had agreed to at the end of October. They didn't. It got worse.

It. Just. Didn't. Make. Sense.

What followed, on January 6, made even less sense. The behaviors exhibited that day and the day before were the most troubling I'd ever seen. Let me put this in context.

Since 1995, as part of my State Department, Department of Defense, and CDC assignments, I've traveled and worked in countries around the world. In some of them, democratic principles were newly in place. In others, the governments preached democracy but practiced authoritarianism. I'd seen free and fair elections contested, both peacefully and violently. In my work in Africa, where violent targeting of candidates, their supporters, and polling places spilled over, catching many in the crossfire, I was often forced to take extra safety precautions pre- and postelection. In other countries where I worked, colleagues involved in suspicious car accidents believed they were targeted due to their support for one politician over another.

I never expected to see a violent uprising, an insurrection, take place on American soil, never expected to see my fellow citizens, individually or in groups, behaving as they did at the U.S. Capitol that horrifying day. Crucially, I never thought I'd see acts of domestic violence watched by the highest official in the land, a man who had sworn to preserve, protect, and defend the Constitution and the founding principles of our democracy, without immediate action and intervention. For all the times over the course of my life I had watched news of unrest from around the world and thought, *That can't happen here*, I was never shaken as badly as I was on January 6, 2021. It *could* happen here. It had.

Periodically, through the summer—particularly in the wake of the George Floyd murder and the rise of the Black Lives Matter movement—I'd seen the perimeter of the White House complex come to resemble a fortress, something you'd normally have seen in a high-threat area like Baghdad's Iraq War–era Green Zone. Security around the White House was always tight, but in the approach to the congressional vote count, these efforts had been redoubled. Barricades were being erected; streets were closed off. Armed troops moved in.

The night before the Capitol riot, in anticipation of leaving the White House for good, I was walking out in the darkness of the city. Because I had spent so much time at the White House, I'd brought many pairs of high heels to change into after walking in in my "reasonable" shoes. On the evening of January 5, I left the office carrying a boxful of them. My usual route, now lengthened due to our being rerouted to an exit nearer the Ellipse, took me through a labyrinth of high chain-link fencing an improvised outdoor corridors. Generators thrummed, and temporary lights on stands illuminated clusters of armed security forces, their shadows splashed darkly across the ground. Walking out the back way, closer to the Ellipse, I could hear the ongoing rally. The crowd there was loud, and although I couldn't see how large it was, the volume of the noise they generated vibrated my chest. From a distance, cries of "Stop the steal!" were carried over the cold breeze. I quickened my pace, the sound of my footfalls competing with the distant, pounding voices.

In that moment, I decided that I would work from home the next day. What was that evening a relatively peaceful protest had an edge to it. What my gut told me was confirmed by the events of the next day. I was glad I trusted my gut; it had saved me many times overseas. *You just never know.*

Still, nothing could prepare me, or the rest of the country, for what transpired on January 6; nor for the risk to life and limb the vice president would face. In conversation with others, I had trouble voicing all the thoughts cascading through my mind, or articulating what the riot at the Capitol meant for me—as a military veteran, a forty-year civil servant, a citizen, and someone who loved her country. In the end, my thoughts returned to the pandemic. I wondered how different its trajectory of loss and pain might have been if all the fury and determination behind the president's words that awful afternoon—"You'll never take back our country with weakness. You have to show strength and you have to be strong"—had been applied to his approach to responding to Covid-19.

Later, I would also wonder how different things might have been if some in the administration and the president hadn't gone on ignoring the pandemic in the days following their insurrection. To his great credit, Matt Pottinger resigned the same afternoon, as protestors stormed the Capitol Building. He had brought me on board, convincing me that my country needed me. It still did, and in the days that followed, I would go on serving the American people, but I would have even fewer allies on whom I could rely.

I'd never found it easy to adjust to the presence of heavily armed police and military, and what I saw every evening in DC was no exception. There were now more police and military troops deployed across the city, and the White House had become an armed camp overnight. Getting to work each day required going through armed checkpoints—as you would see in an a U.S. Embassy in unstable areas around the globe. But this was Washington, DC.

THE EVENTS OF JANUARY 6 disrupted, but did not halt, the orderly transition of power. Meanwhile, I kept my attention focused on one part of that transition: the task force had only a few days left, and there was much to be done to help the incoming administration, and even more urgency.

Back on November 30, I had had my first meeting with members of President-elect Biden's transition team: Vivek Murthy, who would take over as surgeon general; Jeffrey Zients, the Biden administration's coronavirus response coordinator; Zients's deputy coordinator, Natalie Quillian; and a small contingent of transition team members. Given how widespread Covid-19 infections had been in the White House, the team made the wise choice to meet via a Zoom call. David Kessler, the chief science officer for the response, was on the call as well, though neither of us acknowledged that we'd already been regularly communicating and that I'd been sharing data packages for weeks. For roughly ninety minutes, I filled everyone in on all aspects of the response, including the current data and the harrowing rise in slope of new cases.

I went through all the data products from the data streams we had created from scratch, to familiarize them with the picture and see if we needed to create others that better met their needs. I wanted them to have everything they needed to hit the ground running and fully execute the Biden Covid-19 plan. I told them about the governor's reports, my daily data analyses, the White House Covid-19 summary we had created for the Covid Huddles, the hospital database, and the HHS community profile we had quietly posted on a publicly available website so they and the rest of America would have constant access to daily data to aid their decision making. The new team didn't engage me in a lot of back-and-forth, but they asked a few pertinent questions.

I would meet only once after that with representatives of the Biden team, to review the data again. In the first week of January, I met over Zoom with the Biden data team's version of Irum Zaidi, Dr. Cyrus Shahpar. Until our very last day, Irum would keep in close touch with him. Because of the absolute lack of comprehensive national data we had inherited when we first arrived at the White House back in March, we had created a new, integrated system for data collection and analysis, and we wanted to make sure Dr. Shahpar was up to speed on it. Different data analysis people often like their information delivered to them in particular differing ways. Dr. Shahpar would know what the new team needed and wanted. This meeting was very much about the nuts and bolts of the current system.

In the New Year, I became increasingly concerned that a mere ninety minutes with the new administration's response leaders wouldn't be sufficient. I didn't see how, in that brief meeting, they could have gotten much beyond a glimpse of the current situation. The way the transition works is very one-sided: As a part of the outgoing administration, I couldn't contact anyone on the Biden team directly, to ask them if they needed more information, wanted to meet again, or had any questions. This had been the case for both presidential transitions I had already been through. The only way the meeting I desired could take place was if the incoming administration initiated it.

With no one else to turn to, I shared my concerns with Jared. He told me he would see what he could do, ultimately suggesting I work with Pete Gaynor, the head of FEMA. I told Administrator Gaynor my concerns: "They don't have a full grasp of just how many moving parts there are to all this." Pete told me he'd try to find a work-around.

The only public communication coming out of the Biden team was their frustration over the Trump people not being very forthcoming with information. Politics was clearly at play. I didn't like that this impression was out there, but there wasn't much I could do about it publicly or within the White House. No matter what was being said, the outgoing team was still responsible, and we needed to make sure the incoming team had everything they needed.

The political reality was that very few people involved in the Trump administration's response to the pandemic would carry over into the Biden White House. Maintaining continuity in the transition of power is always difficult, but I sensed that this time it would be especially so. After that initial contact, I didn't hear directly from any of the participants on that Zoom call or from their data person. The message was clear: *Thanks. We've got this. We don't need your insights.*

I'm not sure how Administrator Gaynor managed it, but the week of January 11, the outgoing response team gathered in person at FEMA headquarters for a second, comprehensive Covid-19 Zoom call with the incoming response team. It was a productive exchange, but I got the impression the Biden team felt we had failed in our response to the crisis and, therefore, had nothing of real value to offer them. They paid particular attention to, and asked more questions about, the vaccine program. Because of this, I believed the Biden response would be firmly centered primarily on vaccines.

In the summer, I had read the Biden administration's Covid-19 plan. Its emphasis on testing, masking, vaccines, and clear, consistent communication assured me that the new team would finally resolve the long-standing issues over the role of asymptomatic spread and the aggressive testing needed to find the earliest cases. Now, though, the Biden team seemed much more focused on vaccines. The more I saw, and the more I heard from Bob and Tony—who were also meeting with various Biden people—the stronger my disquiet grew. For, as much as their plan relied on increasing testing and enhancing mitigation, their emphasis on vaccines heightened my fears. I wanted to attribute this focus to timing; the vaccine rollout was a recent development. Still, we'd seen how the Trump administration's shift in emphasis from testing to vaccines, in August 2020, had harmed the response and cost so many more lives. We couldn't let the same thing happen here.

Even in January, my early fears about the CDC vaccine guidance were being realized. Their multiple tier distribution plan was too complex; they didn't have a mechanism in place to put their personnel on the ground in the states to help monitor the situation; and without their direct, eyes-on observation and assessment, they couldn't effectively evolve and adapt the guidance to improve the immunization rates of those at the highest risk. Any new people heading into this already problematic situation would be operating at a deficit not of their own making. Taking time to smooth the rollout bumps would further delay getting shots into arms.

Before the winter holidays, I had spent a week in the field, going to five or six states to talk about the current surge, suggest solutions, and do everything possible to simplify state-level vaccination programs. I urged the states to consider a simpler, age-based prioritization of available doses, in place of the more complicated, CDC-endorsed plan. Still, for as much time as had been spent in advance of the vaccines' approval back in November, we were still dealing with evolving issues in January.

Secretary Azar and HHS believed that the Walgreens and CVS drugstores across the country could aid in immunizing the most vulnerable in LTCFs. I was worried about speed. The National Guard was extraordinary at tackling the hard implementation issues we had encountered over the past eleven months, and I lobbied to have Guard troops sent to these facilities to support on-site immunization.

Bob, Tony, Steve, and I brainstormed, coming up with other ideas for working around the kinks in the CDC-approved state plans. With the Janssen/Johnson and Johnson vaccine nearing approval, we factored in how to make best use of this one-shot vaccine. We were drawing a road map to at least one million immunizations before January 20, when President-elect Biden was sworn in. It would be a slow start—all new initiatives are—but we thought if we pushed the system, we could flatten the learning curve and ramp up immunizations, so the new administration could accelerate the response. In the clinical trials, the single-shot Janssen vaccine had been shown to be slightly less effective than Moderna's or Pfizer's shot at preventing disease. Perhaps it would be better suited to inoculating those younger and less vulnerable to severe disease. The FDA would need to gather efficacy data by age group. If we could use the J&J dose for those under age forty-five (including health care workers), we'd preserve the supply of the two-shot mRNA vaccines for the over-seventy group. We even considered administering half doses with the first shot of the Moderna vaccine for the under-forty-fives, as the tested Moderna dose was found to have three times the mRNA of the Pfizer-BioNTech vaccine.

We considered every possible option to provide protection for severe disease, using the right vaccine based on age and vulnerability. We needed to ensure that every dose not utilized for health care workers was immediately made available where it was most needed—in those over seventy. On the day of the insurrection, 230,000 new cases were reported; two days later, the winter surge peaked, at 250,000 cases per day. The ensuing downward trend would continue until mid-March before the Alpha variant emerged, producing a surge in the Upper Midwest. What followed were the predictable peaks and valleys throughout 2021 mirroring 2020's regional surges.

The docs and I agreed that the product of our brainstorming session on the vaccine rollout should be formalized. We decided to prepare a vaccine position paper for the vice president, discussing our concerns and offering solutions. On January 7, Alex Azar learned of our efforts and called me in a huff. We weren't going over his head: days before this I had sent him a draft of the recommendations we hoped to share at the next task force meeting. Still, the secretary was angry—angry that I had taken the rollout issues to the task force for discussion, angry that I had alerted the vice president to mine and the other doctors' concerns. His angry phone call to me was on a par with President Trump's fury back in August. But with the CDC under the umbrella of HHS, he viewed the position paper as our tramping across a clearly established border and into his territory.

I pushed back. If the rollout was not working as planned, I said, we should brainstorm and suggest other solutions. Our paper wasn't a criticism, I told him, but an acknowledgment that the rollout was not going perfectly and that precious doses of the vaccines were sitting around unused.

"That's bullshit, and you know it," the secretary said.

"I don't want to see doses sitting unused," I said, "like we already have in many hospital freezers."

General Perna had shipped enough vaccine doses for 100 percent of all hospital personnel, regardless of their age, but only 60 percent of them had been used on this group, with some very young hospital workers hesitant to be first in line. (There were still questions about the vaccines' safety in early pregnancy in January 2021 due to limited data from the original trials.) In their minds, they weren't at risk for serious Covid-19 disease and could wait until there was more data on the shots. Yet, General Perna was already sending the full complement of second doses now, regardless of how many of the supply of first doses had been administered. To make matters worse, the doses were being sent to the same hospitals in the same quantities, despite their not all being used. Large numbers of unused doses sat in freezers instead of going into the arms of the second prioritized group, those over age seventy. I'd argued and argued that the decision to put health care workers at the head of the line for vaccines was wrong. Now, this major flaw in the rollout plan was coming to light in the middle of a deadly surge.

The secretary and I went back and forth for a good ten minutes. In the end, I toned down the paper based on his comments, and it went to the vice president.

I didn't know then, and I don't know now, if Secretary Azar was aware that even though I'd been told I couldn't visit the states in January, I'd been working the phones every day to check in with them on all aspects of their response, putting special emphasis on vaccines. I prioritized speaking with the governors of states that had been particularly hard hit, with older populations and more vulnerable residents, such as North Dakota, Tennessee, and Arizona, and with governors who had reached out to me personally, as was the case for Governor Andy Beshear of Kentucky and Governor Jim Justice of West Virginia. Governor Justice was being innovative and was prioritizing vaccination rollout by age group. I made sure he was in on the governors' call to convey his practical approach. Age, age, age, he said over and over.

I was gratified when, on January 12, the Executive Office of the President and the Office of Intergovernmental Affairs (the same group that had been so instrumental, particularly Tucker Obenshain, in coordinating Irum's and my state visits) issued an "Analysis of State Vaccination Administration Programs." Prepared by IGA as a guide for Operation Warp Speed to interface with the states, the document identified the ten states that had vaccinated the most citizens per one hundred thousand people. Included on that list were the five states I've just mentioned: Arizona, Kentucky, North Dakota, Tennessee, and West Virginia. Many of IGA's top ten had simplified the CDC guidance and were focusing on getting the doses they had into the arms of those greatest in need. I was particularly heartened that IGA mentioned the best practices these states had used to earn their place among the top ten.

This was why I had been reaching out to governors, state health officials, and others since the summer. Despite so many people telling me it wasn't our role to "tell the governors what to do," for me, it was never about that. It was about learning what was working on the ground. Yes, while in the states, I had said bluntly what I was seeing and how the response could be improved, even when they contradicted what the White House was saying or wanted to believe. But more important, I engaged in productive dialogue, to see what was working and what wasn't. We accounted for the unique characteristics and needs of each state, each tribal nation, each racial and ethnic group, each college and university community, and spoke about how these best approaches could be adapted to those needs. We learned how this virus operated, how it mutated, and we flexed our approaches and our actions to counter it.

• • •

IN THAT LAST WEEK prior to January 20 and the inauguration, I saw the vice president for one of the last times. He called me into his office, thanked me, and told me I should tell the story of the response. Without directly stating it, he was giving me permission to go out and do press hits. I think he understood that what I'd been saying all along, the direction I wanted to take the response, had been the right one. Now, with the transition days away, he and I both wanted to make clear what had worked and what hadn't.

I agreed to participate in a CNN taping and do a spot on CBS's *Face the Nation*. I delivered the same message, expressing hope that past failures would not be repeated and that present successes and solutions would be maximized. I wanted to keep getting the word out. We needed to remain vigilant. We needed to keep learning. We needed to stay in close contact with the states to learn from them.

Unfortunately, the tone of the CNN interview, which wouldn't air until March 28, 2021, was a familiar one: part of the usual pattern in the media of looking to the past to pin blame rather than the future and the corrective action we had taken based on the lessons we'd learned. I had hoped that my appearing on national television could help point the way to a better future. There was plenty of blame to go around, but there were also solutions found during the Trump administration that shouldn't be ignored, that needed to be followed, and new interventions to be discovered.

Along with working the phones to contact governors, I spent the days before Christmas and then up to the New Year speaking with colleagues and friends across the United States, and calling in favors, to get the living former U.S. presidents immunized. The past presidents wanted to use footage from their vaccinations, and their famous faces and voices, as part of a broader pro-vaccine ad campaign. The ad campaign they had planned was great. I wanted to ensure the past presidents and the First Ladies got shots in the arms now. I was particularly concerned about the Bushes, the Clintons, and the aged but vibrant Carters. It was important for the past presidents to be at the pending inauguration, especially in the wake of the insurrection. Having past presidents from both parties standing side by side, united as Americans, was important for the public to see. Having them show their bipartisan support and cooperation in action would be important at this perilous moment in our country's history.

Although Biden's swearing in would take place outdoors, the president-elect and Jill Biden and the former presidents and First Ladies would be gathered indoors prior to the inauguration, where the virus could be present. The Bushes and Clintons, all outgoing and social, would be interacting with many people that day, so I wanted to be sure they were vaccinated with both doses before January 20. The clock was ticking.

Getting the former presidents and First Ladies their jabs proved to be quite difficult, but I had friends in some of the presidents' states whom I had met on my travels, and I had support in Texas (the Bushes) and Georgia (the Carters). In particular, New York (the Clintons) was a problem. I tried three different approaches through my colleagues and friends, but each promising lead took me to a dead end. I had run out of ideas when my go-to Jared told me to contact Adam Boehler, who was taking a few days off with his family. Adam, who seemingly knew every hospital organization's chief executive officer, came through, and within twenty-four hours, days before the New Year, the Clintons were immunized and had added their taped voices to the others'.

This type of behind-the-scenes work always gave me hope. Across the country, I had seen solutions. I had seen red counties and their leadership supporting blue counties, and vice versa. I had seen private-sector leadership focus on solutions, not profits. And I had personally experienced moments when people saw a need and helped.

With the former presidents over age seventy protected from the virus, I took a few hours with my family, cooking a Christmas dinner for less than half of them; an additional son and daughter were in DC, within a few short miles of us. Not letting our guards down, we employed the usual strict practices to protect my daughter, who was eight months pregnant, and my parents, who were in their nineties. There was no communal Christmas dinner for us. Just takeout plates handed outside, just as we'd done at Thanksgiving. Like many Americans, we also opened presents and emptied stockings apart from those who gave them—another family gathering fallen victim to the virus circulating in our community.

TO THE VERY END, I remained inflexible in my determination to do the job I had been assigned. I continued to send out my daily reports, and my team and I sent out our last batch of governor's reports on Monday, January 18, including in it vaccination reports and distribution recommendations. We again emphasized the importance of sharing information across the states—what was working and what wasn't.

It was too soon for the inoculated population to have any great effect on the overall numbers, but county by county and state by state, the nation as a whole was stabilizing through the week of January 11. As before, I remained concerned about the most populous states, California, Texas, and New York—we had to ensure that they were at a true plateau and that cases there were on the decline. Between twenty and thirty million people were infected—likely more, as the numbers would never reflect the number of untested asymptomatic.

In my last week, the pattern of good news/bad news continued. I received some heartfelt thank-you emails from people whose work I valued (Jared Kushner, Tony Fauci, Brad Smith, Tucker Obenshain, and Adam Boehler among them).

I had been including in the governor's reports my concerns about evolving variants expanding here as well as imported variants and to be alert and sequence more. I raised this alert in the governor's report because I couldn't get the CDC to expand sequencing. The CDC was not genetically sequencing enough virus samples, which meant it would be easier for new, more virulent strains of SARS-CoV-2 to arise silently and unseen and to spread. I fielded an inquiry from NBC News to shed light on an Ohio State University study that had isolated and identified a new variant of the virus. I wrote back, off the record, giving my assessment and pointing out that I'd already addressed this in a prior governor's report. I explained that some less-fit mutations made it harder for the virus to replicate, while other, fitter mutations made replication easier and the virus more transmissible. At the time, we didn't know if this variant had a competitive advantage over the previous one. It had been identified and was still being studied.

I was both encouraged and dismayed by this turn of events. It was good news that the Ohio State researchers had stepped up in terms of genetic sequencing. I still couldn't understand why the CDC hadn't: Back in July 2020, we had asked the CDC to expand sequencing and work with Illumina, a private-sector biotechnology company. But the CDC decided that a cooperative effort at the scale I had recommended wouldn't be needed. They could handle it internally, with their own laboratory and through a few contracts with public health and research labs. With our access to the private sector and additional research laboratories, national capacity was within reach, but taking advantage of the private-sector labs at our disposal was never deemed a priority, as it was in other countries. The United Kingdom was a world leader in genetic sequencing and testing (and remains so to this day), sequencing at a rate nearly ten times that of the United States, so we were forced to rely on them to get this important job done. We didn't want to be caught flat-footed again.

Even though the United States had more capacity and reach, I often had to rely on other countries for data and answers that we should have had at our fingertips from the outset—whether it was the original mortality and morbidity data that allowed me to warn the elderly and Americans with comorbidities in March 2020, the breadth of asymptomatic spread at universities, the evolution of new strains, or the durability of our vaccines. I could never push the CDC to do the countrywide data collection, the science, to determine mask effectiveness or better understand vaccine hesitancy and how to overcome it.

MY TIME IN THE White House and as a formal civil servant was at an end. I couldn't at this point even return to my old post at the CDC in any official capacity; I would adapt to my new circumstances.

I was pleased that, in my final daily report on January 18, I was able to say that we were seeing mostly significant improvements. I had wanted to hand over a surge in retreat, and we did. But as I stated earlier in this book, worry is endemic to an epidemiologist's work.

On January 16, a Saturday, I had to remove all my belongings from my old State Department/Global Ambassador's office. The building exterior was completely covered in plywood, with a small plywood door on temporary hinges. I was grateful that the guard at the entrance had allowed me, my daughter, and Irum into the building in the first place. Over the next hours, we packed up and carried box after box out to our cars, parked blocks away. The streets were eerily empty.

I decided to take a drive toward downtown DC. I was concerned. The city was one of the metropolitan areas I had had to report as problematic. In advance of the Biden swearing-in ceremony, National Guard troops had arrived from hot spots around the country and were comingling with DC Metro police officers. These troops huddled close together in tents to eat and, sometimes, sleep. After the inauguration, they would return home, and I worried that, in ushering in the new administration, we would create a superspreader event. In my final act, I contacted several members of the new administration to recommend that all National Guard and active-duty troops in DC be tested for the virus.

In far too many ways, we were all being tested. We couldn't give up hope. We couldn't stop working. We couldn't stop learning. We had to stop repeating the mistakes of the past. I was out of the White House, and would soon be out of federal service, but I was definitely not out of the fight.

Epilogue: Looking Back and Thinking Ahead

As the preceding pages have demonstrated, I'm a big believer in the adage that where there's a will, there's a way. My personal version of that cliché goes on to add, "and comprehensive data will show us the best way." I was no more relentless in tracking data once I was out of the White House than I was while there. The Biden data team was still updating and maintaining the HHS Community Profile site and all the state profiles from the governor's reports that we had quietly put up in December 2020. All Americans continued to be able to track the pandemic's status in their county and across the United States in real time. I used this data and triangulated the global data in my ongoing communications with some on the task force, on webinar after webinar, and in private conversations to help guide friends, family, and anyone else I spoke with about layered protection and pandemic preparedness. The data had shown us the way in 2020 and it was showing us the way in 2021, and it is still showing us the way in early 2022.

While the virus was silent, the numbers spoke clearly to guide us out of what felt to many of us like a recurring bad dream. Many believed: Once Trump is out of office things will get so much better. Once we get vaccinated, we'll be able to go back to doing what we did before. The summer of 2021 would be different. This Thanksgiving will be different. This Christmas will be different. This will all be over soon.

I would never suggest that hope isn't helpful or that we haven't made strides in tackling this pandemic. That is a realistic assessment, but so is the reality that too many people continue to die, with each surge we do a little worse than our colleagues across the ocean, and some have grown so weary of the pandemic that they appear to be willing to accept an excessive number of preventable casualties from this crisis. Hoping we can learn to live with the pandemic no matter the consequences, or accepting that we have to learn to live with it, and figuring out *how* to live with it as safely as we can as a country are very different approaches to this ongoing public health crisis.

I don't *hope* we have the answers to how to live with it. I *know* we do. The data has shown us the way forward using commonsense mitigation. Many obstacles continue to exist that prevent that message from being made clear. Some of them have their roots in politics, some in ineffective communication on the benefits of various public health measures, some in our collective failure to account for the variability in human experience and opinion in a socially and geographically diverse country, and some are the result of the desire to find blame and not offer effective solutions.

Some people might say that in my time serving as the White House Coronavirus Task Force coordinator, I didn't really live the experience that most Americans did from the start of 2020 to the start of 2022. That's wasn't true then and it's not true now. I understand and have felt the frustration and anxiety people have experienced as a result of the government's response to the pandemic.

The Biden administration hit the ground running in January 2021, focusing their efforts on consistent messaging and ensuring access to vaccine supplies across the United States. The Biden Covid-19 plan was thoughtful and comprehensive—espousing the need to continue to expand testing, and the use of proactive testing, and continued expansion of critical therapeutics as well. After the inauguration, the Biden administration turned their full attention to effectively rolling out vaccines, and so did my family and I. We spent weeks getting an appointment for my mother and father, aged 91 and 96 respectively, in the state where they now lived—Maryland. (I could have gone elsewhere and skirted the rules but opted not to.) Frankly I wanted to see what everyone was up against. Unable to successfully navigate the online process myself, my Millennial children took over. They determined that their phones refreshed faster than my laptop and used them to secure those precious slots.

We weren't alone in facing and overcoming that challenge. Nor were

we the only ones overcoming the difficulties of physically assisting aging parents or grandparents, weakened by a year's isolation, to vaccination sites that we had to drive to, getting them in and out of the
car, to face long lines, dozens and dozens of masked people, with few
accommodations being made to account for their diminished mobility.
It was a nightmarish scenario for them, but they managed, knowing, as
so many other vulnerable Americans did, that the vaccine was a potentially lifesaving measure. I remember sitting with them, during their
post-immunization wait time, and wondering about the elderly who
didn't have the extensive support system that my family provided. How
would I manage if I had a condition that caused shortness of breath or
was in memory care and was told to wear a mask?

I also wondered, as did so many of you: Did every part of this have
to be so hard?

The effort and the discomfort were well worth it. But I understood
that vaccinations alone weren't the answer to protect my parents and
other family members. I was worried about my parents' ability to even
mount an effective immune response. None of the rest of us, their care-
givers, were immunized. I also worried, regardless of a person's age,
whether their immunity, if they developed it, was going to wane over
time. At first, resistance to infection would decrease, followed by, in
some, a decline in the immune system to ward off significant disease.
The original vaccine trials, while large, were short, and most volunteers
were only four to six months post two shots. I feared that not everyone
was made aware of and fully understood those concepts, because that
message wasn't conveyed clearly. Ultimately my fears were realized.

Reviewing data from around the world both in the vaccinated and
unvaccinated in late 2020 into early 2021 painted a clearer picture. The
immunity induced after natural infection waned. There was a clear pe-
riodicity to the surges and increasing reports about reinfections with
each surge from South Africa and other countries. With some viruses,
like measles, mumps, and rubella, that initial infection in nature results
in years of protection. Many Baby Boomers will remember measles and
mumps infections—they were once and done occurrences. The vaccines
for measles, mumps, and rubella were created to mimic the long-term
protection provided by natural disease. But there are other viruses
where natural infection does not result in long-term protection against
reinfection. This was clearly the case with this SARS-CoV-2 virus. Nat-

ural infection immunity waned and reinfection was not only visible but increasingly common. Vaccines created to mimic this immunity will also wane, lessening the level of protection against infection. So, even the vaccinated could contribute to community spread. It wasn't just the unvaccinated who were responsible for ongoing transmission of the virus. (Although not as clear, it appeared that there was longer-lived protection against severe disease with prior infection.)

Once again we were slow to act to this global evolving picture. We continued to under-test. We didn't actively look for those mild and asymptomatic "breakthrough" infections in vaccinees. Once again our data didn't keep pace with the virus. We kept talking about herd immunity and percent vaccinated. It became clear, after we'd analyzed the global data (since we didn't collect domestic data on mild and asymptomatic vaccine breakthroughs) that immunity against infection wanes. We were late in recommending boosters because we didn't have a correlate of protection. We didn't communicate effectively why they were so needed. Indeed, late into the fall of 2021 many were still talking about the rarity of breakthrough infections—rare only because we weren't testing and measuring the breakthroughs in this group and not because they were happening infrequently. We were neither telling Americans with prior Covid-19 infection that they were susceptible to becoming reinfected and they may become infected, asymptomatically or mildly and, crucially, nor that they were at risk of infecting others. Fortunately, many colleges continued to require weekly testing of all their college students and they found the breakthrough infections and they were not uncommon, just less symptomatic.

The net effect of this was that the same cycle of silent invasion was recurring.

As was clear in the summer of 2020, and with every surge onward, when there is widespread community viral spread the virus makes its way to the susceptible and vulnerable, both unvaccinated and vaccinated individuals in nursing homes and in the community, resulting in hospitalizations and deaths. Many people, pundits and public health officials alike, focused on the unvaccinated hospitalizations and deaths and didn't address the mild infection rates in those vaccinated who could infect others. Regardless of vaccination status, people needed to hear that the way to protect the vulnerable remained what they'd been doing for the past year or more—getting tested, using masks, and restricting the size of gatherings. And there was a clear difference in the risk that remained to the vulnerable, especially the unvaccinated, and clear commonsense recommendations should have been made in the summer of 2021 to protect those who were unvaccinated to ensure rapid treatment with antivirals. While our understanding of the virus had changed, while the virus itself had mutated into different variants, those fundamental mitigations were still highly effective. What wasn't as effective was the protection the vaccines offered against infections six months later.

Many didn't see this evolution in the pattern, but it was there. In the summer of 2021, when vaccines were still highly effective in those who'd gotten theirs in the spring, the hospitalization data revealed that 99 percent of admitted patients were unvaccinated. Later, that number fell to 85 percent. Yes, far fewer vaccinated individuals got sick, but the highest rate of hospitalizations remained for those over sixty and seventy (over 22 percent of Americans), just as it had during each and every surge. Although due to vaccination, antivirals, monoclonal antibodies, and advances in care many would survive that hospitalization, hospitalization among the elderly significantly impacts future life expectancy. Older Americans are just not as resilient. That trend in hospitalizations was significant, worrisome, and predictable.

Still too many people were surprised that vaccinated people were getting sick. How could that be? they wondered. Without remaining in a higher state of vigilance and employing the basic mitigation measures, it couldn't have gone any other way. Simple moves like requiring every vaccinated person to read and sign a form that spelled out clearly what these vaccines were studied to do and not do, having those administering the shots reinforcing that basic information prior to or after the jab: "We don't know how long the protection against infection will last. We believe that you will be protected from severe disease but we are studying the rest. You could get infected. You could potentially spread the virus to others even if you don't get very sick. We advise you to continue to take the same precautions as before to protect the vulnerable in your household and when you gather."

Doing that would have gone a long way toward preventing what I experienced at a Walmart store in Berlin, Maryland. I saw an older woman in a wheelchair, moving down the aisle unmasked. I politely asked her why she wasn't wearing one. "Oh, I'm vaccinated. I won't get infected." That same scenario played out countless times in different parts of the country as I traveled. People believed they were invincible. Families believed their grandparents were protected and weddings and gatherings increased in frequency across the United States.

Throughout the fall of 2021 on webinar after webinar speaking to workplace groups, I talked about waning immunity against infection and the potential that the vaccinated were a part of the chain of transmission in the workplace. Each time, those on Zoom were shocked. They believed they were protected completely—both from infection AND severe disease—and didn't need to mask anymore during viral surges. The CDC, in a significant public health stumble, told the vaccinated that they no longer needed to wear masks, reinforcing that sense of invincibility. Too late, the CDC altered that message, as it had about waning immunity, but by the time they did so the resulting confusion was already at work, just as the virus was.

Along with that, in some circles the concept of herd immunity was still being promoted. Moving forward, it is crucial that we understand that this phase requires us to be aware of the need to protect from severe disease—among the general population and particularly among the most vulnerable. The number of those susceptible to severe disease and hospitalization due to SAR-CoV-2 has significantly declined. The small but not insignificant group without an adequate immune response to the vaccine and those that remain unvaccinated still need to be protected.

As much as I welcomed the rise in the number of vaccinated individuals, I worried because the continuously rising slope of testing we had created throughout the summer and into early winter of 2020 began to immediately decline. We had succeeded in establishing testing per population rate that was on par with the United Kingdom. But that bottomed out at a record 12-month low in the summer of 2021, falling from a peak of nearly 2 million per day in 2020 to 300,000. Of course case counts declined as a result.

Testing was so low before the summer surge of 2021, the test positivity rate rose rapidly across the Sun Belt as the Delta variant exploited the increased frequency of indoor gathering due to the heat. But it was more than that. Quietly, vaccine protection against infection was waning, as were mitigation efforts. The CDC implied in their messaging that vaccines alone would be enough to protect the vaccinated against any infection and then, by saying that the vaccinated do not need to mask, cemented that concept across the country.

On May 13, when the CDC director announced that the fully vaccinated no longer needed to wear masks, she characterized this as the
moment "we all longed for." She went on speaking about how the
downward trajectory of cases, the vaccine's performance, and the better understanding of how the virus spread all contributed to making
this "unmasking" guidance possible. She was right about the longing,
but wrong about the reasons that underpinned this change in guidance.
This was the exact moment that I had feared would come. I dreaded
the consequences of its arrival. Time after time we got ahead of ourselves with this virus and then suffered the consequences. We believe
we know more than we do. We react too early, and incomplete data
combined with hopeful thinking leads to premature or just plain wrong
decisions, eroding trust in public health overall.

Once again, we didn't have eyes on that early silent invasion that
could only have been made visible through aggressive testing. With
that single CDC announcement on masking, overnight far too many of
the vaccinated and unvaccinated stopped wearing masks indoors right
before the massive spread of the Delta variant in the South—the variant
we could see coming from the UK. Critically, those who had been careful throughout the first fifteen months of the pandemic due to underlying conditions now believed they were invincible if vaccinated. Many
of those careful individuals were the first in line for vaccination in January and February, and as a result, by July, many of the vaccinated,
including those in the South, had waning protection against infection,
while others may never have developed an adequate immune response
from the vaccine. This subset of individuals were not only susceptible
to infection but to severe infection, hospitalization, and death. Yes, the
severe illness in the vaccinated elderly was a small component of the total number of hospitalizations and deaths, but we could have protected
them better. We could have learned how to protect those over seventy
and those with immunosuppression in that summer surge of 2021, and
carried that important knowledge into the fall and winter of 2021. The
same series of events that haunted our response from January 2020 had
been repeated. Essentially, the CDC was saying what President Trump
and other officials were saying then: You are low risk. Drop your guard
(instead of not raising it, as had been the case in 2020) and go about
your business as before. No need to mitigate at all if vaccinated. You
can't get infected, you can't infect others, your grandparents and those

on cancer treatment are safe. Everyone can exhale after eighteen months of holding our collective breath.

I continued to integrate the data and could see from the curves that vaccinated individuals across the South must be contributing to the community spread. I called and wrote to those still in the federal fight and in the media off the record—something I only did rarely and with full knowledge of White House Comms staff while in the White House. With the CDC telling states that they weren't going to collect data on asymptomatic and mild SARS-CoV-2 infections post vaccination, states didn't recommend any testing of vaccinated individuals, which further contributed to community spread. Also, vaccinated individuals, believing that a bad cold had to be something else, waited until they were really sick to come forward, substantially diminishing the effectiveness of our therapeutics. Vaccinated individuals were gathering unmasked indoors, creating superspreader events, thinking all the while Grandma was protected and none of them could possibly be infected. The vaccinated grandma got sick and died.

People presented late because they weren't tested early.

Maskless vaccinated and unvaccinated silent spreaders were back in full circulation, contributing to community spread.

Surveillance testing wasn't in full use and wouldn't find the silent spread.

The virus invaded communities.

Another 135,000 American lives were lost in the summer surge of 2021.

That was the truth.

Through the spring, summer, and fall of 2021, I was again traveling the country. Every place I went I talked with people and they were shocked that so-called "breakthrough" infections were occurring. They were really only seeing the tip of the iceberg, as many of the post-vaccine infections were so mild many thought they had a cold. Others had no symptoms at all. Without testing they never thought they had become infected with SARS-CoV-2, after all, they'd been vaccinated. Their rationale to me was the CDC said the vaccinated do not need to wear masks anymore because we cannot get infected and we cannot pass the virus to others or they wouldn't have told us not to mask.

But no one should have been surprised that protection against infection waned, because they should have been told from the onset that this was not just possible but likely, especially as the predictable, more transmissible variants came to our shores—Delta and then Omicron. The 95 percent efficacy rate of the Pfizer and Moderna vaccines in clinical trials was against symptomatic infections of the original virus. Also, as in many clinical trials, these were biased results from a study of a highly selected group of volunteers, most with intact and robust immune systems. The truly vulnerable—those in nursing homes, those in their eighties and nineties, those in memory care who couldn't wear masks, those with significant immunodeficiencies, those undergoing cancer treatments, and those on immunosuppressants, from steroids to the new biologics, were not in the trials. The older individuals in the trials were "healthier" than many of their contemporaries who were not physically capable of making all of those vaccine trial appointments.

Even with this healthier subgroup in the most highly controlled conditions possible, 5 percent of the recipients did develop significant illness, although all were protected from death. In real-world conditions, that number could be expected to rise as high as 15 to 20 percent in the highest risk group, those over seventy, resulting in hospitalizations and potentially but rarer fatalities. The implication that these vaccines could and would induce sterilizing immunity—prevent the vaccinated from even getting a silent or a mild infection—was not fully explored in the real world in that moment. Some made assumptions. Some public health officials without complete data made that assumption. Down the line, without required weekly testing we didn't measure the asymptomatic breakthroughs that could potentially contribute to community spread.

The harsh reality was that silent spread could occur even among the vaccinated. It is normal and expected that our neutralizing antibodies wane. We can't have our B cells constantly churning out a high concentration of antibodies for every vaccination we have received throughout our lifetime. It's why we have a memory immune system—a sophisticated system that can immediately, within hours to days of encountering that pathogen, regenerate high levels of protective antibodies and initiate a full cellular immune response and clear that virus. That is how it works in most of us, but it may not work that way in the vulnerable aged and immune-suppressed vaccinated. They should have been warned that during viral surges in the community they needed to continue wearing masks and taking other precautions.

Requiring a vaccine to prevent even that first small initial invasion of the virus and initial replication is a very high bar, and if natural infection doesn't produce long-lived sterilizing immunity to reinfection it is very difficult to design a vaccine that does. In the original trials, asymptomatic breakthroughs were not systematically measured. What I had been working so hard to do for so long was to make clear to the Trump administration, and later through back channel exchanges with others, this very message, so it could be effectively communicated to the American people. The vaccines were a cause for celebration because far fewer Americans would get seriously ill and, critically, those with an immune response wouldn't die. But we still needed to balance the joy of personal protection with the tempered reality of what these vaccines could do and not do. Vaccinated people can infect their loved ones. Perhaps next generation vaccines will be created that produce long-lived immunity to any infection; perhaps an intranasal vaccine to produce high levels of mucosal immunity through generation of local IgA will be the development that will aid those who are needle averse.

Moving forward, we have to learn from these mistakes. CDC guidance and other communications must consistently be based on evidence or, in the cases when that evidence base is being developed, the CDC must explicitly alert the public to the provisional nature of its recommendations and clearly state what is known and not known—what data they are still collecting—and share all the data so every American can look at the data for themselves and make decisions based on all the information. Americans can handle nuance. They can understand that at a point in time this is the CDC's best judgment and they are working to get definitive evidence.

The CDC and NIH could have created a curriculum for each grade level so schools had an opportunity to seize a teachable moment about the nature of viruses, the immune system, and vaccines. I am sure some teachers did this on their own, but going forward, teaching these scientific concepts in the context of this pandemic is another way to ensure these important concepts get disseminated among younger people early. As is true with so many things, behavioral change begins with knowledge. We can't continue to rely on people becoming so uncomfortable—due to direct exposure to the devastating effects of trauma, like this one has produced—in order to initiate change in behavior, in how we educate our young people, and in how we as adults respond to new situations that require us to learn. Also, SAMSA and the NIH needed to take the lead on developing comprehensive mental health studies of students across all ages to understand the kinds of mental health support students need to aid them in negotiating a very stressful experience. The same is true of college-age people. So many of us struggled, but understanding the effects on younger people, those with addiction issues, and other vulnerable cohorts of the population is essential to an effective defense against potential collateral damage.

Many of the CDC's communication missteps were due to lack of data or the utilization of incomplete data. They lacked the real-time data from across the country representing all ages, races, ethnicities, and medical conditions. It is not that the employees at the CDC aren't dedicated and hard working, but they needed to ensure that all the possible outcomes and impacts were considered and whether the data was robust enough to make such definitive statements. They lacked the behavioral research.

The CDC must engage in directly, or provide grants to behavioral science experts, including marketing specialists, to develop clear strategies to deal with vaccine hesitancy to increase uptake. Implementation of a plan is where the proverbial rubber meets the road. The CDC must engage in implementation science to a far greater degree than it ever has in the past to ensure that its evidence-based guidance is practical and can be implemented on the ground. They also must develop mechanisms to evaluate the outcomes and impacts of all guidance they issue (using real-time data across America, not isolated substudies out of convenience). Those outcomes must be evaluated not only at the broad, national level. They must be done at the state and local level to best reflect the very real local geographic and demographic differences within our country's regions. A country as vast and as diverse as ours necessitates gathering huge amounts of data that is reflective of that degree of diversity.

TRACKING THE PANDEMIC, DOMESTICALLY, globally, and personally, trying to make sure the rest of my family (except for the three grandchildren under five) was immunized in sequence with the CDC guidelines, took me into late spring. From tracking domestic and global data I learned a lot that kept my family Covid-free, and public health agencies should have been discovering, researching, and appropriately implementing much of what I believed we were both analyzing and projecting. Yes, we worked outside the home, yes, we traveled for work and went into public spaces, yes, we went on vacation, yes the grand-children went back to preschool; but at this writing my family heeded the call for layered protection—just those extra commonsense steps have kept my family Covid-infection free. We had to because we have very vulnerable family members in each and every one of our collective households.

It was also clear that many states, both red and blue, were having difficulty following the complex tiered vaccine guidelines. It was also evident that those with early high immunization rates were the leaders of states willing to adapt and simplify those guidelines. But the massive early rush for vaccines in the highly motivated was hiding the depth of the underlying hesitancy. Well before the vaccines were rolled out, well before the pandemic began, even, the CDC and the FDA and other agen-cies should have been working to gather data on how to improve adult immunization uptake. We didn't know what drove some people to get vaccinated and, critically, what caused others to pass and not get the flu, pneumococcal pneumonia, or other adult vaccines. If we had been armed with the data of who was missing from annual flu vaccination and had developed clear solutions to address those very evident gaps, and removed barriers to vaccine access, and shown progress each and every year in adult immunization rates, we would have had a critical road map to follow to reduce Covid-19 hesitancy and we would have been in a different place in early 2021, with a clear plan, community by community, to roll out the Covid-19 vaccine. Having been out in the states, I believe the underlying and unstudied reasons for not being immunized were as diverse as the composition of our country's popu-lation. This lack of understanding, lack of data, and lack of solutions continues to haunt our pandemic response.

As the months moved on, out of frustration some railed against the vaccine hesitant, pointing fingers at them, demeaning their intelli-gence. With each negative comment that cascaded across the airwaves and social media, unvaccinated Americans were being pushed further to the margins. I was reminded of years of battling HIV, TB, and Ebola. Creating a hostile environment, alienating people who may have clear concerns, not listening, not hearing, and not addressing their concerns leads to less compliance with public health guidance, not more. Blam-ing others for behaviors and choices never changed minds or moved us forward in any pandemic, ever. As before, among the measures I recommend to improve this pandemic response and future ones, I call for a rigorous study of vaccine hesitancy across all demographics, including race and ethnicity, and urban/rural geography. This study and the insights gained must be one part of a multi-faceted revision of the pandemic preparedness plan.

While vaccines were rolling out, no one was out in the states listening and learning what was working and not working. Instead of finding out what was happening in rural America, there was further politicization and polarization of the pandemic and vaccination specifically. The majority (two-thirds) of counties in America are red counties by voting choice. But more important than how one votes, many are rural counties that for decades have not had access to routine primary health care. In public health lingo we say these Americans don't have a "health home" where they receive routine checkups with a primary care physician, nurse practitioner, or physician assistant. I would listen to some of the experts on the nightly news talk about getting information to American's primary (family) doctors. We were decades from the reality of family doctors and the rural local physicians. Rural residents often had to make choices between driving hours to the emergency room or going to work to support their families.

With Covid-19 disease, as with many prior medical issues in their families, they would wait to see "if it improved on its own," delaying care and ensuring many rural residents had a poorer response to Covid-19 infections. Rural counties knew they were far beyond the distance of the "golden hour," that sixty minutes that often makes the difference between life and death with cardiovascular events and traumatic injuries. These were communities that understood they had been left behind in routine health care access. They'd been ignored. Left to their own devices, some consulted with the only "experts" in the field they trusted—traditional and social media, or those in their community they trusted to be more plugged into those outlets than they were themselves. This isolation from routine and preventive health care and from sound information happened not only in our rural counties but also specific isolated urban communities. In my travels in the summer of 2020 I'd witnessed this level of distrust. I'd seen the void in public health education and services in those locales. I'd felt the absence of on-the-ground CDC personnel.

Now residents in these areas were being further marginalized and ostracized. People were pushed further away without dialogue, no one was willing to listen or answer their questions. No one was looking at their unique circumstances. It was spring planting season. Did we make appointments available late Saturday night so that farmers could get inoculated after coming in from the fields? Were they offered the opportunity to get vaccinated on Sunday mornings at their places of worship, so that they could recover from any post-inoculation reaction and be ready for work on Monday morning? Did we fully account for the many diverse needs of our entire population and not just those on the mainstream coasts? Was anyone working with the "flyover" states? No, we could have been and should have been.

Did we go into communities and listen and change our course of action? In many cases the answer was no. And the divide only got wider. I could see on social media that the demeaning comments from public health officials about the "unvaccinated" were being repeated on those platforms, furthering the divide. At times, that vaxxed vs. unvaxxed gap became a rationale—a justification for the overwhelming death rate of the summer surge in 2021. I saw individuals comparing the higher death rates in rural red counties to their presidential voting pattern in the last election. Didn't they understand health care across this country is not equal and the community hospitals in these rural regions may not have the same access to technology that medical centers in major metropolitan areas have? Look where Level 1 trauma sites are—they aren't in our rural areas—they are in our main metros.

The vaccination issue ran so much deeper than how people voted. It was more about access. Some struck the tone, that it was "those people, those people who didn't get vaccinated" that were dying—almost creating a sense that they deserved it. It was the "UNVACCINATED" who were filling our hospitals and dying. But, public health is about the entire public, not just the people who agree with you; it's critical to find a way to convince those who don't agree with you. If you choose to be in public health, you must choose to serve the entire public—you cannot dismiss people—you must find a way to reach people where they are. You don't give up, you don't walk away, you find a way or make one.

I have a solution to resolve these diverse population issues. The CDC has to evolve into a decentralized state presence with strong headquarter coordination and not remain one where 90 to 95 percent of the domestic staff are housed at CDC headquarters in Atlanta. By having more boots on the ground in more areas of the country, the CDC can establish a more continuous, more timely, more real-time data-driven responsive feedback loop between its in-state personnel and those back in Atlanta. They can't rely on fax machines and viewing data solely on fifteen-inch computer screens. They need fact finders on site to ensure the timely development and modification of guidance based on the reality on the ground. The CDC must fully support the states with these long-term staffing initiatives to address this ongoing pandemic and all other current public health issues.

It bears repeating: you can't fix the problems you can't see. Seeing up close and personal what is transpiring in the community is the best way to serve the needs of all of us.

I STATED EARLY ON that all pandemics are political. I hope by now that this distinction is clear: pandemic responses are political but viral infections and the diseases they produce are apolitical. That's the reality that data supports. Politics can't guide the response to an apolitical entity. For the most part what I experienced was true in terms of actions, but not in the rhetoric. Words do matter.

As has become abundantly clear, fundamental issues have transcended a change in presidential leadership. Chronic underlying issues in our federal institutions and deep societal issues contributed to the deadly surges that occurred across both administrations. We lost nearly 510,000 Americans in the twelve months from March 1, 2020, to the end of February 2021. What is so dismaying is that despite all our knowledge, our advanced technology, our vaccine rollout and additional therapeutics we lost over another 430,000 Americans in the next twelve months from March 1, 2021, to the end of February 2022. The 2021 summer surge was more deadly than the summer surge of 2020. And although the winter surge of 2021/2022 will be less deadly than the winter surge of 2020/2021, tens of thousands of American lives will still be lost unnecessarily. We will pass 950,000 American lives lost by March 1, 2022.

Explaining that most were unvaccinated is an excuse, not a solution. We need solutions: solutions to hesitancy; solutions to save the lives of those who are currently unvaccinated; we need to make clear to everyone what vaccines can do and not do, we have to let people know we care about them even if unvaccinated; and for the unvaccinated, we need to ensure their access to tests and ensure they are diagnosed early and have immediate access to effective antivirals. We need antiviral cocktails that reduce the risk for the development of resistance pre-positioned across the country and in the national stockpile so everyone in urban and rural areas who needs them gets them. Everyone must understand their own risk and their options. There are many solutions.

Other economically secure countries were able to blunt hospitalizations and deaths by routinely making tests available to everyone in the public who wanted them at a minimum or no cost. We didn't. We needed those tests available without restriction and we didn't get enough data or tests soon enough. That was true during the Trump administration and the problem has grown worse so far in the Biden White House.

Comprehensive data is unhindered by politics. In my time on the task force, decisions on critical supply distribution were made based on numbers not politics. Projections were critical. We predicted the severity of the winter surge of 2020 in the late spring of 2020, resulting in the Next Generation Stockpile, and the proactive use of the Defense Production Act purchasing all the available Covid antigen tests in the late summer of 2020. These tests were then surged to specific unique needs—BCIUs, tribal nations, nursing homes, and specific states based on where the virus would be the worst.

As much as politics and perceptions filled the social media and mainstream media, at no time did the vice president or any senior leader in the White House tell me or the task force to respond to a state's needs based on the population's political affiliations or whether the governor was a Trump supporter or a Biden supporter. I was told over and over—get the governors and mayors what they need to respond to the pandemic. Although at times we didn't have enough supplies—what we had was aligned to the need. Tests, supplies, and therapeutics were distributed based on need (equity) and vaccines by population. The call lists that went to the vice president were based on where the virus was going and where the response needed to be increased. The visits to the 44 states and over 30 universities that the White House organized were based on need and pushing states to be more proactive and aggressive early. With steers from FEMA and then the Unified Command Group (UCG) under Admiral Abel, the private sector used our data and our requests to deliver tests, supplies, and treatment based on need not based on profits or ability to pay. The innovations I saw at the state and local level were highlighted in the governor's report independent of whether the innovative state was red or blue. The on-the-ground response and the depth and breadth of the response and the mitigation efforts were independent of party affiliation. No one on the task force, none of the doctors ever discussed the pandemic in terms or red or blue counties or states. The vice president made it clear in task force after task force that these were Americans, not Democrats or Republicans.

While the development of tests, from PCR to point-of-care home tests, vaccines, and many therapeutics were produced in record time by the Trump administration, this had to be combined with the Herculean efforts of the Biden administration, enabling millions of Americans to be rapidly immunized. This clearly demonstrates the best of our administrations building on each other for the benefit of Americans. If you allowed yourself independent of party to step back for even a moment, you could see this was only possible as a bipartisan effort across administrations and represented the best of the American spirit.

Unfortunately, all of that got lost in rhetoric. Our most successful global health programs—like PEPFAR that was controlling HIV in Africa without a vaccine—was only possible due to a depth and breadth of commitment that transcended any one political party. Developed and initiated by President Bush, it was supported through both the Obama and the Trump administrations—through nine Congresses independent of the party of the Speaker of the House. This is the road map of public health impact. This will need to be the road map for the next generation pandemic preparedness. Stop the finger pointing, let's lay bare what didn't work, what did work, and let's learn together and find what works so we can be better prepared next time.

But throughout 2021, everything began with red and blue county and state comparisons. But it wasn't just vaccination rates that were responsible for the community spread of the summer surge; rather it was the lack of testing, mitigation, and the needed clarity from the CDC that vaccine protection from infection was dramatically waning, turning the immunized into the silent transmitters. The South has more red counties, the North more populous blue counties, so, of course, the summer surge was more in red counties. But as the winter surge evolves into more populous blue urban counties you will see fatalities rise in blue counties—even blue counties with high vaccination rates, because we still have vulnerable Americans who are not adequately protected. I never talked about red or blue counties. I talked about where the virus was and where it was going and what we needed to do. The virus doesn't know whether that vulnerable American is a Democrat or a Republican.

Throughout the spring, summer, fall, and winter of 2021/2022, I quietly wrote to those I knew who could have an impact. Testing had to be dramatically increased especially among those who were vaccinated, in families with vulnerable members. We needed to know immediately if they were infected to ensure early access to antiviral and monoclonal antibodies. I was watching what Florida was doing to dramatically expand the access to monoclonal antibodies to save more lives and prevent hospitalizations, but it appeared the federal government wasn't willing to recommend the adoption of this promising proactive practice. It seemed to me the reason for this unwillingness was political. Governor DeSantis is a Republican in a state where recent presidential elections have seen razor-thin margins between winning and losing.

In place of that reasoning, I heard some say that the South was using "too much" of the monoclonal antibodies supply and the federal government needed to adjust supplies. In every surge the region with the most hospitalizations and the most serious infections needed the most access to the most therapeutics—not less. The virus was primarily in the Southern United States and the Southern states were the ones needing the therapeutics. At all times, equitable distribution of resources based on data must guide the response to any public health care crisis.

The focus on a perceived imbalance of distribution distracted from the real issue: With the fall coming we should have been stockpiling tests and all therapeutics, including the new oral antivirals, in advance of the coming winter surge, not waiting for the crisis. We spent the summer of 2020 getting ready for the coming winter surge with PPE, tests, and therapeutics. Did we spend the summer of 2021 getting ready for the pending winter surge? We must develop a better balance of near-term and long-term projections of all aspects of the pandemic response. And how will we spend all of our efforts in the lull that will come after this Omicron surge to prepare for a potential summer surge in 2022 and winter surge in 2022/2023? We need a new plan to address vaccine hesitancy and clear guidance on how families with vulnerable members can protect them in surges while the rest of America moves to ensure we are open, working, and all of our students are in school. We have to provide better access to oral antivirals no matter where we live.

We knew that viral variants would occur at any time and that they would do so again and again. Throughout 2020 numerous variants emerged from natural infection immune pressure, and that would continue through 2021 and will continue. That's how viruses survive. Not only that, but, so far, most of these variants were first visible in other countries, providing us with a potential head start to combating them. The Alpha variant made its way across the Atlantic and exploded in the Upper Midwest states in April and May 2021, causing thousands of deaths. However, those cases and deaths were in "blue" states so this wasn't featured prominently on the news. As a result the alert didn't get out loudly enough and early enough to advise enough Americans of our collective susceptibility to variants that caused serious community spread in Europe. They will get here—use the early warning of what was happening in Europe to prepare in the United States.

In late May and early June of 2021, at the county level, the status of the pandemic was much like it had been in 2020. We were seeing those now predictable broad improvements, but there were those early troubling signs of increasing test positivity, coupled with fewer tests being conducted and no active sentinel testing in place. Instead of refocusing on the fundamental public health tools we had collectively developed—testing, masking, reducing friend and family gatherings indoors when you see the early warning signs—we chose to ignore the early warning signals and were once again in full community spread across the Sun Belt by July. Public health pundits and federal leaders blamed it on the unvaccinated, the more contagious Delta variant, the lack of mitigation by governors. Yes, all those contributed, but the reasons didn't matter—what mattered was using data, developing actions, and implementing solutions. What mattered was Americans were sick and dying and we weren't doing enough. The frequency of my warning emails increased, networks asked me to appear, but I knew if I did the only questions would be about my year in the Trump administration not what was happening and not what needed to be done now.

I'm saddened that it took the explosion of the Delta variant to rouse the country from our collective pandemic slumber and fatigue. As we moved through the summer surge the vaccination rates for eligible

Americans remained stalled just shy of 50 percent by the end of July. By January of 2022, it had only risen to 64 percent and 26 percent of them boosted. Did we use those six months to find solutions to the vaccine hesitancy? Did we evaluate granular data to see if any counties were successful in combating hesitancy—did we change the messages, the platforms, the words? Did we bring in more marketing experts to conduct in-depth focus groups across different age, race, ethnicity, and geography? Or did we just map the areas with lower vaccination rates and blame the surges and the resulting fatalities on them?

There needs to be equal emphasis on behavioral science and implementation science research to understand immunization dynamics and how to confront hesitancy and anti-vaxxers. How to communicate risk and how to specifically mitigate that risk based on your family's profile? Ignoring the hesitant and not giving practical and implementation information to Americans has resulted in our current situation. We need to understand the drivers of personal decisions and choices when it comes to vaccines and other mitigations that we will need in the future. The importance of implementation science to develop a better understanding of who should be the lead agency and how funding should be prioritized can't be minimized. I am always struck when I hear public health officials say, "We didn't think the public would do it, so we had to change our approach." In the case of vaccines we knew full well that many wouldn't do it, but little was done to help us better understand why and develop approaches to change the public's mind. Put another way, if we know the product is good but isn't selling, that's not the fault of the product developers. It's a marketing and sales issue. The CDC needs to recognize that and rely on outside help to refine its marketing messages better. That costs money, and those efforts need to be funded.

Masks are another example of this. Some within the CDC said that the KN95 and N95 masks are more "uncomfortable" and Americans won't wear them. What data supports that premise? I actually find those masks to be more comfortable—but that is anecdotal and they needed data. And this is how public health officials get themselves in trouble. We cannot base public health recommendations on our own personal biases. We can't say we are following the science when it's never been studied. We should have studied the acceptability and effectiveness of different masks then and still need to, since we will need them again. We should then present the data transparently to the American people.

Following the science requires us to actually fund the science; for too long we haven't valued implementation science or behavioral science research. We need to fund this area of science now either through the NIH or the CDC and these trials should be rigorous and represent all Americans, not just urban Americans.

The CDC can't continue to issue guidance like it did in December 2021 when it decreased the length of its recommended days of isolation from ten to five without requiring a test—especially with the availability of antigen tests that track with infectivity. Instead of conducting a thorough study and data analysis, using people from different ages, races, ethnicities, and with different underlying health conditions, it relied on a cohort of convenience—those readily available and not chosen randomly to best represent the general populace. We cannot use biased data, just because it's the only data available. We need to fund the research proactively so we can truly tell Americans we are following the science because we have the science and the data, not wait and, out of emergency, use small biased data sets.

By July 2021 the vaccine protection against infection had begun to wane across the country in those immunized first. We didn't let the American people know this was happening. We didn't use the early data out of Israel and the UK to understand this over the summer. We didn't ask Americans to test even though they were vaccinated to understand the depth and breadth of breakthroughs that were evident every day among sport teams and college students in August 2021; we didn't update the data to know if this was occurring. We didn't tell vaccinated Americans that their protection from infection had waned and they were now a potential risk of infecting their vulnerable family members. They also weren't aware that despite vaccination some people didn't develop a robust immune response and weren't protected from severe disease. We didn't ask those over eighty to be tested for antibodies to the spike protein—a readily and commercially available test—even if we didn't know the precise correlation between antibodies in the blood and precise levels of protection. Did we develop the study to define the immune correlates of protection so every American and every American over seventy and every American in a nursing home would know when they were vulnerable to reinfection and potentially significant disease? We have that capability—why didn't we use it so we would know who needed to be boosted when. Who needed extra layers of protection when the surges came? Did we give specific guidance to parents with children under five about what to do to keep them safe? Did we develop clear risk profiles so that every American knew their risk to serious disease, whether they were vaccinated or unvaccinated; previously infected or not? We have the capacity to know this proactively, not by just tracking hospitalizations and deaths when it's too late to save them.

We should have done the studies, but even without those studies, we should have known who didn't have any detectable antibody protection—we would have known who was at significant risk to severe disease. Families would have been informed so they could assess their own family and their risks and make informed decisions.

This would have been a critical information point among the elderly in the community and nursing homes as well as those with significant immunosuppression. They would have known and their families would have known their elders may be at increased risk of severe disease despite vaccination before the holidays. Across this country there are millions of Colin Powells with underlying medical issues who may have a blunted (not fully protective) immune response to the vaccine, or even the vaccine and booster. In my years of work around the globe I have found that more information, even if preliminary, is better than withholding information until all is perfectly understood. When we did surveys across communities in Africa we could see whom we were missing. Who didn't know their HIV status and who hadn't accessed lifesaving treatment. We didn't know why—we didn't have every detail; that would have taken months to define. Instead we immediately treated this as a crisis. We brought communities and public officials to brainstorm and come up with solutions and immediately implemented those solutions following the data for whether it improved testing uptake and treatment uptake. We didn't wait for the perfect, we acted. If there's a change, then explain the rationale behind that change clearly.

In the meantime, I was confused by the fact that while this repeated pattern was detectable, we were allowing this cycle to spin us all in another frustrating revolution that had so many wishing and hoping that it would just stop. We didn't have all the tools to stop the virus completely, but we have the means to recognize at what point and in what part of the country and at what time we need to be on highest alert. We know when we need to increase our level of mitigation, where and when to mask up, where and when to increase efforts to get the symptomatic and the asymptomatic, the vaccinated and the unvaccinated, tested regularly if they are in contact with vulnerable family members. As humans, it is difficult to be in a constant state of high vigilance. We need to use data down to the county level to let people know when there is a threat in their community—just like our immune system can't be made to stay at a constant level of vigilance, we can't stay in that state of constant anxiety and worry. We don't have to. When our hearts and minds are under attack our mental health suffers, some resort to addictive behaviors, some see suicide as the last resort.

We are now, due to how much time has lapsed and how many of the repeated patterns we've been through, better able to see how surges work and how long they last. This is based on what we've seen here and elsewhere in the world. In parallel to our experience with vaccines, global supplies were even more limited and additionally very slow to get shots into arms. It was clear from following the curves of the South African data that two things were happening in parallel. They had experienced high rates of general population infections with each surge. Due to the population having fewer comorbidities and the overall youth of their populations, more than a decade younger than the USA, their hospitalization and fatalities, in relationship to community spread and cases, were lower and continued to decline. But there was also evolving real-life data and clarity on durability of natural infection immunity—each surge was five to six months trough to trough and peak to peak and each surge was with a different primary variant in South Africa, where they were carefully and regularly sequencing the virus to detect the presence of variants very quickly.

Along with looking at global numbers, I was examining weather data and other figures to triangulate an analysis of what to expect. By September 2021, I was very worried about the coming fall/winter surge. I heard pundits stating we "were through the Delta surge" and all would be well in November. I knew we were only through the summer and early fall part of the Delta surge—not the winter part. Many who were comparing timelines of when the fall surge occurred last year were reassured when we made it into October without rising cases. They weren't accounting for the fact that this fall had a different weather pattern than the previous year. By the end of September 2020, the temperatures were consistently around freezing across the Northern Plains. This year was the warmest October on record. I went to the Northern Plains states in October—people were still eating and drinking outside—it was clear we weren't *through* the Delta surge across the North—it hadn't yet *started*. Late enough that we could have alerted all Americans to get boosted and tested and make it clear there were Americans without an optimal immune response that may be vulnerable to severe disease.

I sent out my family alert. We had vulnerable family members in each family cohort. Increase vigilance and increase testing. Even if you don't have symptoms. We sent the grandchildren back to preschool. We understood the risks and critical rewards of the grandchildren socializing with other children. We made decisions based on data and used available tools to mitigate the risks. We helped the preschool improve its indoor air quality, as we knew two-year-olds couldn't consistently mask all the time. We worked with the school on practical implementable solutions. We bought at home antigen tests in lots of ten because we could afford them.

Why wasn't the government distributing these to those who couldn't? Why weren't we testing young people in community colleges, which provided a wonderful opportunity to see the earliest community spread in cohorts that were together in one place and easily accessible? That's where you would find the variants first. Early warnings with high levels of testing change the community spread—we knew this from nursing homes, colleges, and countries like the UK. They kept their hospitalizations and fatalities much lower than ours through aggressive testing with minimal mitigation. We should have tested aggressively early on to identify the first instances of asymptomatic and mild infections in communities. Knowing as soon as possible where and when those infections were taking place would have helped us alert vulnerable Americans and those in contact with them that they needed to mask up and test to make sure they weren't infected and hadn't passed the virus to their family members. If we'd tested early, empowered families with the knowledge they needed to protect their family members, and mitigated early, we could have prevented 30–40 percent of the fatalities in each surge in 2020 and 2021. It wasn't about changing your lifestyle, it was about testing before you visited Grandma. It was about making sure Grandma had access to testing so she could immediately access the effective treatments available, but they needed to be used immediately not after Grandma became so seriously ill that she needed to be hospitalized.

But we still weren't testing strategically, we still weren't collecting enough data in real time, and we still weren't communicating all the information effectively to the American people. I was talking with some in the media off the record. I kept asking them why this wasn't happening—they felt that no one wanted to discourage people from being immunized so they didn't want to talk about the potential frequency of breakthrough infections—that would discourage people. Then there were masks. Some were great about telling people to move to KN95 or N95 masks that were now readily available on Amazon all through 2021—but what were we doing for those who couldn't afford home tests and better masks? This persisted through the most significant spread of the Omicron variant until late into January when the federal government finally acted, expanding tests and mask access.

Finally, the "happy talk" of the early fall met the reality of cooling across the northern United States with the Delta surge, initially in the Northern Plains and Rocky Mountain states, and the predictable arrival of the more infectious Omicron variant in the Northeast that made its way across the country. There were finally some actions on testing and the importance of boosting but we were again in reactive vs. proactive mode. Instead of being fully prepared and sending a clear alert to the dangers that lay ahead, the holiday message was to celebrate with families, celebrate the victory over this virus. These holidays will be different. If you are vaccinated gather together! Don't worry! As vaccinated families and friends gathered with others of unknown status or those who hadn't developed an effective immune response, around those dining room tables we would infect others. Many would get sick and some would die, fewer than last winter but still tens of thousands of Americans would lose their lives. From the middle of October to the end of January 160,000 lives have been lost, and by the time the Omicron surge ends and deaths decline, we will have lost nearly 220,000 Americans in the winter surge of 2021/2022. Yes, it was 90,000 less than the winter surge of 2020/2021, when we lost 300,000 Americans, but we have vaccines and additional therapeutics. This 100,000 fewer deaths should not be viewed as a success but as a red flag that we need to do better and be better prepared.

Perhaps since a more "normal" Thanksgiving was promised, the CDC and the Biden administration felt they had to deliver that message. But we knew from 2020 that Thanksgiving proved to be a superspreader holiday. As a family we knew what was coming and we hunkered down again, severely limiting gathering, still traveling, still working in public, but now we had the advantage of testing to ensure we weren't exposing others. I worried about those without an effective immune response to the vaccine who believed they were fully protected from severe disease. In the months and years to come we will see precisely who was lost and why. It's what I told every governor in 2020—it will be clear in the next few years, the impact of their actions or inactions. It won't be cryptic. Messaging matters, information matters, and warning people of actions and risks saves lives.

We need to confront our blind spots and our misperceptions. Yes, even scientists have beliefs and perceptions that color their interpretation of data. Yes, we could see the Delta variant AND the Omicron variant coming. Every variant that made it to the UK and Europe and caused a surge there has made it to the United States. Alpha, Delta, and Omicron all caused surges in Europe and made it to the United States. Not all variants from South America made it to the U.S., but ALL of the European surge variants made it to the United States—predictably.

We cannot ignore predictable patterns. We need to learn from these patterns. We know the regional pattern of the summer surge; we know the massive spread of Thanksgiving and holiday gatherings. We should be using this human behavior knowledge and data to ensure that Americans are empowered with the information they need to adapt to the reality of SARS-CoV-2 now and in the future. Not out of fear but out of clear knowledge. What is clear is had we been testing and sequencing at the same level as the UK, we would have seen that the highly contagious variant was already here. The UK had dramatically lower hospitalizations and deaths with both their Delta and Omicron surge compared to the United States. In January 2021, we were on par with population-adjusted Covid-19 deaths with Europe. Since then we have 20 percent more fatalities than the UK. Why? They ensured the public had the most important tool to prevent spreading during gatherings: testing. The UK made tests available and is testing at nearly five times the rate as the United States—their test positivity despite Omicron didn't go above 10 percent whereas ours in January 2022 was well over 30 percent. The summer, fall, and winter 2021 recurrent pandemic community spread continues to be largely driven by the asymptomatic unvaccinated, the pre-symptomatic unvaccinated, and also the asymptomatic and mild symptoms vaccinated—to find these infections you need to TEST. Paid leave if you are positive is critical so people don't have to choose between their own health and family well-being. Many hourly workers don't want to test because they don't want to know their status and potentially miss their daily wage. In immigrant communities, rural communities, uninsured, unvaccinated, we need public health tools that meet their specific needs.

What if, instead of delivering that okay to gather, preemptively beginning in late September or early October, the federal government sent all families a pre-holiday "gift." What if we spent the summer buying and building enough tests for our stockpile and could have stockpiled enough masks to send or to make available free of charge KN95 and N95 masks to every household in the country. Along with that, a simple set of bullet point recommendations/reminders of best practices for social gatherings could have been included on a one-sheet or a card. Better yet, instead of mailing them out to every address, we could have used our vast data collection and aggregation tools to deliver them to the residences (private and communal like nursing homes and memory care facilities) of people over the age of sixty-five who might not have had the ability or desire to go out and who were more likely to be receptive to this "gift." We could have also utilized schools to distribute those packages to parents of K–12 students. For the age cohort that covers college age to thirty-five and possibly to sixty-five the packages could have been made available at places that age group frequented—grocery stores, at drive-through restaurants, coffee shops, churches, community centers. Since the private sector was so eager to contribute, places like Starbucks, for example, could have had them available at their locations.

I could go on, but the point is clear. The government had to overcome any possible barrier through clear and effective communication, age and platform specific—making the risk clear to specific Americans and the need to mask up as effectively as possible and to test prior to hosting or attending gatherings that included vulnerable family members. Many people were doing this anyway, but for those who didn't, not having to go out and secure these supplies, seeing masks and testing normalized to a greater degree than before could have stimulated them to do what was best to protect themselves and their loved ones. We need to address social pressures occurring at the community level that decrease vaccine uptake. My colleague heard from people in South Dakota who didn't tell their friends they were vaccinated because of peer pressure—adult men not telling their friends they are vaccinated for fear of how it would be perceived . . . we can and must change these societal pressures.

Again, I witnessed local leadership. Five blocks from my house I saw long lines snaking up the street. Many people respectfully waiting in line—the young, the old, at least four feet apart, outside waiting for hours. I wondered what was happening and investigated. The mayor of the District of Columbia was making at-home tests available to its residents at local libraries so everyone could test before gathering with others. That proactive effort by the mayor decreased Covid-19 spread during the holidays and protected the vulnerable. She did this before the holidays—proactively. But what it also showed me was people were willing to do the right thing to protect others if we reduce barriers and make mitigation like testing available. Imagine if you could walk to your local library everywhere across the country and pick up free masks and tests. If clearly explained, everyone would understand their personal risk and what situations were most likely to lead to potential exposure, and I know they would have been willing to protect their grandparents with testing and masking. I have seen this across the country.

Or, we could take advantage of the corporate/private sector's desire to be involved and to contribute. Fast food restaurants, retail businesses, and others could become distribution centers for free masks and tests. The government had the ability to provide stimulus funds to millions of Americans. What if we had been given the choice to opt out, and instead had those funds go to the purchase of mitigation supplies to be delivered to those in need? Clearly, many Americans continue to be generous throughout this pandemic, and a program that would allow for funds, or the supplies themselves, to be diverted to charitable organizations, community organizations, already doing outreach in the community, could help overcome some of the wariness of the federal government many people I met in different states around the country felt. If the government showed that those people were seen and heard, they'd be more willing to listen, I believe, when the CDC, for example, had something to tell them. They'd better understand that the government understood them, and, at the very least, acknowledged they existed.

The Biden administration has just enacted a similar program to send free tests to every household. However, it requires people interested in receiving them to go to a website to enroll. While I hope the program works, I also hope that this current administration and those supervising this program remain flexible, and move from reactive to proactive. It's February, and in many areas the cases are waning and many haven't

Epilogue

received their tests yet. Also, many of us have had bad experiences while trying to register to get vaccinated via online systems. Some of the people who most need these free supplies may not have the infrastructure or the time to navigate a system that gets too easily overwhelmed. There are multiple barriers to entry the government has to account for when providing services that they believe most everyone will want simply because they are free. As we all know, all programs like this need to take into account that they have to be flexible in devising distribution channels that will meet the diverse needs of ethnically, economically, age, and attitude variable groups.

One of the most important takeaways here is to not wait until the surge or the holidays are upon us. We have learned that there are lulls in the outbreaks. During these interludes, we needed to expand testing access. We should have used that time to prepare for the next surge by getting as many Americans as possible to buy into making critical behavioral changes needed to protect their vulnerable family members. Many communities have done similar things on their own, and with more CDC personnel on the ground in those communities they can work with mayors, governors, and public health officials to make these efforts happen, evaluate their effectiveness, and make changes as necessary.

To put it another way, the CDC must become more customer-friendly as they deliver services to the American people. The trickle-down effect of their limited research and generic guidance is a starting point—but working with states and local communities to water the grass roots efforts that currently exist and providing people with what they need, not what the CDC believes they need. They can do both, but only if they engage in dialogue. As we've all likely experienced at one time or another, top-down authoritative management has its place in some crises, but not at all times and in all places. The CDC is a part of public health, but it also needs to be part of evidence-based community health as well. As before, words matter, and though "Centers for Community Disease Control and Prevention" might be more of a mouthful, it can serve as a reminder to us and to the CDC what its true mission is.

I have seen the CDCers do this elsewhere and in other contexts. For the last twenty years they have been in the trenches working alongside Ministries of Health officials and communities to serve those most in need of both prevention and treatment services to combat the HIV pandemic. They use weekly and monthly comprehensive data to engage in data-driven decision making week by week to improve access and quality of services down to the local clinic and community. They define gaps and find solutions. Through the data we could all see the tyranny of averages—not just looking at country-level data and applauding our progress but using demographics and geographic location to see who we were missing, young men and women and, critically, the most vulnerable key population groups. Hidden marginalized people lost in the averages. Together the CDC worked in deep partnership alongside those they served. Working with key populations—people who inject drugs, men who have sex with men, transgender women, and young women and men—allowed the CDC and PEPFAR to address the cultural and policy barriers to access. This is the CDC I have seen across the globe and this is the CDC we need domestically. A CDC that works with everyone to increase vaccine uptake through listening rather than solely mandating their uses. A CDC that works to create commonsense guidance that Americans utilize. A CDC that recognizes not everyone will choose to get vaccinated and some will need specific guidance linking the level of viral community spread to recommendations of enhanced testing and immediate access to antiviral and effective cocktails or monoclonal antibodies to save their lives without hospitalization. A CDC that recognizes guidance needs to evolve with the science and the community cultures and be adaptable. CDC guidance must be unambiguous to bring communities together and not drive them apart.

An evolved CDC would have an enhanced mission to ensure effective services at the community level. Intertwined with that shift in perspective is a greater need to hold itself accountable, and for us as those receiving those services to hold it accountable, for what it promises and what it delivers. In my work at PEPFAR, the conference I was attending while deciding to come to the White House served as a model of accountability. Annually, each country's program was subject to review and analysis of progress, holding each of us accountable, from the funders to the program implementers, to the communities in need of the services.

We need a CDC that is practical and focused on changing the public health of this country—focusing on decreasing health inequities and addressing the social determinants of health and disease. Our fatality rates are higher than most European countries because we are unhealthy in comparison. By using big data and continuous data analysis to define in real time improved outcomes and the impact of interventions, we can not only continuously make improvements in the health of Americans and develop the implementation science platform that will serve us in pandemics and pandemic preparedness but can also improve the nation's health by addressing both chronic diseases (comorbidities) and infectious diseases. By diving into and addressing the social determinants of health, ensuring progress in health equity year after year. We learned from HIV/AIDS you need to be embedded in communities, to understand the core structural barriers to access, to address concerns and be in continuous dialogue to understand not only access issues but issues around vaccine hesitancy. We have clarity on what isn't working and now we need to develop the evidence and in-depth behavioral science studies to understand what does work. Transforming the CDC into a proactive partner in implementation makes sense and is adoptable and adaptable community by community. This requires a new model—a model of being in the states, being flexible, and learning and gathering evidence through population-based studies and not isolated cohorts of convenience. We need to address our overall health with specific funding and focus on areas of greatest need, including tribal nations. There are solutions—we saw them and we need to fund them.

Much of what we need to do to be better prepared for the next pandemic begins with the routine definitive laboratory diagnosis of viral diseases. We do that for bacterial diseases, including mapping antibiotic resistance. We need to definitively diagnosis by a lab test each and every viral disease at all levels of health-service providers—urgent care, doctor offices, emergency rooms, and hospitals. Importantly, we already have the technology to do it. To show how important this is, health insurance companies and Medicare and Medicaid must demand that a confirmed laboratory diagnosis (test) has to be performed and coded in the database any time a patient comes in with a possible viral infection or for treatment of a viral disease. If CMS makes this one rule change, we will always know what viral respiratory diseases are circulating in the community. We need a national database with reporting in real time that collects both the laboratory and medical codes for community-acquired infections. Not creating a parallel system but using the electronic systems that already exist, collated together. I believe hospitals and clinics would be happy to provide this electronic data stripped of personal identifiers, with ages blurred by creation of age bands but including sex, race, and ethnicity. This will provide our baseline so we can "see" in the data any disruption in the pattern. This will allow us to see when something different is out there but it will also allow us to develop vaccines and treatments for current viral diseases as more rapid clinical trials can be done once you know who is infected. We have been flying blind or using sentinel sites and collecting syndromes rather than definitive laboratory diagnoses. This national database should be publicly available in an easy to interpret form down to the county or zip code so parents can know what's in the community, and assess the risk to their children, and know what is in the community when their own child gets sick. Too many viral diseases present similarly, and without a lab diagnosis you are making assumptions.

The CDC should have access to all the data to bring the full strength of their data analysts and experts to the table, but so should the private sector and state and local officials, along with the public. It starts with the data, and data gets better when it's used. The CDC should work with each state to decide if any other critical public health data should eventually be added. Imagine if we were using real-time data to understand our opioid epidemic, suicide attempts, and childhood obesity, and held ourselves, the state and local governments, and the CDC accountable to improve step by step the health of our nation. It's possible with data and using the data to chart outcomes and impact in real time, expanding the solutions that work and stopping those that don't. It's not about more money; it's about using the money we have more effectively. We did this in PEPFAR. In a flat budget for twelve years, we tripled the number on lifesaving treatment and we rolled out or increased the most effective prevention programming each quarter, evaluating outcomes and impact and holding ourselves jointly accountable. You can't manage what you don't measure, and we should not accept monies going to programs or states if not linked to mutually defined outcomes and impacts. States need cooperative agreements with the federal government and not grants. The CDC needs a permanent presence in states—a team, not just one individual, so they have a deep understanding of the issues that need to be addressed at the local level, the performance of the programs they are funding at the state and local level based on objective outcomes and impact, the effectiveness of their federal guidance, and direct funding of and support of populations-based implementation science.

The pandemic response must evolve with the evidence—which requires systemic collection of data—big data across the country and utilization of these data sets for decision making. The decisions made in March 2020 were based on the evidence that the virus was extremely deadly for those over seventy, with case fatality rates of near 30 percent, and we didn't have a proven therapeutic to blunt the deadly virus and needed to buy time to protect the highly populated cities. The virus was isolated to fewer than ten cities of more than one million residents. With the availability of remdesivir and vastly improved clinical practice in the care of patients, the fatality rate in those over seventy dropped to under 10 percent. Still too high and requiring additional layers of protection, but throughout the summer surge of 2020 we learned how to maintain retail, restaurants, and schools fully open with masking and testing, and decrease in home gatherings during community surges. With the development of monoclonal antibodies, the fatality rates in those over seventy dropped first to under 5 percent and then continued to fall through the surges of 2021 with vastly expanded treatment options. Now with vaccines, oral antivirals, and all of the other treatment advances, there is a clear pathway to living with Covid-19 with dramatically decreased fatalities. But this will require more durable vaccines with improved immunogenicity in the elderly, routinely available home tests for those that are still vulnerable, and rapid access to antivirals. We have learned who is susceptible to severe disease and we can define those without a protective immune response despite vaccination. These individuals need to proactively test during a community surge in their location—these community surges are predictable: fall and winter for the Northern United States, and for the Southern United States potentially summer and late November to February. With proactive access to Covid-19 tests, with potentially weekly testing during community surges and immediate linkage to aggressive antiviral and immune therapy, lives can be saved without disruption to everyday activities. During these clear periods of high community spread—readily identified by a color alert system at the county level—multigenerational households, households with immunosuppressed family members, households with children not eligible for vaccination, and unvaccinated Americans with risk factors will have to take additional precautions centered on regular testing while the majority of Americans can continue normally. It is estimated that nearly forty million Americans are in this category of high vulnerability and will need clear guidance on testing and access to immediate treatment. It is more difficult for workplaces and schools, as they need to protect the most vulnerable workers and students during this time, with limited community-spreading events. Other issues to consider or rethink are in-person meetings, air quality, and direct support for high-risk students and workers. It sounds complicated, but we have twenty-first-century technology that can be deployed to improve safety in public spaces, including active viral inactivation in the air in a safe manner.

All universities receiving federal research dollars must be required to immediately engage in a pandemic response, bringing all of its personnel and equipment to the fight. This alone would allow us to do 250,000 to 500,000 additional nucleic tests per day. But it's more than equipment; it is the ability to do behavioral and implementation science research.

There needs to be a national pandemic team that is constantly reviewing the available data and includes the private sector, universities, and the federal government, including the military and National Guard—not tabletop exercises but use of real-time viral disease data to chart responses to current viral seasons and outbreaks. Ensuring everyone is used to reviewing the data, understanding patterns, and reacting to deviations from baseline in a proactive rather than reactive manner.

Why there was no central clearinghouse organized for states to share best practices, worst practices to be avoided, and all areas of the distribution and inoculations is inexcusable. We can't repeat that mistake again. Fortunately, through various health agencies, through public health officials, and with some governors communicating with one another, some of that valuable information was exchanged. It should not have been a catch-as-catch-can effort; instead, if there had been a more proactive approach to every phase of the response, including vaccination and how to best facilitate it, more doses would have been administered sooner.

There needs to be equal emphasis on behavioral science and behavioral science research to understand immunization dynamics and how to confront hesitancy and anti-vaxxers. Ignoring them and this issue has resulted in our current situation. We need to understand the drivers of personal decisions and choices when it comes to vaccines and other mitigations that we will need in the future. The importance of implementation science is encapsulated in the questions *Who is the lead agency and will funding be prioritized?* I am always struck when I hear public health officials say, "We didn't think the public would do it." Like recommending more effective masks. To say the KN95 and

N95 masks are more "uncomfortable" and Americans won't wear them. What data supports that premise? And this is how public health officials get themselves in trouble. We cannot base public health recommendations on perception. We can't say we are following the science when it's never been studied. We should have studied the acceptability and effectiveness of different masks, as we will need them again, and we should present the data transparently to the American people. Following the science requires us to actually fund the science; for too long we haven't valued implementation science or behavioral science research. We need to fund this area of science now, either through the NIH or through the CDC, and these trials should be rigorous and represent all Americans. Not a cohort of convenience. The data on the change from ten days to five days of isolation could have been done weeks ago in Americans of different ages, races, ethnicities, and underlying health conditions. We cannot use biased data just because it's the only data available—we need to fund the research so we can truly tell Americans we are following the science because we have the science and data.

This pandemic laid bare the gaps we have in the United States, and it's not about funding more of what resulted in these gaps. First and foremost, we are a country of health disparities and devastating social determinants of health, where zip code determines the quality and access to health care. Second, we don't have the national databases we need to respond to this or the next pandemic, which would need to include 100 percent of the community-acquired infectious diseases, whether presenting at urgent care, a clinic, or a hospital emergency room, and 100 percent of these community-acquired infectious diseases must be definitively diagnosed by a laboratory assay or a point of care test. Every approved vaccine must have research to define the correlate of protection that is measurable or a surrogate of the correlate—such as this antibody titer to spike protein that correlates with neutralizing antibody titers. We need to ensure a vibrant biotech industry in this country and move critical essential medicine and PPE production back to the United States and back to the Northern Hemisphere. We cannot ignore the lessons learned and must ensure they are part of pandemic preparedness in the future. Core to this is seamless integration of the public and private sector into the response. Our health care delivery and our entire supply chain for everything the hospitals use and need are through the private sector. Let's for a minute use testing as an illustration. During 2020, the federal government, through research as well as financial incentives to the private sector in the form of massive purchases of tests given to the states to use, supported the research and development of next generation testing through the congressionally-funded RADx. These incentives stopped in 2021 and tests no longer purchased by the federal government, especially antigen tests, overflowed in warehouses because the federal government had no plan to stockpile tests. There was no plan for viral surges of Covid-19 in 2021 and supplies were not purchased to combat these surges. We have an economy based on supply and demand and the federal government needs to stockpile supplies for when the demand occurs. This requires prioritization and discussion. Tests could be part of the national stockpile and be provided to states during surges or in between surges to support routine cohort screening.

In early February of 2022, media outlets reported on a bipartisan group of senators expressing support for an independent commission to investigate the origins of the pandemic and the federal response to it across both administrations. Modeled after the National Commission on Terrorist Attacks Upon the United States (aka the 9/11 Commission), this pandemic investigation would be comprised of a twelve-member panel with subpoena power. Two members of the Senate Health Committee, Patty Murray (D-WA) and Richard M. Burr (R-NC) were working on a draft of legislation to make this a reality. I applaud their efforts and urge readers to contact their senators to pledge their support of this commission. Much like this book, its job would be two-fold—to investigate how the response was handled and to offer recommendations for moving forward.

Among those recommendations, I hope they will include a provision to establish an office of National Pandemic Preparedness—an independent office whose oversight would supersede HHS and its preparedness plan. It would bring together not just the federal agencies and state and local governments, but the private sector and our research institutions that receive federal dollars. It should also include representatives from members of the community. Not only would this group oversee preparedness for future novel viral outbreaks, but by using national data tracking systems, it would also take on other current health pandemics—obesity, diabetes, cardiovascular diseases, and opioid and other addictions.

If 9/11 permanently altered the intelligence community, Covid-19 must similarly force a reckoning at HHS and other related agencies. The federal institutions must evolve to address the failures that have occurred so that they are better prepared to embrace the kinds of approaches that will protect us in the future. And often these kinds of commissions are the only way to instigate that kind of broad, dramatic change when it comes to institutions as entrenched as these.

It is my greatest hope that 2022 will see our response to this pandemic, individually and collectively, adapt to the evolving nature of the outbreaks. We have to hold on to what has been successful, adapt, and understand that the cyclical nature of the surges demands different things from us. Just as our immune systems can break down over time, so does our ability to remain vigilant. A constant state of hypervigilance isn't healthy, possible, or needed. Using the right tools at the right time, we can learn how to live with this virus, while we continue to evolve and develop even better tools. Science and medicine have an amazing track record of success in eradicating some diseases, and treating and managing those that were once considered death sentences, and I'm confident that a similar story arc is one we can expect with Covid-19. As much as I am a worrier, I am also a believer. I don't just believe that we have to do better; I know that we will.

Appendix

Summary of Issues to Be Addressed with Proposed Solutions that Will Require Legislative Commitment and Funding

Critical issues today that require addressing for this pandemic and future pandemics:

1. **Command and control**—the addition of ASPR to HHS has caused confusion about the division of labor and roles and responsibilities. There needs to be clarity between the CDC and ASPR specific roles. This separation causes more work and confusion for the states.
2. **Lack of definitive laboratory diagnosis of viral diseases— need definitive laboratory diagnosis of all significant respiratory diseases.**
   a. Flu is tracked and diagnosed by a symptom complex rather than definitive laboratory testing with either nucleic acid testing or antigen testing.
   b. Many viral respiratory infections, including RSV, parainfluenza, and adenovirus, are diagnosed as a process of elimination or assumptions based on symptom complex.
   c. Lack of critical sequencing of SARS-CoV-2 and other respiratory viral infectious diseases at a level that is consistent with the pandemic or real-time data collection.
   d. **Solution:** all respiratory diseases should be definitively diagnosed in the twenty-first century to drive local testing capacity, routine reporting, and development of enhanced antivirals; accountability to Congress with regular reporting on testing and sequencing completed with full analysis.
      i. Before doctors can prescribe any antivirals or code for specific respiratory infections for reimbursement, insurance companies, Medicaid, and Medicare should require that patients have a definitive laboratory diagnosis of the viral infection. This would incentivize physician's offices, urgent care centers, and all hospitals to have local lab plat-

form capability, thus driving up the number of facilities capable of making these diagnoses in a timely manner.

   ii. This would also create a baseline of known viral disease so new infectious agents could be easily identified.

  iii. This would create the critical laboratory capacity at all levels that will be useful between and for pandemics.

  iv. Expand access to quality home rapid testing to empower every American with the knowledge they need to protect themselves and their families.

3. **Data**—there are no comprehensive links of public health data and clinical data and these are two separate and partially overlapping systems and often require duplicate data entry.

  a. The CDC's hospital and ER data was primarily modelled from sentinel reporting sites.

  b. All reporting was voluntary and sporadic. Specific regions in the United States were either over- or underrepresented with inadequate visibility into all U.S. counties.

  c. Incomplete: It did not include daily hospital admission nor distinguish on a patient basis confirmed or suspected Covid-19—just generic Covid. It did include an absolute bed count of available ICU and regular hospital beds but again in a modelled methodology.

  d. Data are in siloed systems across the CDC without a single common system.

  e. **Solution:**

    i. Work with all six thousand U.S. hospitals to establish routine, regular, and timely reporting from hospitals that is modular, adaptable, flexible, and electronic, not paper based and that transcends any specific infectious disease but can rapidly build out new modules based on any pandemic.

    ii. Required regular reporting of already collected codes for specific community-acquired infectious diseases with age bands, race, and ethnicity appropriately blurred at the county level or combined counties to ensure HIPAA compliance.

   iii. All data, including lab, hospitals, case, mortality, should be integrated at the community level and available internally and externally.

iv. Use current technology and set up adaptive systems as technology innovates.

4. **Guidance without definitive evidence base creates confusion.**

   a. The CDC must ensure all guidance is accompanied by or linked to the evidence base or clearly state that the evidence base is being developed.

      i. Masks: when the CDC recommended cloth masks in April 2020, stating it was solely to ensure infected individuals were not spreading the virus this led to significant confusion across the country, as Americans could not understand that cloth fabric only blocked droplets in one direction. Despite repeated requests for the CDC to conduct or commission the conduct of these simple experiments, the CDC waited until the fall of 2020 to brief on the bi-directional protection of cloth masks.

   b. **Solution:** The CDC must have mechanisms or internal capacity to investigate and provide the proof of all the elements included in guidance in real time.

5. **Guidance without evidence base of implementation, mechanisms to conduct implementation science in real time, and continual feedback loops on outcomes and impact of the recommended public health actions.**

   a. The CDC should have been conducting behavioral research into flu vaccine uptake over the past decade to understand adult vaccine uptake and have developed clear strategies to increase vaccine uptake and show their efficacy, and these would have informed the initial vaccination strategies for states with continuous behavioral implementation research to improve and evolve messaging.

   b. **Solution:** The CDC must engage in all aspects of timely implementation science in partnership with states to ensure guidance is optimized for execution and with mechanisms to evaluate outcomes and impacts of all guidance and state-level public health funding and continuously improve the outcomes and impact of their public health guidance.

6. **Must increase the speed of published data to support evolutions of policy in the *MMWR* or other public health science journals, or publish in real time within the guidance documents.**

a. Recent experience in Marin County on the spread of the SARS-CoV-2 in the school situation would have been critical information to school boards; however the incident that occurred in May 2021 was not published until the end of August 2021, long after the Southern schools had made decisions and the students were fully back in school.

7. **The CDC must fully support states with long-term staffing to address both current public health issues and for the next pandemics.**

   a. 90–95 percent of the CDC domestic staff are in Atlanta and not in the field, where the information dissemination and implementation are occurring.

   b. 85 percent of the CDC staff worked remotely over the past 2 months.

   c. Public health involves the public, and their customers are the states and populations in those states.

   d. **Solution:** The CDC must evolve into a decentralized state presence with continuous feedback loops between the in-state personnel and HQ to ensure timely development and modification of guidance based on the reality on the ground.

8. **Accountability**

   a. There is no accountability of federal dollars to policies, outcomes, and impacts at the federal or state level.

   b. The CDC does not have specific annual milestones for their performance or state performance of grant monies.

   c. There are no annual granular county- and state-level assessments of the health of the country especially among major public health issues facing America.

   d. The CDC often relies on delayed reporting and delayed publication on a 3- to 5-year basis that doesn't align with the annual funding to ensure continuous public health improvement.

   e. **Solution:** CDC must develop timely reporting, implementation results, and outcomes linked to the major public health issues of the country translated down to each and every state with timely data and reporting to ensure continuous program improvement for obesity, hypertension, diabetes mellitus, and community-acquired infections in partnership with states and territories, tribal nations and communities to ensure culturally appropriate and highly impactful programming for dollars invested that looks at incremental improvements through annual reporting and trend lines.

9. **Overarching—we don't need one set of standards and processes for improving the public health of the country and another for pandemic preparedness—they need to be integrated and utilized between pandemics to improve the health of the nation: Next Generation Pandemic Preparedness** as part of our public health response to existing public health issues—health disparities, obesity, diabetes, cardiovascular disease—with data and implementing science.

   a. **We will be better prepared if** we develop definitive diagnosis of viral diseases and there are preexisting equipment, trained technicians, and shared laboratory information systems ready to detect new infectious agents.

   b. **We will be better prepared if** we continuously conduct behavioral research on the uptake of adult vaccines and year over year address the structural and perception issues that limit vaccine uptake and show year over year progress with programming.

   c. **We will be better prepared if** we work to address the social determinants of health and the health disparities through trusted partnerships between community and federal, state, and local partners using real-time data to show improved outcomes and impact year over year, not just once or twice a decade.

   d. **We need to stop just observing the problem and begin addressing the problem.**

   e. **We will be better prepared if** we listen and actually hear and in deep-partnership address the paternalistic and culturally insensitive manner of service delivery to our tribal nations.

10. **Deep dive into what held us back from a rapid response at the FDA and NIH and what actually increased responsiveness.**

    a. Testing—
       i. **good**—Laboratory Designed Assays (LDA), EUA
       ii. **bad**—limiting EUA and therefore testing to symptomatic disease—inhibited the aggressive asymptomatic testing

needed to prevent community spread and willingness to try new ways of reporting from the new devices and tests.

b. Treatment—

    i. **Good**—expanded compassion use, rapid review, and EUA—further streamline regular processes.

    ii. **Needs improvement**—lack of pre IRB approved generic protocols for early stage hospitalized and late stage patients, lack of access to these critical study agents with controlled trials at all hospitals across the United States rather than solely the currently established research hospitals—there should have been a national CRO that would surge to states and territories when we first saw an increase in test positivity to ensure all hospitals had access to controlled trial agents and we would have gotten the answers in days to weeks and not months later—this would have also facilitated the objective trial of agents that some proposed, from hydroxychloroquine to ivermectin, in the regions where they were interested in studying efficacy. Community hospitals and rural hospitals must be eligible for research activities.

c. Vaccines Development

    i. **Good**—rapid movement from Phase I to Phase III trials as warranted, adding community-based research sites to established NIH network of sites, parallel GMP production at risk to ensure immediate access to new and effective antivirals, additional therapeutics, and next generation vaccines.

    ii. **Needs improvement**—inadequate education during the summer of 2020 to continually update the American people on the vaccines being developed with Town Halls, children's books for all households, a chat line to answer questions about the science proactively, education at all levels from K–12, higher education, adults through community centers and churches.

    iii. **Enhanced vaccine protection**—increased vaccine induced durability of protection against infection; potential for cross variant vaccine boosting and intranasal vaccines for durable IgA mucosal immunity.

11. **Bring essential medicines and PPE manufacturing back to the United States.**
    a. The United States ran out of not only protective equipment but essential medication and this is an emergency and needs to be addressed.
12. **Must ensure that full vaccine production and surge fill and finishing capacity exist in the United States to decrease dependence on international facilities.**
13. **Ensure a robust biotech industry for rapid development of new vaccines and viral and fungal treatments.**
14. **Ensure private sector is at the table for all pandemic preparedness planning and response—they can move faster and take risks the federal/state governments won't.**
15. **Ensure the state leaders are in constant communication with the White House and the federal agencies so that their lesson learned can be immediately highlighted to all the states—solutions to many of these issues exist at the state level and we all need to learn from them.**

Acknowledgments

We all have moments in our lives, moments that are clear inflection points personally and professionally, but they aren't moments we share with others around the globe, but this pandemic was.

First, I am grateful to Jeff Carneal who reached out in cold February 2021 and said I should write a book. He thought it was important to document what actually happened—what went well and what would need to change. Not people's perceptions of what happened but what *actually* happened. To date, I have only written technical papers and book chapters and I am grateful to the team at HarperCollins who believed a scientist could write a book that was understandable. To Lisa Sharkey and Matt Harper, I am grateful for your patience and your wisdom, as you guided a newbie through the complex and a bit overwhelming process of writing a book. I appreciate your willingness to support a book that was different from the headlines and your willingness to listen without preconceived ideas and prejudices. I am grateful for your finding Gary Brozek, who wrote beside me month after month, chapter after chapter, and sentence after sentence. Gary, you have a depth of patience that is remarkable. You were unflappable and tolerated my harping on the numbers. You know more about the science of the pandemic than any other writer. Thank you for believing that this virus was bigger than who was in the White House. You understood the stories needed to be told. You had an open and inquisitive mind combined with brilliant writing skills. You are a true partner. Thanks also to the entire team at HarperCollins who shepherded this book through the production process. I can't name you all, but in particular I want to thank Jenna Dolan for her copyediting, Frieda Duggan and Pam Rehm for their keen eyes as proofreaders, and senior production editor Rebecca Holland for bringing that team together and overseeing all facets of the process. Thanks also to Tina Andreadis and Theresa Dooley from the publicity department for their work in helping spread the word about this book. Trina Hunn did an exemplary job on the legal review of these pages, and I'm indebted to her for her diligence and editorial suggestions.

But this book has its roots in decades prior. I was fortunate to be a child of the sixties and seventies—a time of boundless discovery in the sciences. As a child to see my country, America, commit to going to the moon when in class we were still using slide rules—there were no cell phones, desktop computers, or calculators—still seems audacious. There wasn't all the needed technology lined up and ready to go. Yet our country believed we could make what on the surface seemed impossible, possible. There was a promise that became a goal and together—not as red states and blue states but as a United States—over the decade we transformed President Kennedy's vision into reality. I was born into a math and science family, where dinner conversations were not focused on politics and the news of the day but how transistors would replace tubes, a revolution that would change the world. Everything could be miniaturized—huge tabletop radios became transistor radios that were portable, small, and battery-operated. Long before STEM, I had two teachers at my small school, Delaware County Christian School, Dr. Roberts and Dr. Borse, who saw me—in all my quirky nerdiness—and my strange abilities to create pictures out of numbers in my head. I remembered numbers when states and capitol names would fail me. Although these were the upper-school science teachers, they allowed me into the afternoon science club while in elementary school. And the students in that afternoon club embraced me and treated me as normal. The teachers took us to watch total eclipses of the sun, on weeks-long summer field trips to explain rock formations as the geologic history of our planet, and they treated me as any other student male or female— they saw my mind not my gender. To Kristina, my friend from second grade onward, your friendship through the decades has been an anchor for me. The sixties and seventies were the days of local and international science and technology fairs. After I won my hometown state fair, the local newspaper, the *Patriot News* supported me, and I went to the International Science and Engineering Fair. That experience was about so much more than the prizes and awards. I was with hundreds of people like me. I am grateful for each of these opportunities and all of my teachers who understood that science and math were meant to be creative. They fueled my ability rather than confining me to a written "it's not done that way."

To my brothers, Dannie, and Donnie: you were smarter than me in the hard sciences but lacked all common sense and let your baby sister tag along on all your adventurers and misadventures. My parents, Don and Adele, let us blow up the house with our chemistry experiments and never stifled our creativity or sense of adventure. We just had more baking soda and fire extinguishers than the average family. We were set free every day armed with quarters and bicycles and allowed to explore until dinner. To my grandmother Helen, who encouraged my fierce independence and gave me my voice. My undaunted cousins Peggy, Kim, and Beth who shared those idyllic summers with me throughout my childhood in rural Maryland: thank you for your continued support and friendship.

This book would not exist without my close friends and family who weren't swayed by headlines, and who, throughout that year in the White House and since, have supported me, unconditionally. My family is amazing. My daughters, Danielle and Devynn, my brother, Don, and his family Lin, Christy, Sierra, Amy, and Donnie who sacrificed to ensure all of our vulnerable family members across the country were protected. Thank you for listening. I am grateful for my family's diligence in protecting one another and following mitigation guidance so no one became infected with Covid despite all of us working full-time. Danielle, you protected your children, your unborn child, my parents, while creating the most incredible supply chain of everything we needed—I am grateful. Devynn, nothing was too big or too small for you to do in support of all of us, from your sewing skills and cooking, to pick ups and drop offs, you kept us moving forward with masks and food. And to your respective husbands, Mike and Anjan, nieces and nephew and their husbands, Dustin and Brian, my husband, Paige, and his sons, Dylan and Taylor, and their partners, Phoebe and Sarah—thank you. I know throughout those darkest times you had my back, and I know you defended me. You should never have been put in that position.

This should have never become personal and you should never have had to endure the threatening text messages and mail. Paige, you were an unmovable rock with unconditional support. You fed me, drove me, listened to me, and kept me connected for months on end.

  This experience taught me a lot about real friendship, those with integrity and those without. To Sandy Thurman and Sarah Schlesinger for the constant and unquestioning support. You remembered who I was—not who the headlines said I was. To Irum Zaidi who instantly understood what was needed—data. Thank you also for agreeing that we needed to go to the states in the middle of their vast community spread across the country, county by county, to see the diversity, and listen to public and community leaders to ensure that we understood the local context. Federal guidance that can't be understood or implemented is just words on the page and doesn't help anyone. You understood that data isn't collected for the sake of publication but collected to save lives. Data needs to be used and translated so everyone is empowered with the knowledge they need to make the right decisions for their health and the health of their families. Thank you, Irum, for driving 25,000 miles across the country week after week—to 42 of the 44 states visited. Thank you for going to all the fast-food drive-throughs in town to create that one perfect meal, for reading the book and making sure the numbers were right. To the Bush Institute, Kenneth Hersh and Holly Kuzmich, thank you for supporting me while I was in the White House and in writing this book. Thank you to the Generals Lester Martinez and Eric Schoomaker, who showed me what real leadership looks like. Even in the hardest moments you both always did the right thing despite the consequences. I'm indebted to the community of HIV/AIDS advocates and the young women of the DREAMS program across the globe who took the time to teach me and trust me to support them. Together we ensured your voices were heard and issues addressed as we tackled the pandemic of HIV and TB/HIV across the globe. Together addressing the systemic and structural barriers to access of needed services.

  To President and Mrs. Bush: You saw a different yet very deadly global pandemic. You saw past the numbers to the people of the world who were suffering and committed themselves and the American people to saving the globe from the ravages of the HIV/AIDS pandemic. You believed there could be a different future and that dreams are being realized because of your decision to start PEPFAR nearly twenty years ago. Today, the HIV/AIDS pandemic is coming under control—even before a vaccine. You believed in people and understood it was people who make programs not just dollars. Thank you for your vision and making what many thought impossible—possible. To our ten Congresses and presidents since 2003, each of you continued to support PEPFAR across administrations and across party affiliation. This continuity of support showed me that we can do big things, hard things, things others thought were too difficult, when we work across the aisle and support programs independent of party to ensure they have the impact we all want. We build on the success of those before to achieve the long-term goal. This is the source of my hope and my belief that we can be more.

This book is possible because I was in the White House in that moment with a task force led by Vice President Pence and task force members who worked around the clock with dedication and sacrifice to save as many lives as possible. They all worked together in that moment to ensure a different future for every American through vaccines, treatment, tests, and supplies. To the amazing data team led by Irum Zaidi and supported by the USDS and especially Amy Gleason of the USDS. To Steve Redd, Sean Cavanaugh, Chuck Vitek, and Daniel Gastfriend, your advice was invaluable and your support made the reports, data integration, and analysis possible. To Vice President Pence, Jared, Adam, and Brad: you believed the numbers and the projections and moved mountains based on those projections to ensure states had as much of what we could mobilize to meet their needs. The White House Comms teams always wanted the facts and figures and supported me behind the scenes for months. Seema Verma looked at the global data with me and understood the implication for Americans and was unrelentingly focused on protecting the vulnerable. Seema and the people at CMS mobilized and responded in a proactive manner that saved lives. I am particularly grateful for continuous dialogue with Steven Hahn, Bob Redfield, and Tony Fauci who were always there and always a sounding board. To Jerome Adams who was innovative and dedicated and who, along with Brett, Steve, and Bob, went to states to listen, learn, and help when there were just too many for me to get to for the second and third time. To Yen and Matt Pottinger for your enduring friendship and support. To Secretaries Purdue, Carson, Mnuchin, and Scalia—no one will know how many crises you prevented—how close we came to running out of all protein sources and how you rearranged the entire food supply chain within days, understood the needs of the vulnerable and working Americans, and created those critical economic safety nets. To the private sector and the Trump administration who worked uniquely in deep dialogue and true partnership. The Trump administration engaged the private sector in a unique and transparent manner throughout the crisis and together delivered to the American people the dramatic expansion of tests, a spectrum of different tests including rapid antigen tests, to therapeutics, and the vaccines. I am grateful to the large medical suppliers across the United States who utilized our data and predictions to align scarce resources to those who needed them the most, ensuring equity independent of profit and ability to pay.

To all the governors, mayors, county commissioners, community members, and health care leaders who took the time to talk with me and share their insights while in the midst of their own state's pandemic crisis: you changed federal policies and recommendations with your valuable insights. To the tribal chairmen and the community of tribal nations who showed me in their words and actions that community and human life is sacred. To the university presidents and boards of trustees who understood their schools were more than institutions of higher learning and believed in their students and opened their doors in the fall of 2020 to all of their students on campus. These university and college leaders found a way or made one. They were committed and brave and I learned from them what to do to keep faculty and students safe in the midst of this pandemic. We translated your road map into a country road map that we took to every governor. You didn't close—you opened. You used technology. You believed your students and faculty to be a community who would protect one another. You understood that students needed to be on campus because, in many cases, that was their bed, their food, their employment to support both their bodies as well as their minds. To the students who truly saw one another, supported one another independent of political affiliations, who saw a need and met it, who created student food banks: through your eyes I am reminded of the passion of discovery but also the deep compassion for others. Many thanks to the National Guard. You filled every gap from supply delivery to vaccine roll out. In addition to our amazing health care providers and all of those who supported them, the men and women of the National Guard were the backbone of the pandemic response.

Finally, thanks to the federal workforce who work year after year with each new administration across the political divide. You understand that no matter the personal impact you do the right thing to the best of your ability inside of the Republican and Democratic administrations because you serve the American people. I know the amazing people at the CDC and throughout the federal government. If they could finally have access to comprehensive national data and become embedded in the states, then the federal agencies will have the information they will need for truly data- and science-driven decision making. They could then empower the American people with knowledge and not threats. They'd be able to help people access better health care and not make everything about red or blue counties. There is a difference in red and blue counties because health care access is different, but our biology is shared. We are one of the wealthiest but most unhealthy nations in the world. We have ignored the often poor access to health care in our rural communities and we need to address the barriers to health care that exist in our urban areas. We haven't tackled our underlying epidemics of obesity, diabetes, or heart disease; we have ignored the high maternal and child mortality, the mental health of our children, and were late in responding to the opioid epidemic. If we tackle these epidemics with data-driven decision making and hold each other accountable for incremental progress year after year, we can change this as we did with HIV on continents far away. We are succeeding when many thought it was impossible. When we work together in a shared goal across borders and cultures in deep partnership with governments and community—success is possible.

Index

About the Author

Dr. Deborah Birx served in public health in the U.S. government for more than forty years, working for multiple presidential administrations in that time. A former colonel in the U.S. Army, Dr. Birx spent much of her career researching HIV and working around the globe combatting the pandemics of HIV, TB, and malaria. She has held leadership positions at Walter Reed Army Medical Center and Institute of Research, the Centers for Disease Control and Prevention, and the State Department as Ambassador-at-Large and U.S. Global AIDS Coordinator as part of the President's Emergency Plan for AIDS Relief (PEPFAR). She is currently a senior fellow at the Bush Institute. She lives in Washington, DC.